METHODS IN MOLECULAR BIOLOGY

For further volumes:
http://www.springer.com/series/7651

For over 35 years, biological scientists have come to rely on the research protocols and methodologies in the critically acclaimed Methods in Molecular Biology series. The series was the first to introduce the step-by-step protocols approach that has become the standard in all biomedical protocol publishing. Each protocol is provided in readily-reproducible step-by step fashion, opening with an introductory overview, a list of the materials and reagents needed to complete the experiment, and followed by a detailed procedure that is supported with a helpful notes section offering tips and tricks of the trade as well as troubleshooting advice. These hallmark features were introduced by series editor Dr. John Walker and constitute the key ingredient in each and every volume of the Methods in Molecular Biology series. Tested and trusted, comprehensive and reliable, all protocols from the series are indexed in PubMed.

Manuscript Instructions https://drive.google.com/file/d/1K4Az5vpAFVzYadJ3jKx2-QSo7Dqa_SNA/view?usp=sharing

Chikungunya Virus

Methods and Protocols

Edited by

Claudia Filomatori

Universidad de Buenos Aires, Facultad de Farmacia y Bioquímica, Departamento de Química Biológica, Buenos Aires, Argentina; Consejo Nacional de Investigaciones Científicas y Técnicas (CONICET), Universidad de Buenos Aires, Instituto de Química y Fisicoquímica Biológicas "Prof. Alejandro C. Paladini" (IQUIFIB), Buenos Aires, Argentina; Universidad de Buenos Aires, Facultad de Odontología, Cátedra de Biofísica y Bioestadística, Buenos Aires, Argentina

Eliana F. Castro

Instituto de Investigaciones Biotecnológicas, Universidad Nacional de San Martín (UNSAM)—Consejo Nacional de Investigaciones Científicas y Técnicas (CONICET), Buenos Aires, Argentina

Humana Press

Editors
Claudia Filomatori
Universidad de Buenos Aires
Facultad de Farmacia y Bioquímica
Departamento de Química Biológica
Buenos Aires, Argentina

Consejo Nacional de Investigaciones Científicas y Técnicas (CONICET)
Universidad de Buenos Aires, Instituto de Química y Fisicoquímica Biológicas "Prof. Alejandro C. Paladini" (IQUIFIB)
Buenos Aires, Argentina

Universidad de Buenos Aires
Facultad de Odontología
Cátedra de Biofísica y Bioestadística
Buenos Aires, Argentina

Eliana F. Castro
Instituto de Investigaciones Biotecnológicas
Universidad Nacional de San Martín (UNSAM)—
Consejo Nacional de Investigaciones Científicas y Técnicas (CONICET)
Buenos Aires, Argentina

ISSN 1064-3745 ISSN 1940-6029 (electronic)
Methods in Molecular Biology
ISBN 978-1-0716-5495-8 ISBN 978-1-0716-5496-5 (eBook)
https://doi.org/10.1007/978-1-0716-5496-5

This Humana imprint is published by the registered company Springer Science+Business Media, LLC, part of Springer Nature.
The registered company address is: 1 New York Plaza, New York, NY 10004, U.S.A.

Preface

With global climate change and the emergence of arboviruses on the rise, there is an urgent need to address the understudied aspects of Chikungunya virus (CHIKV) biology. CHIKV is a mosquito-borne virus that has re-emerged and spread globally since 2006, with reported transmission in 115 countries worldwide. CHIKV is endemic in Latin America, Africa, and Asia, causing millions of infections worldwide with a significant proportion of patients developing chronic and debilitating sequelae. Despite multiple efforts and preclinical development, there are currently no licensed antivirals. In 2025, two chikungunya vaccines—a live-attenuated formulation (IXCHIQ, Valneva) and a virus-like particle vaccine (Vimkunya, Bavarian Nordic)—received accelerated approval from major regulatory agencies; however, the U.S. FDA subsequently suspended IXCHIQ's license due to significant safety concerns, while Vimkunya was licensed only for use in non-endemic regions to protect high-risk travelers. Consequently, both vaccine effectiveness and global availability remain insufficient, limiting the impact of vaccination as a public-health tool for controlling chikungunya transmission, particularly in endemic countries. On the other hand, effective mitigation of CHIKV and other arboviral pathogens necessitates rigorous entomological surveillance and vector control interventions.

This book provides a comprehensive collection of laboratory protocols designed to support the study of CHIKV across multiple dimensions. The chapters detail experimental approaches for investigating viral RNA and protein functions, virus–host interactions, pathogenesis, immune responses, diagnostic methodologies, quantification of viral particles, viral genome sequencing, structural biology, vector competence, field surveillance, and antiviral discovery. Together, these protocols aim to empower researchers with practical, reproducible tools to accelerate fundamental discoveries.

Buenos Aires, Argentina — *Claudia Filomatori*
Buenos Aires, Argentina — *Eliana F. Castro*

Contents

Contributors

TALITA ÉMILE RIBEIRO ADELINO • *Central Public Health Laboratory of Minas Gerais (Lacen-MG), Ezequiel Dias Foundation – Funed, Belo Horizonte, Minas Gerais, Brazil; Instituto René Rachou – Fiocruz Minas, Belo, Horizonte, Minas Gerais, Brazil*

DIEGO ALVAREZ • *Instituto de Investigaciones Biotecnológicas, Universidad Nacional de San Martín (UNSAM)—Consejo Nacional de Investigaciones Científicas y Técnicas (CONICET), Buenos Aires, Argentina; Escuela de Bio y Nanotecnologías (EByN), Universidad Nacional de San Martín, Buenos Aires, Argentina*

CECÍLIA ARTICO BANHO • *Laboratório de Pesquisas em Virologia, Faculdade de Medicina de São José do Rio Preto, São José do Rio Preto, So Paulo, Brazil*

EUGENIA S. BARDOSSY • *Institut Pasteur, Université Paris Cité, CNRS UMR3569, Viruses and RNA Interference Unit, Paris, France; Viruses and RNA Interference Unit, Institut Pasteur, Université Paris Cité, Paris, France*

JESÚS J. BARRAZA SÁNCHEZ • *Universidad de Buenos Aires, Facultad de Farmacia y Bioquímica, Departamento de Química Biológica, Buenos Aires, Argentina; Consejo Nacional de Investigaciones Científicas y Técnicas (CONICET) – Universidad de Buenos Aires, Instituto de Química y Fisicoquímica Biológicas "Prof. Alejandro C. Paladini" (IQUIFIB), Buenos Aires, Argentina*

VIVIANE LIMA BATISTA • *Department of Microbiology, Institute of Biological Sciences, Universidade Federal de Minas Gerais, Belo Horizonte, Brazil*

LEANDRO BATTINI • *Structural biology of infectious diseases G5+ unit, Institut Pasteur, Université Paris Cité, Paris, France*

POOJA BHARDWAJ • *ICMR-Regional Medical Research Centre Gorakhpur, BRD Medical College Campus, Gorakhpur, India*

FRANCESCO BRANDA • *Unit of Medical Statistics and Molecular Epidemiology, Università Campus Bio-Medico di Roma, Rome, Italy*

TOMMASO CABANI • *Department of Microbiology, Icahn School of Medicine at Mount Sinai, New York, NY, USA; Graduate School of Biomedical Sciences, Icahn School of Medicine at Mount Sinai, New York, NY, USA*

ELIANA F. CASTRO • *Instituto de Investigaciones Biotecnológicas, Universidad Nacional de San Martín (UNSAM)—Consejo Nacional de Investigaciones Científicas y Técnicas (CONICET), Buenos Aires, Argentina*

NICOLE T. CERF • *Universidad de Buenos Aires, Facultad de Farmacia y Bioquímica, Departamento de Química Biológica, Buenos Aires, Argentina; Consejo Nacional de Investigaciones Científicas y Técnicas (CONICET), Universidad de Buenos Aires, Instituto de Química y Fisicoquímica Biológicas "Prof. Alejandro C. Paladini" (IQUIFIB), Buenos Aires, Argentina*

LIMIN CHEN • *The Joint Laboratory on Transfusion-Transmitted Diseases (TTD) Between Institite of Blood Transfusion, Chinese Academy of Medical Sciences and Nanning Blood Center, Nanning Blood Center, Nanning, China; Institute of Blood Transfusion, Chinese Academy of Medical Sciences and Peking Union Medical College, Chengdu, China; The Hospital of Xidian Group, Xi'an, China*

MASSIMO CICCOZZI • *Unit of Medical Statistics and Molecular Epidemiology, Università Campus Bio-Medico di Roma, Rome, Italy*
THAIS ALKIFELES COSTA • *Laboratório de Vírus, Instituto de Ciências Biológicas, Universidade Federal de Minas Gerais, Belo Horizonte, Minas Gerais, Brazil*
VIVIAN VASCONCELOS COSTA • *Department of Morphology, Institute of Biological Sciences, Universidade Federal de Minas Gerais, Belo Horizonte, Brazil*
BETÂNIA PAIVA DRUMOND • *Laboratório de Vírus, Instituto de Ciências Biológicas, Universidade Federal de Minas Gerais, Belo Horizonte, Minas Gerais, Brazil*
GAURAV RAJ DWIVEDI • *ICMR-Regional Medical Research Centre Gorakhpur, BRD Medical College Campus, Gorakhpur, India*
AKINO EMI • *Laboratory of Biosafety Research, Faculty of Medicine, Osaka Medical and Pharmaceutical University, Takatsuki, Japan*
SANTIAGO E. FARAJ • *Universidad de Buenos Aires, Facultad de Farmacia y Bioquímica, Departamento de Química Biológica, Buenos Aires, Argentina; Consejo Nacional de Investigaciones Científicas y Técnicas (CONICET) – Universidad de Buenos Aires, Instituto de Química y Fisicoquímica Biológicas "Prof. Alejandro C. Paladini" (IQUIFIB), Buenos Aires, Argentina;*
CLAUDIA FILOMATORI • *Universidad de Buenos Aires, Facultad de Farmacia y Bioquímica, Departamento de Química Biológica, Buenos Aires, Argentina; Consejo Nacional de Investigaciones Científicas y Técnicas (CONICET), Universidad de Buenos Aires, Instituto de Química y Fisicoquímica Biológicas "Prof. Alejandro C. Paladini" (IQUIFIB), Buenos Aires, Argentina; Universidad de Buenos Aires, Facultad de Odontología, Cátedra de Biofísica y Bioestadística, Buenos Aires, Argentina*
VAGNER FONSECA • *Instituto René Rachou – Fiocruz Minas, Belo, Horizonte, Minas Gerais, Brazil; Department of Exact and Earth Sciences, University of the State of Bahia, Salvador, Brazil; Centre for Epidemic Response and Innovation (CERI), School for Data Science and Computational Thinking, Stellenbosch University, Stellenbosch, South Africa*
ALEXANDER W. E. FRANZ • *Department of Pathobiology & Integrative Biomedical Sciences, University of Missouri, Columbia, Missouri, USA*
MARTA GIOVANETTI • *Department of Science and Bio-Technology, Universita Campus Bio-Medico di, Roma, Italy; Oswaldo Cruz Institute, Oswaldo Cruz Foundation, Rio de Janeiro, Brazil*
JEANNE A. HARDY • *Department of Chemistry, University of Massachusetts, Amherst, MA, USA; Center for Bioactive Delivery in the Institute for Applied Life Sciences, University of Massachusetts, Amherst, MA, USA; Models to Medicine Center in the Institute for Applied Life Sciences, University of Massachusetts, Amherst, MA, USA*
HARI SHANKER JOSHI • *ICMR-Regional Medical Research Centre Gorakhpur, BRD Medical College Campus, Gorakhpur, India*
MELODY LI • *Department of Microbiology, Immunology and Molecular Genetics, University of California, Los Angeles, CA, USA; Molecular Biology Institute, University of California, Los Angeles, CA, USA; Eli and Edythe Broad Center of Regenerative Medicine and Stem Cell Research, University of California, Los Angeles, CA, USA*
SPARSH MAKHAIK • *Department of Chemistry, University of Massachusetts, Amherst, MA, USA*
JENNIFFER RAMOS MARTINS • *Department of Biochemistry and Immunology, Institute of Biological Sciences, Universidade Federal de Minas Gerais, Belo Horizonte, Brazil*

MÓNICA R. MONTES • *Universidad de Buenos Aires, Facultad de Farmacia y Bioquímica, Departamento de Química Biológica, Buenos Aires, Argentina; Consejo Nacional de Investigaciones Científicas y Técnicas (CONICET)—Universidad de Buenos Aires, Instituto de Química y Fisicoquímica Biológicas "Prof. Alejandro C. Paladini" (IQUIFIB), Buenos Aires, Argentina*

OLIVIA BORGHI NASCIMENTO • *Laboratório de Pesquisas em Virologia, Faculdade de Medicina de São José do Rio Preto, São José do Rio Preto, So Paulo, Brazil*

MAURICIO LACERDA NOGUEIRA • *Laboratório de Pesquisas em Virologia, Faculdade de Medicina de São José do Rio Preto, São José do Rio Preto, So Paulo, Brazil; Department of Pathology, University of Texas Medical Branch, Galveston, TX, USA*

MARIA GABRIELA NOVAL • *Department of Microbiology, Icahn School of Medicine at Mount Sinai, New York, NY, USA*

GRISELDA INES ORIA • *Área de Entomología, Instituto de Medicina Regional, Universidad Nacional del Nordeste, Resistencia, Argentina*

CELSO M. QUEIROZ-JUNIOR • *Department of Morphology, Institute of Biological Sciences, Universidade Federal de Minas Gerais, Belo Horizonte, Brazil*

KÍSSILA RABELO • *Laboratory of Ultrastructure and Tissue Biology, Institute of Biology, Universidade do Estado do Rio de Janeiro, Niteroi, Rio de Janeiro, Brazil*

MUHAMMAD ARSLAN RAHAT • *Department of Chemistry, University of Massachusetts, Amherst, MA, USA*

RUCHI RANI • *Department of Biosciences and Bioengineering, Indian Institute of Technology Roorkee, Roorkee, Uttarakhand, India*

LÍVIA SACCHETTO • *Departamento de Genética, Evolução, Microbiologia e Imunologia, Universidade Estadual de Campinas, Campinas, So Paulo, Brazil*

NATÁLIA G. SALOMÃO • *Department of Anatomical Pathology, Hospital Universitário Antônio Pedro, Universidade Federal Fluminense (UFF), Niteroi, Rio de Janeiro, Brazil*

FABIO SCARPA • *Department of Biomedical Sciences, University of Sassari, Sassari, Italy*

RAJEEV SINGH • *ICMR-Regional Medical Research Centre Gorakhpur, BRD Medical College Campus, Gorakhpur, India*

JUJUN SUN • *The Hospital of Xidian Group, Xi'an, China*

YOUICHI SUZUKI • *Laboratory of Biosafety Research, Faculty of Medicine, Osaka Medical and Pharmaceutical University, Takatsuki, Japan*

MAURO MARTINS TEIXEIRA • *Department of Biochemistry and Immunology, Institute of Biological Sciences, Universidade Federal de Minas Gerais, Belo Horizonte, Brazil*

SARA ANN THANNICKAL • *Department of Microbiology, Icahn School of Medicine at Mount Sinai, New York, NY, USA*

SHAILLY TOMAR • *Department of Biosciences and Bioengineering, Indian Institute of Technology Roorkee, Roorkee, Uttarakhand, India*

ANUSHKA UPADHYAY • *Department of Biosciences and Bioengineering, Indian Institute of Technology Roorkee, Roorkee, Uttarakhand, India*

NIKOS VASILAKIS • *Department of Pathology, University of Texas Medical Branch, Galveston, TX, USA; Center for Vector-Borne and Zoonotic Diseases, University of Texas Medical Branch, Galveston, TX, USA; Institute for Human Infection and Immunity, University of Texas Medical Branch, Galveston, TX, USA*

SEBASTIANO VOLPE • *Departamento de Química Biológica, Facultad de Farmacia y Bioquímica, Universidad de Buenos Aires, Buenos Aires, Argentina; Consejo Nacional de Investigaciones Científicas y Técnicas (CONICET), Instituto de Química y Fisicoquímica Biológicas "Prof. Alejandro C. Paladini" (IQUIFIB), Universidad de Buenos Aires, Buenos Aires, Argentina; Escuela de Bio y Nanotecnología, Universidad de San Martín - CONICET, Buenos Aires, Argentina*

JIAYI WANG • *Department of Pathogen Biology, School of Clinical and Basic Medical Sciences, Shandong First Medical University & Shandong Academy of Medical Sciences, Jinan, Shandong, China*

CHENYANG WU • *Department of Microbiology, Icahn School of Medicine at Mount Sinai, New York, NY, USA*

JOILSON XAVIER • *Institut für Virologie, Freie Universität Berlin, Berlin, Germany*

HE XIE • *The Hospital of Xidian Group, Xi'an, China*

XIPENG YAN • *The Joint Laboratory on Transfusion-Transmitted Diseases (TTD) Between Institite of Blood Transfusion, Chinese Academy of Medical Sciences and Nanning Blood Center, Nanning Blood Center, Nanning, China; Institute of Blood Transfusion, Chinese Academy of Medical Sciences and Peking Union Medical College, Chengdu, China*

ZHENLAN YAO • *Department of Microbiology, Immunology and Molecular Genetics, University of California, Los Angeles, CA, USA*

LEILIANG ZHANG • *Department of Pathogen Biology, School of Clinical and Basic Medical Sciences, Shandong First Medical University & Shandong Academy of Medical Sciences, Jinan, Shandong, China*

JUNXIANG ZHAO • *Department of Pathogen Biology, School of Clinical and Basic Medical Sciences, Shandong First Medical University & Shandong Academy of Medical Sciences, Jinan, Shandong, China*

Check for updates

Chapter 1

Overview of Chikungunya Virus Epidemiology, Biology, and Pathogenesis

Eliana F. Castro and Claudia Filomatori

Abstract

Chikungunya virus (CHIKV), an arthropod-borne alphavirus within the *Togaviridae* family, is transmitted primarily by *Aedes aegypti* and *Aedes albopictus*. The virus causes an acute febrile illness characterized by severe, often bilateral polyarthralgia, with potential progression to chronic musculoskeletal pain and rare systemic complications involving cardiovascular and neurological systems. CHIKV exhibits a spherical, enveloped virion (~70 nm) with $T = 4$ icosahedral symmetry, incorporating E1/E2 glycoprotein heterodimers that mediate receptor binding and membrane fusion. Its positive-sense RNA genome (~11.8 kb) encodes nonstructural proteins for replication and structural proteins for virion assembly. Replication occurs in cytoplasmic spherules, involving synthesis of genomic and subgenomic RNAs, followed by glycoprotein maturation and budding at the plasma membrane. Epidemiologically, CHIKV has expanded beyond Africa and Asia, with major outbreaks driven by adaptive mutations enhancing transmission via *A. albopictus*. Since introduction to the Americas in 2013, the global incidence remains high, with >180,000 confirmed cases reported in 2025. Preventive strategies rely on vector control and vaccination; VLP-based vaccines (e.g., Vimkunya) show promise, while live-attenuated formulations face safety concerns. No licensed antivirals exist; current management is supportive, though investigational therapies targeting viral replication and immune modulation are under development.

Keywords Chikungunya virus, Epidemiology, CHIKV genome, CHIKV virions, CHIKV life cycle, CHIKV pathogenesis

1 Viral Disease

Chikungunya virus (CHIKV) belongs to the family *Togaviridae*, genus *Alphavirus*, which includes several other medically significant arthropod-borne viruses (arboviruses), such as Sindbis virus, Semliki Forest virus, and Eastern, Western, and Venezuelan equine encephalitis viruses [1, 2]. CHIKV was isolated in 1952 during a dengue outbreak in Tanzania [3]. The term *chikungunya*, derived from the Makonde language spoken in the region, translates to "bent" or "stooped posture," reflecting the severe and often persistent joint pain that characterizes the infection. The virus is primarily transmitted

Claudia Filomatori and Eliana F. Castro (eds.), *Chikungunya Virus*, Methods in Molecular Biology, vol. 3067,
https://doi.org/10.1007/978-1-0716-5496-5_1,

by mosquitoes of the *Aedes aegypti* and *Aedes albopictus* species. Following an incubation period of approximately 1–12 days, patients usually report acute symptoms, such as high fever (often exceeding 102°F/39°C), joint pain (arthralgia), headache, fatigue, and muscle pain (myalgia), all of which can be both debilitating and distressing [4–6]. Joint pain, the hallmark feature of CHIKV disease, is frequently bilateral and symmetric, affecting corresponding joints on both sides of the body, including hands, feet, knees, and wrists. Although the disease is generally self-limiting, a substantial proportion of patients develop chronic and debilitating sequelae, most notably persistent joint pain and other long-term complications that can severely impair quality of life [7, 8]. In some cases, CHIKV infection is associated with severe atypical manifestations involving multiple organ systems, including the cardiovascular and central nervous systems, which may result in life-threatening outcomes or death [8]. In addition to its clinical impact, CHIKV outbreaks impose a considerable economic burden on affected regions, reflected in increased hospitalization rates, intensified outbreak response and vector control efforts, and disruption of social and economic activities [9, 10].

2 Epidemiology

CHIKV is a re-emergent arbovirus [11–14]. For approximately six decades, CHIKV circulation was largely restricted to Africa and Asia. Three major CHIKV lineages have been defined based on their geographic distribution: West African, East/Central/South African (ECSA), and Asian. The ECSA and Asian strains have disseminated globally during the last decade, while the West African remained confined to a sylvatic circulation in sparsely populated regions of Africa. In 2005, a CHIKV strain derived from ECSA lineage caused a major epidemic in the Indian Ocean islands and then spread to Europe. This outbreak was driven by viral adaptations that enabled efficient transmission by *Aedes albopictus*, in addition to the primary vector *Aedes aegypti* [15]. Key determinants of this expanded vector competence were later mapped to the viral envelope glycoproteins. The widespread distribution of *A. albopictus* in these regions has raised concerns about the further expansion of CHIKV-endemic areas.

CHIKV was introduced into the Americas in 2013, and since then, both Asian- and ECSA-derived isolates have been reported. Overall, since its re-emergence, both autochthonous and imported CHIKV cases have been documented in the Americas and Europe [16, 17]. According to available data from January to September 2025, a total of 263,592 suspected cases, 181,679 confirmed cases of CHIKV disease, and 155 CHIKV-associated deaths have been reported globally [18]. The Region of the Americas has reported the highest number of cases, followed by the European region, which primarily comprises cases reported from French Overseas Departments in the Indian Ocean.

3 Virion Architecture

Alphaviruses have a common virion structure, consisting of a roughly spherical, enveloped particle approximately 65–70 nm in diameter with a highly ordered icosahedral symmetry ($T=4$) [19] (Fig. 1a). The outer layer comprises a host-derived lipid bilayer into which heterodimers of the viral envelope glycoproteins E1 and E2 are embedded, forming 80 trimeric spikes on the virion surface. The E2 glycoprotein, generated by furin cleavage of the p62 precursor into E3 and E2, is primarily responsible for receptor binding. In turn, the E1 glycoprotein is a class II fusion protein that mediates membrane fusion during viral entry. Beneath the envelope, a single copy of the viral genomic RNA is packaged with hundreds of copies of the capsid (C) protein, together forming the viral nucleocapsid (NC). Although little is known about the detailed organization of the virion interior and the mechanisms governing genome packaging, it has been proposed that the highly basic capsid protein interacts with

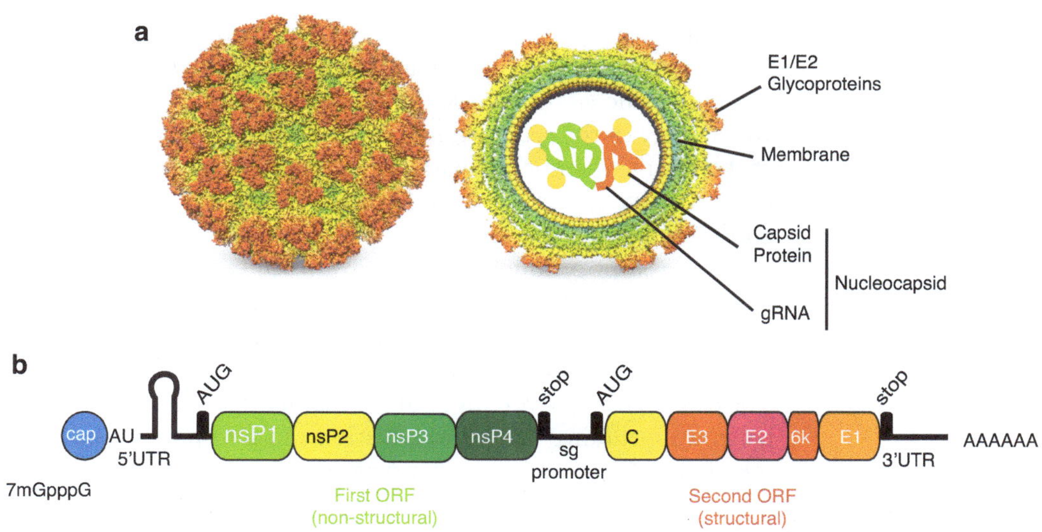

Fig. 1 Chikungunya virion and genome. (**a**) The virion is composed of the capsid (C) protein, which together with the viral genome forms the nucleocapsid, and the E1/E2 glycoproteins embedded in the lipid envelope. Mature virions display a smooth surface formed by E1/E2 heterodimers arranged in a $T=4$ icosahedral-like lattice. (**b**) The viral genome is a single-stranded, positive-sense RNA molecule bearing a 5′ cap and a 3′ poly(A) tail. It comprises two open reading frames: the first encodes the four nonstructural proteins (nsP1–4), and the second encodes the structural proteins (C/E3/E2/6 K/E1). The genome is flanked by 5′ and 3′ untranslated regions (UTRs)

the viral genome through electrostatic forces at multiple packaging signals [20, 21].

4 Organization of the Chikungunya Virus Genome and Replication Cycle

The CHIKV genome is a single-stranded, positive-sense RNA of approximately 11.8 kb, capped at the 5′ end and polyadenylated at the 3′ end, resembling a cellular mRNA [22, 23] (Fig. 1b). It contains two major open reading frames (ORFs): the first ORF encodes the nonstructural proteins and the second ORF encodes the structural proteins [2]. Flanking the ORFs are the 5′ and 3′ untranslated regions (UTRs) of the viral genome, and there is an additional UTR between both ORFs, which contains the subgenomic promoter of the second ORF. Although these regions do not code for proteins, they contain cis-acting elements that interact with viral and host factors, regulating the steps of the viral life cycle [1, 24, 25].

The viral life cycle begins with the attachment of the viral E2 glycoprotein to the host cell surface receptors, followed by internalization through endocytosis [26] (Fig. 2). Acidification of the endosomal compartment induces dissociation of the E2–E1 heterodimer and E1 rearrangement into fusogenic homotrimers that induce fusion of viral and endosomal membranes [27]. This process enables the release of the nucleocapsid into the cytoplasm, where disassembly occurs and the viral genomic RNA (gRNA) is made available for translation. The incoming RNA is translated into the P1234 nonstructural polyprotein. Proteolytic processing of this polyprotein by the viral protease nsP2 regulates the temporal assembly and activity of the replication complex [28]. RNA synthesis occurs at the plasma membrane, within membrane-invaginated structures known as spherules [29]. Early replication complexes synthesize a full-length negative-strand RNA intermediate, which serves as a template for the production of new gRNA and a shorter subgenomic RNA (sgRNA) [30]. The sgRNA encodes the structural polyprotein, which is translated at the endoplasmic reticulum and subsequently cleaved into the capsid protein and the envelope glycoproteins [31]. The envelope glycoproteins undergo maturation within the secretory pathway, including heterodimerization and transport to the plasma membrane. Concurrently, newly synthesized genomic RNA associates with capsid proteins to form nucleocapsids in the cytoplasm, which then diffuse to the plasma membrane. There, nucleocapsids interact with membrane-embedded glycoproteins, leading to virus assembly, budding, and the release of mature infectious virions.

Overall, the chikungunya virus replication cycle relies on precisely regulated molecular interactions between viral components and host factors at each stage, highlighting critical steps that may be exploited for the development of targeted antiviral strategies.

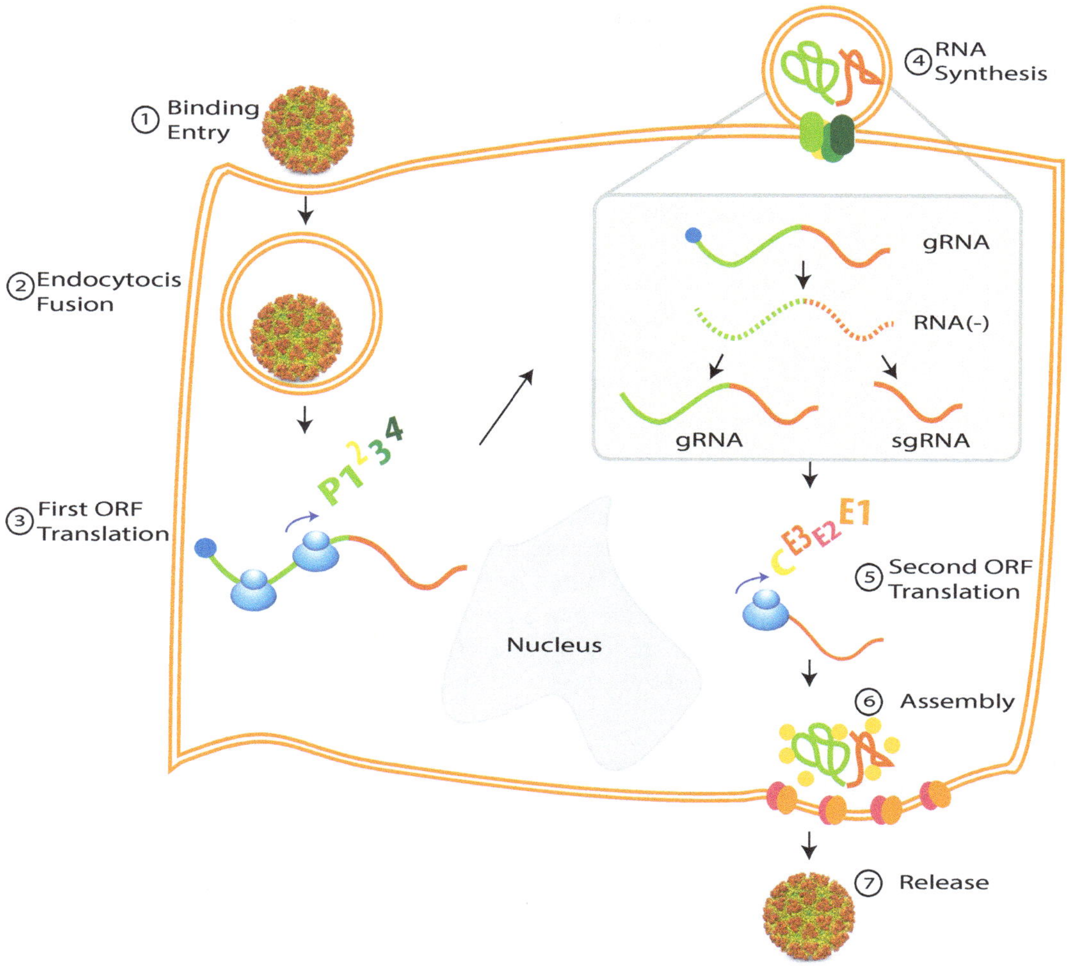

Fig. 2 Chikungunya virus replicative cycle. The replication cycle begins with virion attachment to cell receptors and internalization by clathrin-mediated endocytosis (1). Endosomal acidification triggers conformational changes in the E proteins, promoting fusion of the viral envelope with the endosomal membrane and release of the genome into the cytoplasm (2). The genomic RNA is translated to produce the nonstructural polyprotein, which is processed into the mature replicase complex (3). Viral RNA synthesis occurs at plasma membrane–derived spherules, where a minus-strand intermediate is generated and used to synthesize genomic and subgenomic RNAs (4). The subgenomic RNA directs the production of structural proteins (5), leading to virion assembly (6) and release (7)

5 Pathogenesis

CHIKV pathogenesis involves a multifaceted interplay between the viral agent, host immune responses, and environmental determinants, collectively driving both the acute inflammatory cascade and the persistent nociceptive symptoms characteristic of chronic disease (reviewed in [32]). During the blood meal of an infected mosquito, CHIKV is inoculated either directly into the bloodstream or deposited within dermal tissue alongside vector saliva [33–35]. Within the dermis, viral replication primarily occurs in fibroblasts, which serve as initial replication niches [34, 36–39]. Additional host cell populations—such as dendritic cells, macrophages, keratinocytes, melanocytes, and endothelial and epithelial cells—are also considered permissive to infection [40, 41]. Engagement with antigen-presenting cells in the skin facilitates viral dissemination via lymphatic drainage into the systemic circulation. CHIKV exhibits marked tropism for musculoskeletal compartments (including synoviocytes, fibroblasts, and myocytes) and neural tissues, establishing secondary replication sites. In these localizations, virus induces cytopathic effects and elicits robust immune activation, culminating in hallmark clinical manifestations: high-grade fever, severe polyarthralgia, and systemic inflammation [32, 42, 43]. Pain pathophysiology in CHIKV infection reflects a complex convergence of direct tissue injury, immune-mediated processes, and neural sensitization [44]. Despite significant progress, the mechanisms underlying chronic pain remain incompletely elucidated; sustained production of proinflammatory cytokines is hypothesized to play a pivotal role. Humans function as amplifying hosts, generating high-level viremia essential for subsequent acquisition and transmission by hematophagous mosquitoes.

6 Preventive Measures

Effective mitigation of CHIKV and other arboviral pathogens necessitates rigorous entomological surveillance and vector control interventions, as these constitute the cornerstone for reducing transmission dynamics and safeguarding public health. Continuous monitoring of *Aedes* spp. populations within endemic regions provides critical epidemiological intelligence for forecasting and preparedness against potential CHIKV outbreaks. Population-based serosurveys and seroprevalence assessments further delineate the magnitude of viral exposure within communities, thereby informing evidence-based public health strategies. Vector control programs primarily aim to suppress populations of competent mosquito vectors, thereby interrupting arbovirus transmission cycles. Core interventions encompass the application of adulticidal and larvicidal agents, coupled with environmental management strategies designed to eliminate or modify breeding habitats. Integrated approaches combining chemical, biological, and ecological measures remain essential for sustainable vector suppression and outbreak prevention [45–47].

Particularly for CHIKV, Ixchiq (Valneva) [48], a live-attenuated vaccine, has received regulatory approval in multiple jurisdictions, including the United States, Canada, Europe, and Brazil. This vaccine leverages an attenuated CHIKV strain to elicit robust humoral and cellular immune responses; however, its clinical benefit in large-scale populations remains under evaluation. Notably, in August 2025, the U.S. Food and Drug Administration (FDA) Center for Biologics Evaluation and Research (CBER) has suspended the biologics license for Valneva Austria GmbH's Ixchiq based on serious safety concerns [49]. In parallel, virus-like particle (VLP)-based platforms have demonstrated promising immunogenicity profiles without the replication-associated risks inherent to live-attenuated formulations. Vimkunya (Bavarian Nordic), a VLP vaccine mimicking the structural architecture of CHIKV without containing viral genetic material, recently achieved accelerated approval from major regulatory bodies, including the FDA, the European Medicines Agency (EMA), and the UK Medicines and Healthcare products Regulatory Agency (MHRA) [50]. Overall, the integration of vaccination programs with vector control strategies and epidemiological surveillance will be essential to achieve sustained reduction in CHIKV transmission.

7 Therapeutic Strategies

From both scientific and clinical standpoints, Chikungunya virus (CHIKV) presents substantial challenges, primarily due to the absence of virus-specific therapeutics. Current management strategies are limited to supportive care, focusing on symptomatic relief through analgesics and anti-inflammatory agents. During the acute viremic phase, recommended interventions include rest, adequate hydration, and administration of antipyretic and analgesic drugs—acetaminophen being the first-line agent—to alleviate pyrexia and musculoskeletal pain. In cases of refractory pain, escalation to more potent analgesics may be warranted [32, 51]. Emerging therapeutic approaches aim to disrupt viral replication, modulate host immune pathways, or mitigate chronic inflammatory sequelae [52]. Investigational modalities encompass direct-acting antivirals, polyclonal and monoclonal antibody therapies, and therapeutic vaccine platforms, most of which remain in preclinical development or early-phase clinical trials. Despite significant progress in antiviral drug discovery for CHIKV over the past decade [53], no licensed antiviral agents currently exist for CHIKV infection [54].

Ultimately, advancing our understanding of CHIKV at molecular, clinical, and epidemiological levels remains crucial for the development of effective strategies to control disease and prevent future outbreaks.

References

1. Barraza Sánchez JJ, Volpe S, Faraj SE, et al (2025) Functional RNA elements in alphavirus genomes. Curr Microbiol 82(8):364.
2. Strauss JH, Strauss EG (1994) The alphaviruses: gene expression, replication, and evolution. Microbiol Rev 58:491–562

3. Robinson MC (1955) An epidemic of virus disease in Southern Province, Tanganyika Territory, in 1952-53. I. Clinical features. Trans R Soc Trop Med Hyg 49:28–32
4. Mourad O, Makhani L, Chen LH (2022) Chikungunya: an emerging public health concern. Curr Infect Dis Rep 24:217–228
5. Nkoghe D, Kassa RF, Caron M, et al (2012) Clinical forms of chikungunya in Gabon, 2010. PLoS Negl Trop Dis 6(2), e1517.
6. Taubitz W, Cramer JP, Kapaun A, et al (2007) Chikungunya fever in travelers: clinical presentation and course. Clin Infect Dis 45(1), e1-4.
7. Paixão ES, Rodrigues LC, Costa M, da CN et al (2018) Chikungunya chronic disease: a systematic review and meta-analysis. Trans R Soc Trop Med Hyg 112:301–316
8. Puntasecca CJ, King CH, Labeaud AD (2021) Measuring the global burden of chikungunya and Zika viruses: a systematic review. PLoS Negl Trop Dis 15(3), e0009055.
9. Feldstein LR, Ellis EM, Rowhani-Rahbar A, et al (2019) Estimating the cost of illness and burden of disease associated with the 2014-2015 chikungunya outbreak in the U.S. Virgin Islands. PLoS Negl Trop Dis 13(7), e0007563.
10. Soumahoro MK, Boelle PY, Gaüzere BA, et al (2011) The Chikungunya epidemic on La Réunion Island in 2005-2006: a cost-of-illness study. PLoS Negl Trop Dis 5(6), e1197.
11. Thaikruea L, Charearnsook O, Reanphumkarnkit S et al (1997) Chikungunya in Thailand: a re-emerging disease?. Southeast Asian J Trop Med Public Health 28:359–364
12. Lanciotti RS, Valadere AM (2014) Transcontinental movement of Asian genotype chikungunya virus. Emerg Infect Dis 20:1400–1402
13. Fredericks AC, Fernandez-Sesma A (2014) The burden of dengue and chikungunya worldwide: implications for the southern United States and California. Ann Glob Heal 80:466–475
14. Weaver SC (2014) Arrival of chikungunya virus in the new world: prospects for spread and impact on public health. PLoS Negl Trop Dis 8(6), e2921.
15. Tsetsarkin KA, Vanlandingham DL, McGee CE et al (2007) A single mutation in chikungunya virus affects vector specificity and epidemic potential. PLoS Pathog 3:1895–1906
16. Stapleford KA, Moratorio G, Henningsson R, et al (2016) Whole-genome sequencing analysis from the chikungunya virus Caribbean outbreak reveals novel evolutionary genomic elements. PLoS Negl Trop Dis 10(1), e0004402.
17. Liu Q, Shen H, Gu L, et al (2025) Chikungunya virus in Europe: a retrospective epidemiology study from 2007 to 2023. PLoS Negl Trop Dis 19(3), e0012904.
18. Organización Mundial de la Salud, https://www.who.int/es
19. Voss JE, Vaney MC, Duquerroy S et al (2010) Glycoprotein organization of Chikungunya virus particles revealed by X-ray crystallography. Nature 468:709–712
20. Brown RS, Anastasakis DG, Hafner M, et al (2020) Multiple capsid protein binding sites mediate selective packaging of the alphavirus genomic RNA. Nat Commun 11(1):4693.
21. Kim DY, Firth AE, Atasheva S et al (2011) Conservation of a packaging signal and the viral genome RNA packaging mechanism in alphavirus evolution. J Virol 85:8022–8036
22. Khan AH, Morita K, Carmen Parquet Mdel et al (2002) Complete nucleotide sequence of chikungunya virus and evidence for an internal polyadenylation site. J Gen Virol 83:3075–3084
23. Vasiljeva L, Merits A, Auvinen P et al (2000) Identification of a novel function of the alphavirus capping apparatus. RNA 5'-triphosphatase activity of Nsp2. J Biol Chem 275:17281–17287
24. Hyde JL, Gardner CL, Kimura T et al (2014) A viral RNA structural element alters host recognition of nonself RNA. Science 343:783–787
25. Filomatori CV, Bardossy ES, Merwaiss F, et al (2019) RNA recombination at Chikungunya virus 3'UTR as an evolutionary mechanism that provides adaptability. PLoS Pathog 15(4), e1007706.
26. van Duijl-Richter MKS, Hoornweg TE, Rodenhuis-Zybert IA, et al (2015) Early events in Chikungunya virus infection—from virus cell binding to membrane fusion. Viruses 7:3647–3674.
27. Kielian M, Helenius A (1985) pH-induced alterations in the fusogenic spike protein of Semliki Forest virus. J Cell Biol 101:2284–2291
28. Lemm JA, Rice CM (1993) Roles of nonstructural polyproteins and cleavage products in regulating Sindbis virus RNA replication and transcription. J Virol 67:1916–1926
29. Tan YB, Chmielewski D, Law MCY, et al (2022) Molecular architecture of the Chikungunya virus replication complex. Sci Adv 8(48), eadd2536.
30. Rupp JC, Sokoloski KJ, Gebhart NN et al (2015) Alphavirus RNA synthesis and non-structural protein functions. J Gen Virol 96:2483–2500
31. Solignat M, Gay B, Higgs S et al (2009) Replication cycle of chikungunya: a re-emerging arbovirus. Virology 393:183–197
32. Freppel W, Silva LA, Stapleford KA, et al (2024) Pathogenicity and virulence of chikungunya virus. Virulence 15(1), 2396484.

33. Agarwal A, Joshi G, Nagar DP et al (2016) Mosquito saliva induced cutaneous events augment Chikungunya virus replication and disease progression. Infect Genet Evol 40:126–135
34. Wichit S, Diop F, Hamel R et al (2017) *Aedes aegypti* saliva enhances chikungunya virus replication in human skin fibroblasts via inhibition of the type I interferon signaling pathway. Infect Genet Evol 55:68–70
35. Fong S-W, Kini RM, Ng LFP (2018) Mosquito saliva reshapes alphavirus infection and immunopathogenesis. J Virol 92(12), e01004-17.
36. Sourisseau M, Schilte C, Casartelli N et al (2007) Characterization of reemerging chikungunya virus. PLOS Pathog 3:e89
37. Young AR, Locke MC, Cook LE, et al (2019) Dermal and muscle fibroblasts and skeletal myofibers survive chikungunya virus infection and harbor persistent RNA. PLoS Pathog 15(8), e1007993.
38. Ekchariyawat P, Hamel R, Bernard E et al (2015) Inflammasome signaling pathways exert antiviral effect against Chikungunya virus in human dermal fibroblasts. Infect Genet Evol 32:401–408
39. Couderc T, Chrétien F, Schilte C, et al (2008) A mouse model for Chikungunya: young age and inefficient type-I interferon signaling are risk factors for severe disease. PLoS Pathog 4(2), e29.
40. Labadie K, Larcher T, Joubert C et al (2010) Chikungunya disease in nonhuman primates involves long-term viral persistence in macrophages. J Clin Invest 120:894–906
41. Gasque P, Jaffar-Bandjee MC (2015) The immunology and inflammatory responses of human melanocytes in infectious diseases. J Infect 71:413–421
42. Zhang R, Kim AS, Fox JM et al (2018) Mxra8 is a receptor for multiple arthritogenic alphaviruses. Nature 557:570–574
43. Lidbury BA, Rulli NE, Suhrbier A et al (2008) Macrophage-derived proinflammatory factors contribute to the development of arthritis and myositis after infection with an arthrogenic alphavirus. J Infect Dis 197:1585–1593
44. Sun W, Shi S, Liao S et al (2025) Chikungunya fever: pathogenesis and mechanisms underlying pain symptoms. Front Immunol 16:1679385
45. Skalinski LM, Santos AES, Paixão E, et al (2023) Chikungunya seroprevalence in population-based studies: a systematic review and meta-analysis. Arch Public Heal 81(1):80.
46. Bustos Carrillo F, Collado D, Sanchez N et al (2019) Epidemiological evidence for lineage-specific differences in the risk of inapparent chikungunya virus infection. J Virol 93:1622–1640
47. Pinto SB, Riback TIS, Sylvestre G, et al (2021) Effectiveness of Wolbachia-infected mosquito deployments in reducing the incidence of dengue and other Aedes-borne diseases in Niterói, Brazil: a quasi-experimental study. PLoS Negl Trop Dis 15(7), e0009556.
48. Valneva. Valneva receives first marketing authorization for ixchiq® in a chikungunya endemic country. Published April 14, 2025. Accessed July 22, 2026. https://valneva.com/press-release/valneva-receives-first-marketing-authorization-for-ixchiq-in-a-chikungunya-endemic-country/
49. FDA update on the safety of Ixchiq (Chikungunya Vaccine, Live). FDA suspends biologics license: FDA safety communication | FDA, Published August 22, 2025. Accessed July 22, 2026. https://www.fda.gov/safety/medical-product-safety-information/fda-update-safety-ixchiq-chikungunya-vaccine-live-fda-suspends-biologics-license-fda-safety
50. VIMKUNYATM (Recombinant) | A chikungunya vaccine | For travelers, Published 2025. Accessed July 22, 2026. https://vimkunya.com/
51. Hua C, Combe B (2017) Chikungunya virus-associated disease. Curr Rheumatol Rep 19(11):69.
52. Ghildiyal R, Gabrani R (2020) Antiviral therapeutics for chikungunya virus. Expert Opin Ther Pat 30:467–480
53. Wang M, Wang L, Leng P et al (2024) Drugs targeting structural and nonstructural proteins of the chikungunya virus: a review. Int J Biol Macromol 262:129949
54. Nizi MG, Massari S, Tabarrini O, et al (2025) A medicinal chemistry overview of direct-acting antivirals approved in 2013–2024 by US FDA. Eur J Med Chem 300, 118105.

Chapter 2

Naked-Eye Visual Loop-Mediated Isothermal Amplification Method for the Rapid Detection of Chikungunya Virus

Jiayi Wang, Junxiang Zhao and Leiliang Zhang

Abstract

Chikungunya virus (CHIKV) represents a significant global health threat due to its recent resurgence. This protocol highlights the use of capillary-modified visual loop-mediated isothermal amplification (LAMP) technology for the rapid detection of CHIKV. We employed Vero cells for virus propagation and utilized the MolPure Viral DNA/RNA Kit for nucleic acid extraction, successfully amplifying extracted viral RNA at 63°C for 25 min. The presence of CHIKV was indicated by a distinct yellowish-green color change, demonstrating the effectiveness and simplicity of this protocol. Our protocol underscores the potential of LAMP technology as a rapid diagnostic tool for CHIKV, enabling timely public health interventions.

Keywords LAMP, CHIKV, Naked-eye visual, Rapid detection, Nucleic acid

1 Introduction

The chikungunya virus (CHIKV) has recently re-emerged on a large scale, impacting multiple countries across Africa, Asia, and Europe [1, 2]. CHIKV is an alphavirus transmitted by *Aedes albopictus* and *Aedes aegypti* [3]. It is a positive-sense single-stranded RNA virus with a genome of 11.8 kb, encoding six structural proteins and four nonstructural proteins (nsP1–nsP4). Since its identification in Tanzania in 1952, CHIKV has spread to various countries and regions, primarily concentrated in tropical areas [1]. A significant outbreak occurred following mutations in the 2000s, and the virus has now spread into subtropical regions, becoming an infectious disease of considerable concern [3]. Its main symptoms include high fever, rash, and joint pain, which can potentially lead to long-lasting arthritis [3]. Individuals infected with CHIKV produce IgM and IgG antibodies as part of their immune response, which can be utilized for serological and immune adsorption testing. Reverse transcription polymerase chain reaction (RT-PCR) is the most commonly used molecular detection method during the early stages of infection.

Claudia Filomatori and Eliana F. Castro (eds.), *Chikungunya Virus*, Methods in Molecular Biology, vol. 3067,
https://doi.org/10.1007/978-1-0716-5496-5_2,

Loop-mediated isothermal amplification (LAMP) technology, unlike PCR, does not depend on temperature cycling; instead, it amplifies DNA or RNA at a constant temperature of 60°C–65°C [4]. This method, developed by Notomi et al in 2000 [5], employs a reaction system composed of deoxynucleotide triphosphates (dNTPs), Bst polymerase, fluorescent dye, primers, and nucleic acid templates. The essence of the LAMP reaction lies in its ability to identify different regions of the target nucleic acid using Bst DNA polymerase and multiple specific primers, facilitating efficient amplification. The results of the LAMP reaction are primarily evaluated through agarose gel electrophoresis, turbidity detection, staining, and fluorescence detection. Characterized by its simplicity, rapid amplification, and minimal equipment requirements, LAMP has been widely applied for the rapid detection of pathogenic microorganisms in recent years, including the Zika virus [6], Ebola virus [7], CHIKV[8], mpox virus [9], and SARS-CoV-2 [10]. Capillary-modified visual LAMP further simplifies equipment requirements while enhancing amplification efficiency [11]. In this protocol, capillary-modified visual LAMP technology was employed to detect CHIKV, utilizing Vero cells for virus propagation and the MolPure Viral DNA/RNA Kit for nucleic acid extraction (Fig. 1). The extracted viral RNA was amplified at 63°C for 25 min, with positive results indicated by a yellowish-green color (Fig. 1).

2 Materials

2.1 Reagents

1. Primers targeted to the highly conserved CHIKV nsP1:
   CHIKV nsP1-F3: ATCCTTGCTACAGAAGTCA
   CHIKV nsP1-B3: AGCAGCAGGTTAGTGTT
   CHIKV nsP1-FIP: CATGGTGTTCGTGTTCCGTGGCT-GAACCAGAGGATAG
   CHIKV nsP1-BIP: ACCTACTTCCCGTGGTCGTCTTC-CATGTCCTTCCG
   CHIKV nsP1-LoopF: GCGTTCTGCCGTTAACCA
   CHIKV nsP1-LoopB: AAGTGGGCAAAGGAGTGC
2. Vero Cells
3. CHIKV 181/25 strain
4. High-glucose Dulbecco's Modified Eagle Medium (DMEM) (1×)
5. Fetal bovine serum (FBS)
6. Trypsin
7. Penicillin–Streptomycin double antibiotics solution
8. Anhydrous ethanol

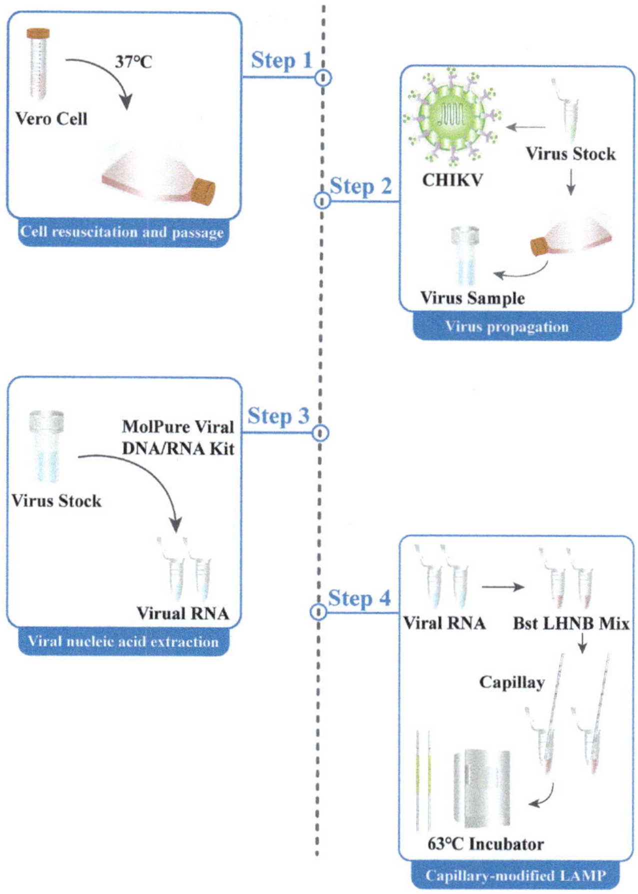

Fig. 1 Experimental procedures of capillary-modified visual LAMP technology for detecting CHIKV. Step 1: Cell thawing and passage. Vero cells are thawed and cultured at 37°C to ensure optimal growth conditions. Step 2: Virus propagation. CHIKV from the virus stock is inoculated into the Vero cells, allowing the infection to propagate and yield a virus sample for subsequent analysis. Step 3: Viral nucleic acid extraction. Viral RNA is isolated from the virus stock using the MolPure Viral DNA/RNA Kit. Step 4: Capillary-modified LAMP reaction. The extracted viral RNA is mixed with the Bst 4.0 LHNB Mix and subjected to amplification in capillaries at 63°C for 25 min using an incubator. The presence of a positive result is indicated by a yellowish-green color

9. Phosphate-buffered saline (PBS)
10. DNase/RNase-free centrifuge tubes
11. MolPure® Viral DNA/RNA Kit
12. HotStartBst 4.0 LHNB Lyo Mix
13. Bst 4.0 LS MasterMix

2.2 Equipment

1. CO_2 incubators
2. Biosafety cabinet class II
3. Liquid nitrogen container
4. Vacuum liquid suction pump
5. High-speed freezing centrifuge
6. Handheld centrifuge
7. Pipettes
8. Autoclave
9. Metal bath
10. Water bath
11. Electrothermal constant temperature incubator
12. Vortex oscillator
13. −80°C freezer

3 Methods

3.1 Cell Thawing

1. Thaw the frozen cells stored in a liquid nitrogen container using a water bath set to 37°C.
2. Once fully thawed, mix the cells with 1 mL of high-glucose DMEM containing 4.5 g/L glucose in biosafety cabinet class II, transfer the mixture to a suitable centrifuge tube, and centrifuge at 1000 ×*g* for 5 min at room temperature. Carefully discard the cell cryopreservation solution.
3. Resuspend the cell pellet in 1 mL of fresh high-glucose DMEM and then add it to the prepared fresh high-glucose DMEM. Mix the cells and incubate them in a 37°C, 5% CO_2 incubator.

3.2 Cell Passage

1. Monitor the cell confluence and passage when it reaches 90%. Vero cells proliferate rapidly, often reaching 100% confluence within 2–3 days; therefore, the typical passage ratio is generally 1:8.
2. Discard the used culture medium, wash the cells three times with PBS preheated to 37°C, remove the PBS, and add 1 mL of trypsin. Incubate the Vero cells in a 37°C incubator for 1–2 min to facilitate digestion.
3. Discard the trypsin, add 2 mL of high-glucose DMEM, and gently pipette to detach all cells from the bottom of the dish, ensuring thorough resuspension.
4. Transfer the resuspended cells into a new high-glucose DMEM according to the desired passage ratio. Mix the cells evenly and incubate them in a 37°C, 5% CO_2 incubator. Regularly monitor cell growth.

3.3 Virus Propagation

1. Culture Vero cells in high-glucose DMEM containing 10% FBS and 1% penicillin–streptomycin to assess cell growth status in biosafety cabinet class II.
2. When cell confluence reaches 80%–90%, add CHIKV 181/25 strain with a multiplicity of infection (MOI) of 1 and culture the cells until cytopathic effect (CPE) is observed under an inverted microscope (*see* **Note 1**).
3. Collect the supernatant 72 h post-infection.
4. Dislodge the infected cells with a small amount of fresh high-glucose DMEM and store them in a sterile centrifuge tube. Freeze overnight at −80°C. After freezing, transfer the tube to a 37°C water bath to thaw.
5. Repeat this freeze–thaw cycle three times.
6. Store the samples in a −80°C freezer, clearly labeling the virus type, infection time, type of cells, and the number of freeze–thaw cycles (*see* **Note 2**).

3.4 Viral Nucleic Acid Extraction using MolPure® Viral DNA/RNA Kit

1. Transfer 200 μL of the virus stock to a 1.5-mL DNase/RNase-free centrifuge tube. If the volume is less than 200 μL, fill it to 200 μL with PBS. Add 400 μL of lysis buffer (LB) provided in the kit and immediately vortex for 15 s using a vortex oscillator. Incubate at room temperature for 10 min.
2. Add 450 μL of ethanol and vortex immediately for 15 s to mix thoroughly.
3. Place RNA adsorption column V1 in a 2-mL collection tube, add the pre-treatment mixture, centrifuge at 12,000 × *g* for 1 min, and discard the waste liquid. Repeat until all the mixture has been processed.
4. Add 500 μL of deproteinizing solution PL, centrifuge at 12,000 × *g* for 30 s, and discard the liquid.
5. Place RNA adsorption column V1 back into the collection tube, add 500 μL of rinsing solution W, centrifuge at 12,000 × *g* for 30 s at room temperature, and discard the liquid.
6. Repeat Step 5.
7. Place RNA adsorption column V1 back into the collection tube and centrifuge at 12,000 × *g* for 2 min at room temperature to remove residual rinsing solution W.
8. Place RNA adsorption column V1 in a new enzyme-free 1.5-mL centrifuge tube, add 30–50 μL of RNase-free H_2O pre-heated to 65°C to the center of the column, and let it stand at room temperature for 1 min.
9. Centrifuge at 12,000 × *g* for 1 min, collecting the filtrate as the RNA sample (*see* **Note 3**).

3.5 Preparation of 25× LAMP Primer

1. Mix 25 μM FIP/BIP primers, 4–8 μM LoopF/B primers and 5 μM F3/B3 primers. Use ddH_2O to dilute and bring the total volume to 0.8 μL.

3.6 Reaction System Configuration

The leuco-hydroxynaphthol blue (LHNB) dye binds to Mg^{2+}, exhibiting a red color initially. As the reaction progresses, positive samples produce PPi, which binds to Mg^{2+} in the system, resulting in a yellow-green color for the LHNB [12]. Mix in a 10-μL total volume: 0.5 μL Bst 4.0 LyoMix A, 3.75 μL Bst 4.0 LyoMix B, 0.4 μL of 25× LAMP Primer Mix (0.12 μL FIP/BIP, 0.02 μL F3/B3, 0.04 μL LoopF/LoopB), 0.2 μL of 50× LHNB Dye, 5 μL of nucleic acid template (*see* **Note 4**).

3.7 Loop-Mediated Isothermal Amplification

1. Extract a capillary tube containing the pre-mixed enzyme reaction system and draw up the sample to be tested (*see* **Note 5**).
2. Apply mineral oil to both ends of the capillary tube and seal it with a capillary glue tip.
3. Place the capillary tube in a constant-temperature incubator at 63°C for 25 min (*see* **Note 6**), observing the color change of the capillary tube during the reaction. HotStart Bst 4.0 LHNB Lyo Mix will change color from violet (negative) to green (positive), while Bst 4.0 LS MasterMix will change from red to yellow.

4 Notes

1. The infected cells reach the CPE stage, characterized by a round, contracted morphology that aggregates into grape-like clusters when observed under an inverted microscope. Ultimately, the single-layer cells rupture and detach, while the high-glucose DMEM turns yellow.
2. The infected cells should be aliquoted and stored after releasing the virus through repeated freeze–thaw cycles at −80°C. Prior to each use, ultrasonic vibration should be performed in an ice-water bath to ensure uniform dispersion of the virus particles.
3. RNA samples extracted from CHIKV should be stored at −20°C for short-term use and −80°C for long-term storage.
4. Aspirate the reactants into the capillary tube, then quickly apply mineral oil to both ends of the tube and seal it with a rubber stopper to prevent aerosol contamination.
5. The optimal ratio of internal to external primers is 1:7, which achieves the highest yield under the same conditions. This ratio also demonstrates good repeatability and specificity.

6. The reaction temperature and time should be precisely controlled at 63°C for 25 min. Excessive temperatures can impact the specificity and yield of the reaction. Higher temperatures may lead to instability in chain binding, reduced Bst enzyme activity, and weakened chain replacement, ultimately resulting in lower yields.

Acknowledgments

This work was supported by grants from Shandong Provincial Natural Science Foundation (ZR2024MH017) and Taishan Scholars Program (tstp20221142).

References

1. Ribeiro Dos Santos G et al (2025) Global burden of chikungunya virus infections and the potential benefit of vaccination campaigns. Nat Med 31(7):2342–2349. https://doi.org/10.1038/s41591-025-03703-w
2. Wang J, Zhang L (2025) The chikungunya virus outbreak in Foshan, China: a rising public health threat in tropical and subtropical regions. J Infect 91(3):106591. https://doi.org/10.1016/j.jinf.2025.106591
3. de Souza WM, Lecuit M, Weaver SC (2025) Chikungunya virus and other emerging arthritogenic alphaviruses. Nat Rev Microbiol 23(9):585–601. https://doi.org/10.1038/s41579-025-01177-8
4. Soroka M, Wasowicz B, Rymaszewska A (2021) Loop-mediated isothermal amplification (LAMP): the better sibling of PCR? Cells 10(8):1931. https://doi.org/10.3390/cells10081931
5. Notomi T et al (2000) Loop-mediated isothermal amplification of DNA. Nucleic acids research 28(12):E63. https://doi.org/10.1093/nar/28.12.e63
6. Song J et al (2016) Instrument-free point-of-care molecular detection of Zika virus. Anal Chem 88(14):7289–7294. https://doi.org/10.1021/acs.analchem.6b01632
7. Kurosaki Y et al (2007) Rapid and simple detection of Ebola virus by reverse transcription-loop-mediated isothermal amplification. J Virol Methods 141(1):78–83. https://doi.org/10.1016/j.jviromet.2006.11.031
8. Parida MM et al (2007) Rapid and real-time detection of Chikungunya virus by reverse transcription loop-mediated isothermal amplification assay. J Clin Microbiol 45(2):351–357. https://doi.org/10.1128/jcm.01734-06
9. Feng J et al (2022) Development of a loop-mediated isothermal amplification method for rapid and visual detection of monkeypox virus. Microbiol Spectr 10(5):e0271422. https://doi.org/10.1128/spectrum.02714-22
10. Baek YH et al (2020) Development of a reverse transcription-loop-mediated isothermal amplification as a rapid early-detection method for novel SARS-CoV-2. Emerg Microbes Infect 9(1):998–1007. https://doi.org/10.1080/22221751.2020.1756698
11. Luan J et al (2025) Development of a capillary-modified naked-eye visual loop-mediated isothermal amplification method for the rapid detection of mpox virus and chikungunya virus. Front Microbiol 16:1615132. https://doi.org/10.3389/fmicb.2025.1615132
12. Zaczek-Moczydłowska MA, et al (2020) A single-tube HNB-based loop-mediated isothermal amplification for the robust detection of the Ostreid herpesvirus 1. Int J Mol Sci 21(18):6605. https://doi.org/10.3390/ijms21186605

Chapter 3

RT-RPA-Assisted CRISPR/Cas12a-Based Isothermal Detection of Chikungunya Virus

Pooja Bhardwaj, Gaurav Raj Dwivedi, Hari Shanker Joshi and Rajeev Singh

Abstract

Chikungunya virus (CHIKV) is transmitted through the bite of *Aedes* mosquitoes, specifically *A. aegypti* and *A. albopictus*. CHIKV belongs to the alphavirus with a positive-sense ssRNA genome of 11–12 kb size. The virus has been reported from various geographical regions across the globe. Chikungunya fever is an acute febrile illness, which, if left untreated, may develop into chronic arthralgia that may persist for several months or acute encephalitis syndrome. Therefore, early diagnosis of CHIKV is crucial to initiate prompt supportive treatment. Laboratory diagnosis of CHIKV typically relies on serological tests such as IgM antigen capture ELISA and molecular methods including RT-PCR or qRT-PCR. However, both these methods are not viable in peripheral settings. This chapter highlights recent advancements in molecular detection techniques for CHIKV, specifically isothermal detection methods that eliminate the requirement for complex instruments. The detection is facilitated by RT-RPA and CRISPR/Cas12a endonuclease. The assay offers advantages over existing methods such as rapid and early detection, and eliminates cross-over contamination, ultra-sensitivity, high specificity, and ease of result interpretation.

Keywords Chikungunya virus, Molecular detection, Isothermal amplification, CRISPR/Cas12a, RT-RPA, Biosensors

1 Introduction

Chikungunya (CHIK) fever caused by chikungunya virus (CHIKV) is a disease of global health concern. It is a mosquito-borne viral disease, transmitted through the bites of infected female mosquitoes. The most common vectors are *Aedes* mosquitoes, including *A. aegypti* and *A. albopictus*. These two vectors also transmit other mosquito-borne viruses, such as dengue virus (DENV) and zika virus (ZKV). CHIKV is one of the silent attackers that caused sudden outbreaks after three decades of protracted periods of silence [1, 2]. Since 2004, there has been an increase in the frequency and geographical extension of CHIK outbreaks, primarily due to viral modifications that facilitate easier transmission of the virus by *A. albopictus* mosquitoes [3]. As

Claudia Filomatori and Eliana F. Castro (eds.), *Chikungunya Virus*, Methods in Molecular Biology, vol. 3067, https://doi.org/10.1007/978-1-0716-5496-5_3,

of November 2024, approx. 480,000 cases of CHIK were reported worldwide across 23 countries [4].

The clinical symptoms of CHIK overlap with those of other viral infections (dengue and zika), and hence, differential diagnosis of CHIK is crucial where arboviruses are in co-circulation [5, 6]. The undiagnosed CHIK can result in high morbidity caused due to arthralgia and arthritis. The laboratory diagnosis of CHIK, including virus isolation, serological methods such as enzyme-linked immunosorbent assay (ELISA), and molecular methods such as reverse transcription-polymerase chain reaction (RT-PCR) or real-time RT-PCR (qRT-PCR), consists of time-consuming procedures requiring sophisticated instrument, good lab infrastructure, and personnel with expertise and hence restricting their usage to specialized laboratory only [7, 8]. Further, CHIK-specific IgM antibody detection is not suitable for acute phase samples (<7 days after the onset of illness) [9, 10]. The lack of molecular point-of-care tests contributes to underdiagnosis and delayed treatment of CHIK. Therefore, developing a rapid, sensitive, and easy-to-interpret test is crucial for early detection and timely treatment of CHIKV in clinical cases.

Isothermal amplification methods such as recombinase polymerase amplification (RPA), loop-mediated isothermal amplification (LAMP), and helicase-dependent amplification (HDA) have provided a viable option for field diagnostics [11]. RPA introduced in 2006, enables nucleic acid amplification at a constant temperature (37°C–42°C) within 15–20 min, without the need for sophisticated equipment [12]. Further, the RPA can be modified to RT-RPA with reverse transcriptase for detection of pathogens having RNA genome [13, 14].

The Clustered Regularly Interspaced Short Palindromic Repeat/CRISPR-associated endonuclease (CRISPR/Cas) system, an immune defense system in bacteria and archaea, is being repurposed for nucleic acid detection. When integrated with RT-RPA, CRISPR/Cas12a can detect low genome copies at slightly raised ambient temperatures [11, 15, 16]. CRISPR/Cas12a recognizes targets directed by crRNA and a specific PAM sequence (5′-TTTV), leading to cleavage of both double-stranded DNA (dsDNA) target and single-stranded nontarget DNA (labelled ssDNA reporters) [15]. This cleavage produces a detectable fluorescent signal or lateral flow strip readout, eliminating the requirement for complex instruments like qPCR or PCR, expanding CRISPR/Cas technology's potential for rapid diagnostics. This chapter describes the use of RT-RPA and CRISPR/Cas12a technology for the diagnosis of CHIKV using primers and crRNA targeting the E1 structural gene.

2 Materials

Treat all biological specimens as potentially infectious, ensuring they are handled in line with the relevant biosafety standards. Use ultrapure nuclease-free water and molecular-grade reagents for preparing all solutions.

2.1 Viral RNA Extraction

1. For detection of chikungunya virus in clinical samples including serum, blood, or cerebrospinal fluid (CSF), we recommend using Purelink or Qiagen Viral RNA extraction kit or equivalent. The manufacturer's instructions for the above-mentioned extraction kits are sufficient.
2. RNaseZap Wipes or spray and DNAZap.

2.2 Oligonucleotide Sequences

1. RPA primer sequences:
   - CHIKV forward primer: 5′-GTAACTGCCTATGCAAACG GCGACCATGC-3′.
   - CHIKV reverse primer: 5′-CCAGTTGTGTATTAGC ATAGACGTC-3′ (*see* **Note 1**).
2. crRNA: 5′-UAAUUUCUACUAAGUGUAGAUGCGAUAU CCAAAGUCGCACAC-3′ (*see* **Note 2**).
3. ssDNA reporter: 5′-6-FAM-TTATTATT-BHQ-3′ (*see* **Note 3**).

2.3 Synthesis of RNA by In Vitro Transcription

1. T7 High Yield Transcription Kit (Thermo Scientific, USA) or any commercially available *in vitro* transcription (IVT) kit.
2. Qubit4 fluorometer.
3. Recombinant plasmid containing target gene or PCR product of CHIKV.
4. 5 M ammonium acetate.
5. 100% ethanol.
6. 70% ethanol.
7. Nuclease-free water.

2.4 Recombinase Polymerase Amplification

1. TwistAmp® Basic Recombinase Polymerase Amplification Kit (TwistdX, USA) or any available RPA for isothermal amplification.
2. Reverse transcriptase (RT) enzymes; any Moloney Murine Leukemia Virus (MmLV) variant.
3. Sterile 0.2 mL tubes.
4. Water bath/dry bath, 39°C.
5. Positive control (PC) synthetic CHIKV RNA (*see* **Note 4**).
6. Negative control (NC) (*see* **Note 4**).

7. Patients or clinical samples.

2.5 CRISPR/Cas12a-Based Detection

1. LbCas12a enzyme (NEB) or any other commercially available (*see* **Note 5**).
2. 10× r2.1 NEB buffer.
3. crRNA.
4. ssDNA reporters.
5. HybriDetect Lateral flow detection strips (MGHD1) (Milenia Biotec, Germany) or any other compatible strips or in-house developed lateral flow detection strips.
6. UV-transilluminator.

2.6 Other Commonly Required Material

1. Micropipettes.
2. Sterile micropipette tips.
3. Sterile tubes.
4. Microfuge.
5. Mini cooler: 0°C.

3 Methods

3.1 Sample Preparation

All procedures should be carried out at room temperature unless otherwise specified following biosafety guidelines handling clinical specimens [17]. Clean the working platform and micropipettes with RNaseZap Wipes and DNAZap to remove all potential RNase and DNA contamination.

1. For detection of Chikungunya virus in clinical samples (serum, blood, or cerebrospinal fluid), we recommend Purelink or Qiagen Viral RNA extraction kit or equivalent. The manufacturer's instructions for the above-mentioned extractions are sufficient.
2. Follow the method as described in the kit insert (*see* **Notes 6 and 7**).
3. Store extracted RNA at −80°C (*see* **Note 8**).

3.2 Preparation of Positive RNA Control with In Vitro Transcription

1. Linearize the cloned plasmid using the appropriate restriction enzyme at 37°C overnight.
2. Analyze the digested linearized plasmid in 0.8% agarose gel electrophoresis at 5 V/cm and excise the desired band from the gel under the UV transilluminator.
3. Purify the PCR product by using the QIAquick® Gel Purification Kit (Qiagen) or any other equivalent kit. Follow the method as described in the kit insert.

4. Anneal the linearized plasmid to T7 primer in a 1:1 ratio by heating at 90°C for 5 min and then slowly cooling down the reaction to room temperature.
5. Perform the IVT steps by using the T7 High Yield Transcription Kit (Thermoscientific, USA). Follow the method as described in the kit insert.
6. Add annealed template:T7 primer complex to transcription reaction mixture containing 4 μL of 5× TranscriptAid reaction buffer, 8 μL of rNTPs (10 mM ATP, CTP, GTP, UTP), 2 μL of TranscriptAid enzyme mix, and nuclease-free water to total volume of 20 μL.
7. Mix thoroughly, spin briefly to collect all drops, and incubate at 37°C overnight.
8. Purify the generated transcripts using RNA Clean & Concentrator™-5 (Zymo Research, USA) or any other compatible kit according to the manufacturer's protocol (*see* **Note 9**).
9. Quantify the RNA concentration with a Qubit4 fluorometer (Thermo Scientific, USA).
10. Adjust the concentrations of template IVT-RNA for RT-RPA reactions to 100 ng/μL with DEPC-treated water and store at −80°C for long-term use.

3.3 CRISPR/Cas12a:CrRNA Hybrid Reaction Setup

1. Thaw all frozen reaction components, briefly centrifuge to collect all drops, and keep them on ice.
2. Keep the LbCas12a enzyme on ice.
3. Keep the ssDNA reporter covered with foil to protect it from light exposure.
4. Prepare the master mix by adding the components as described in Table 1 to a 1.5-mL tube in top-down order (with reporters in the tube lid) (Fig. 1). The master mix given in Table 1 is enough for one sample. Scale up proportionally for the number of samples (*see* **Note 10**).
5. Allow it to incubate at room temperature (25°C–32°C) or 25°C for 10–15 min (*see* **Note 11**).
6. Briefly spin, vortex, and quickly spin down (using a microfuge for 3–5 s at room temperature).

3.4 Isothermal Amplification with RT-RPA

1. Prepare the pre-amplification area by wiping down the work surface and pipettors with RNase Away (*see* **Notes 12 and 13**).
2. On the ice, prepare the master mix as described in Table 2 (sufficient for ten replicates) in a 0.5-/1.5-mL tube.
3. Add 37.5 μL of master mix to a single lyophilized pellet aliquot and carefully resuspend the mixture on ice. Transfer the entire

Table 1
Master mix A (at room temperature)

S. No.	Components	Volume	Final concentration
1	10× 2.1 buffer	1.0 μL	2×
2	1 μM LbCas12a	0.5 μL	0.2 μM
3	crRNA	0.75 μL	20 ng/μL
4	50 μM ssDNA reporters	0.4 μL	4 μM
5	Nuclease-free water	Up to 5.00 μL	

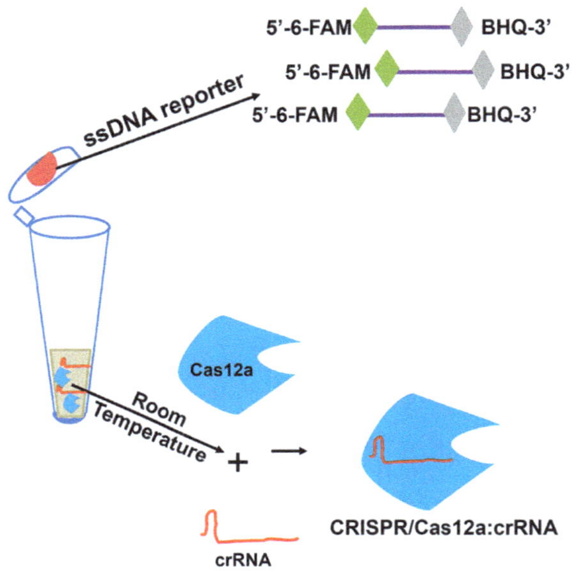

Fig. 1 Assembly of CRISPR/Cas12a:crRNA hybrid

reconstituted reaction back to the initial Eppendorf or PCR master mix tube (*see* **Note 14**).

4. Prepare aliquots of 3.75 μL of the reconstituted reaction in PCR strip tubes or PCR plates, or PCR tubes (*see* **Note 15**).
5. Add 0.3 μL of 280 mM MgOAc to the sidewall of the reaction tube (*see* **Note 16**).
6. Add 1 μL of template RNA/extracted RNA from clinical samples to the reaction mix and carefully mix by pipetting up and down.
7. Vortex briefly and quickly spin down (using a microfuge for 3–5 s at room temperature).

Table 2
Master mix B (at 4°C)

S. No.	Components	Volume
1	Rehydration buffer	29.5 μL
2	10 μM forward primer	2.4 μL
3	10 μM reverse primer	2.4 μL
4	RT enzyme	1.0 μL
5	Nuclease-free water	2.2 μL
	Total	37.5 μL
Add this to a lyophilized pellet or resuspend the pellet into the mix. Distribute 3.75 μL to each tube (× 10) at the bottom		
6	280 mM MgOAc	0.3 μL
7	Template	1.0 μL

8. Carefully open the RT-RPA reaction tubes to avoid aerosol generation and load 5 μL of CRISPR/cas12a:crRNA preassembled complex in the tube lid (*see* **Note 17**).
9. Allow the reaction to incubate for 20 min at 39°C in a PCR thermocycler/incubator/waterbath/dry bath.
10. Intermittent pause.
11. Briefly centrifuge, tap mix or vortex, and quickly spin down to ensure both Mix A and Mix B are mixed to remove potential bubbles and initiate CRISPR/Cas12a:crRNA-directed fluorescence detection of CHIKV (Fig. 2).
12. Resume further incubation at 39°C for another 10–15 min (*see* **Notes 18 and 19**).
13. After the completion of incubation time, observe the tubes under the UV transilluminator and capture the resulting photograph (Fig. 3).

3.5 Result Interpretation

The obtained cumulative fluorescence is analyzed and determined according to the following criteria (*see* **Note 20**):

1. Negative determination criteria: The amount of fluorescence is less than or equal to the amount of negative control fluorescence.
2. Positive determination standard: The amount of fluorescence is greater than the amount of negative control fluorescence.

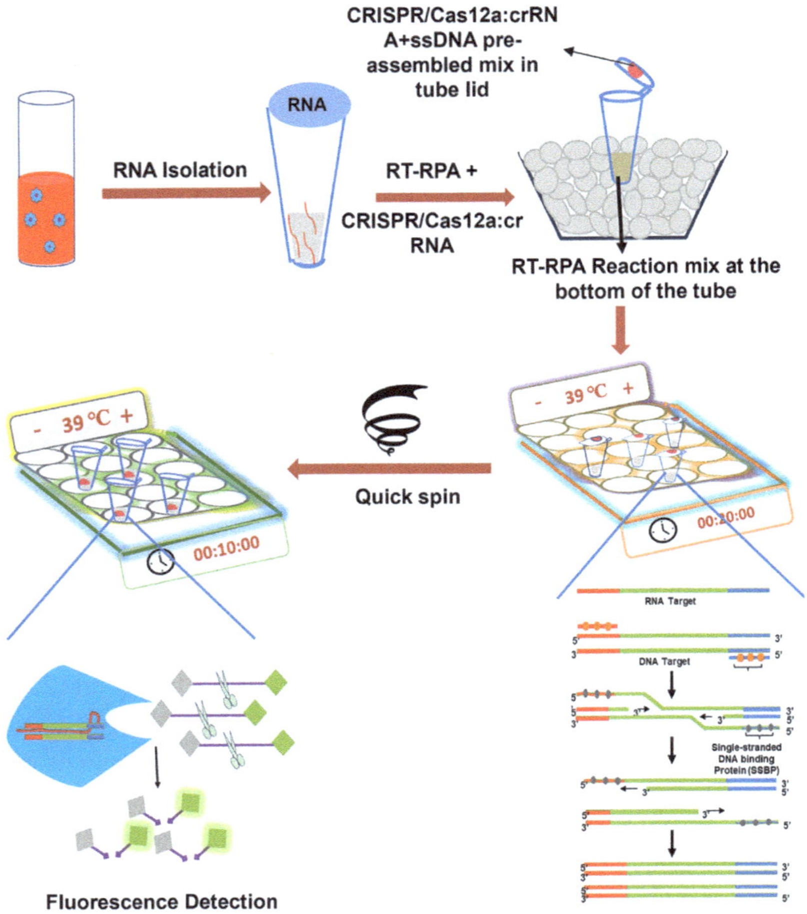

Fig. 2 Flow diagram showing setup of RT-RPA reaction mix and CRISPR/Cas12a:crRNA-assisted detection of CHIKV target.

3. Invalid results: The amount of fluorescence in negative control is greater or equal to the amount of positive control fluorescence.

4 Notes

1. The complete E1 gene sequences of CHIKV genotypes (ECSA, West African, and Asian) are used for RPA primer and crRNA design. The potential primers are selected by identifying highly conserved regions among different CHIKV genotypes by aligning their E1 gene sequences using the ClustalW program in MEGA-X software [18]. BLAST analysis is used to assess the specificity of the selected primers

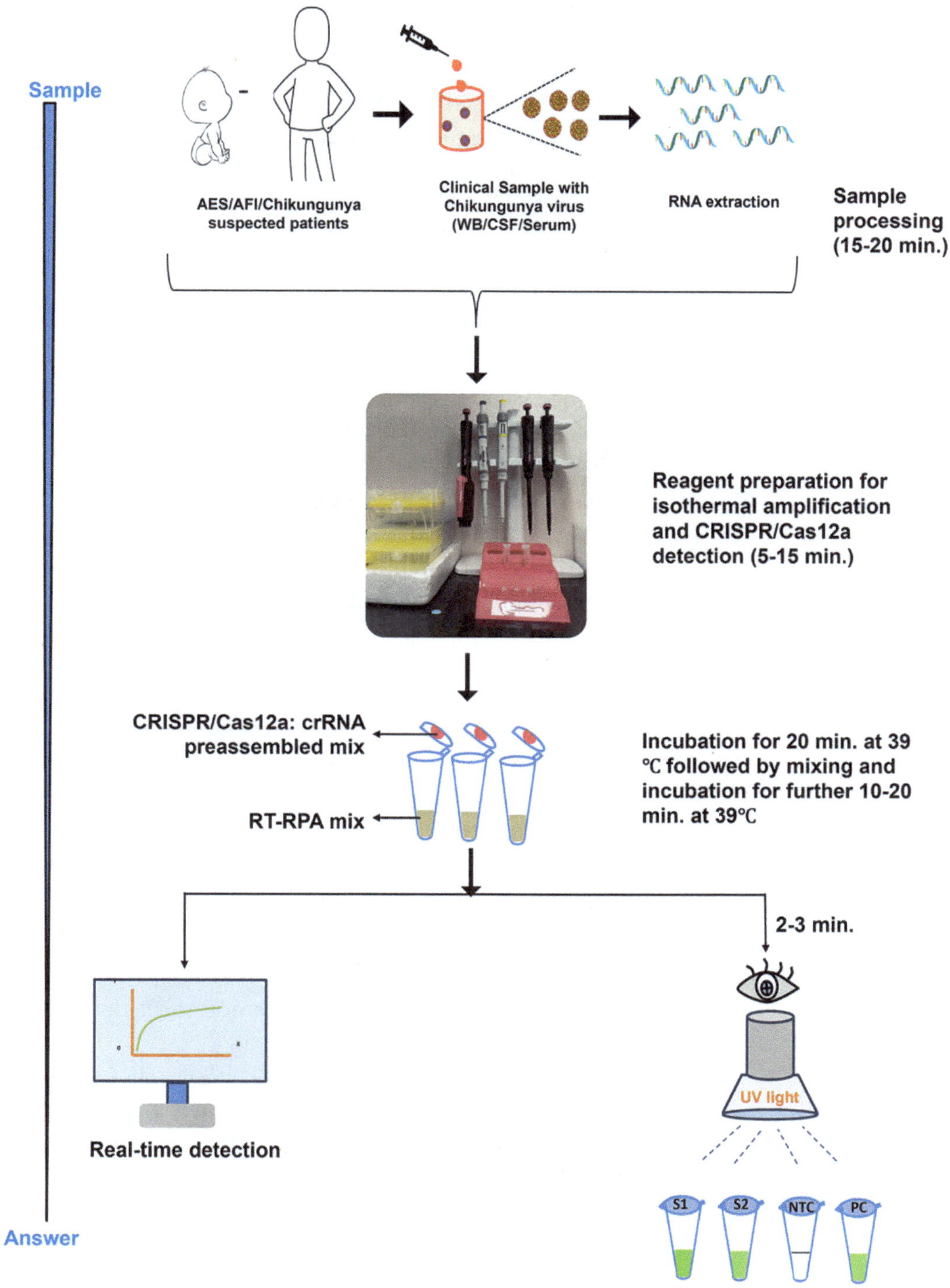

Fig. 3 Flow diagram showing turnaround time for RT-RPA-assisted CRISPR/Cas12a:crRNA-based fluorescent detection of CHIKV target

and crRNA by comparing them to the NCBI nucleotide database. Researchers can opt to design their own RPA primer sequences. Ideally, an amplicon length for RPA should be 200–400 bp for optimal isothermal amplification.

2. crRNA was designed corresponding to the target region amplified with RT-RPA using the designed primer pairs with online Benchling software [19]. The designed crRNAs were commercially synthesized as ssRNA.
3. ssDNA reporters for detection of CHIKV were commercially synthesized and HPLC purified for fluorescence-based detection (5′-FAM-TTATTAT-BHQ-3′).
4. Positive control (PC) consists of RNA that has been *in vitro* transcribed using a recombinant plasmid containing complete CDS of the target E1 gene of CHIKV. Negative control (NC) is DNA extracted from a person not exposed to CHIKV.
5. LbCas12a can also be expressed and purified in-house [20].
6. For the spin-column-based RNA extraction kit, incubation at 56°C in a water bath after elution buffer addition for 5 min before elution helps increase the product yields.
7. DNA contaminants can be checked in RNA transcripts by subjecting it to RT-RPA analysis without reverse transcriptase enzyme. RNA transcripts containing DNA will yield PCR product, whereas pure IVT RNA transcripts will only yield PCR product in the presence of reverse transcriptase enzyme.
8. Quantify RNA immediately after purification. Multiple freeze–thawing cycles may degrade RNA.
9. Alternatively, phenol (pH 4.7): chloroform extraction and ethanol precipitation of RNA transcripts can also be performed [21].
10. Handle gently to avoid mixing reporters in the mix.
11. Exceeding the incubation time does not affect the resulting outcome.
12. RT-RPA is an isothermal amplification technology that produces large quantities of amplified products within minutes. This could result in cross-over contamination of both reagents and the work area, potentially causing false-positive signals. To prevent this, it is strongly advised to establish separate work areas for pre- and post-amplification. Laminar flow hoods, commonly used in standard tissue culture, offer an ideal environment for pre-amplification work.
13. The use of primer aliquots, rather than stocks, is recommended as another measure to prevent contamination spread.
14. One pellet yields approximately ten individual RPA reactions. If a larger number of reactions are needed, scale up the

master mix volume and pellets accordingly. When doing so, it is essential not to switch pipettes between transfers of resuspended pellets, as the sticky nature of the RT-RPA reaction components could lead to significant loss of reaction components if multiple pipette tips are used.

15. We recommend using a PCR strip with an attached cap for multiple reactions or PCR tubes for the low number of reactions to avoid crossover contamination.

16. The addition of 280 mM MgOAc to the reconstituted pellet activates the enzymes and may lead to strong primer–dimer formation. It is therefore important to either quickly proceed to sample addition or mix after template addition.

17. Gently close the cap to avoid mixing the two reactions A and B.

18. Alternatively, the plate can be incubated into a 39°C preheated BioTek plate reader, or equivalent for on-time screening of results by monitoring fluorescence wavelength at excitation (λ_{ex}) and emission wavelength (λ_{em}) set to 495 and 520 nm respectively (over 30 min at 39°C with a 1-min interval between well-data).

19. In this case, following plate preparation, quickly place it into the fluorescence plate reader because the CRISPR/Cas12a reaction may begin as soon as mixed with the RT-RPA reaction mix.

20. Here, we provide a protocol for single-target identification using a two-step method. For one-pot chikungunya detection, follow the instructions provided. We also recommend strictly performing positive and negative controls alongside tested samples. A negative control could include a water-only input sample for RT-RPA or a sample in which the target molecule is known to be absent. Positive controls include synthetic RNA or RNA in which the target is known to be present.

References

1. Lahariya C, Pradhan SK (2006) Emergence of chikungunya virus in Indian subcontinent after 32 years: a review. J Vector Borne Dis 43(4):151–160
2. Bartholomeeusen K, Daniel M, LaBeaud DA et al (2023) Chikungunya fever. Nat Rev Dis Primer 9:17. https://doi.org/10.1038/s41572-023-00429-2
3. Chikungunya facts. https://www.who.int/news-room/fact-sheets/detail/chikungunya#:~:text=Chikungunya%20virus%20is%20transmitted%20by,bite%20primarily%20during%20daylight%20hours. Accessed 7 Jun 2023
4. Chikungunya virus disease case notification rate per 100 000 population, December 2023-November 2024. https://www.ecdc.europa.eu/en/publications-data/chikungunya-virus-disease-case-notification-rate-100-000-population-december-2023. Accessed 20 Dec 2024

5. Alves ADR, Raposo JV, De Sousa RMP et al (2022) Beyond arboviruses: a multicenter study to evaluate differential diagnosis of rash diseases and acute febrile illness cases in Rio de Janeiro, Brazil. PLOS ONE 17:e0271758. https://doi.org/10.1371/journal.pone.0271758
6. Mori A, Pomari E, Deiana M et al (2021) Molecular techniques for the genomic viral RNA detection of West Nile, Dengue, Zika and Chikungunya arboviruses: a narrative review. Expert Rev Mol Diagn 21:591–612. https://doi.org/10.1080/14737159.2021.1924059
7. Natrajan MS, Rojas A, Waggoner JJ (2019) Beyond fever and pain: diagnostic methods for chikungunya virus. J Clin Microbiol 57:e00350-19. https://doi.org/10.1128/JCM.00350-19
8. Pezzi L, Reusken CB, Weaver SC et al (2019) GloPID-R report on Chikungunya, O'nyong-nyong and Mayaro virus, part I: Biological diagnostics. Antiviral Res 166:66–81. https://doi.org/10.1016/j.antiviral.2019.03.009
9. Lanciotti RS, Kosoy OL, Laven JJ et al (2007) Chikungunya virus in US travelers returning from India, 2006. Emerg Infect Dis 13:764–767. https://doi.org/10.3201/eid1305.070015
10. Andrew A, Navien TN, Yeoh TS et al (2022) Diagnostic accuracy of serological tests for the diagnosis of Chikungunya virus infection: a systematic review and meta-analysis. PLoS Negl Trop Dis 16:e0010152. https://doi.org/10.1371/journal.pntd.0010152
11. Bhardwaj P, Kant R, Behera SP et al (2022) Next-generation diagnostic with CRISPR/Cas: beyond nucleic acid detection. Int J Mol Sci 23:6052. https://doi.org/10.3390/ijms23116052
12. Piepenburg O, Williams CH, Stemple DL, Armes NA (2006) DNA detection using recombination proteins. PLoS Biol 4:e204. https://doi.org/10.1371/journal.pbio.0040204
13. Bhardwaj P, Gulafshan S, Singh R (2024) A rapid, specific and ultrasensitive detection of the Chikungunya virus based on RT-RPA:CRISPR/Cas12a one-pot dual mode end-point detection system. Anal Chim Acta 343221. https://doi.org/10.1016/j.aca.2024.343221
14. Bhardwaj P, Dhangur P, Kalichamy A, Singh R (2025) RT-RPA assisted CRISPR/Cas12a based one-pot rapid and visual detection of the pan-dengue virus. J Med Virol 97:e70219. https://doi.org/10.1002/jmv.70219
15. Chen JS, Ma E, Harrington LB et al (2018) CRISPR-Cas12a target binding unleashes indiscriminate single-stranded DNase activity. Science 360:436–439. https://doi.org/10.1126/science.aar6245
16. Kellner MJ, Koob JG, Gootenberg JS et al (2019) SHERLOCK: nucleic acid detection with CRISPR nucleases. Nat Protoc 14:2986–3012. https://doi.org/10.1038/s41596-019-0210-2
17. Bayot ML, Zubair M, Limaiem F (2025) Biosafety guidelines. In: StatPearls. StatPearls Publishing, Treasure Island (FL)
18. Kumar S, Stecher G, Li M et al (2018) MEGA X: molecular evolutionary genetics analysis across computing platforms. Mol Biol Evol 35:1547–1549. https://doi.org/10.1093/molbev/msy096
19. Benchling, https://www.benchling.com/crispr
20. Mendoza-Rojas G, Sarabia-Vega V, Sanchez-Castro A et al (2021) A low-cost and open-source protocol to produce key enzymes for molecular detection assays. STAR Protoc 2:100899. https://doi.org/10.1016/j.xpro.2021.100899
21. TranscriptAid_T7_High_Yield_Transcriptionkit, https://assets.thermofisher.com/TFS-Assets/LSG/manuals/MAN0012652_TranscriptAid_T7_High_Yield_Transcription_UG.pdf. Accessed 14 Feb 2025

Chapter 4

Quantification of Chikungunya Virus RNA by Real-Time RT-PCR Assay

Xipeng Yan, He Xie, Jujun Sun and Limin Chen

Abstract

RT-qPCR is characterized by high sensitivity, strong specificity, excellent reproducibility, and operational simplicity, enabling rapid quantification of viral RNA within a short timeframe. The real-time PCR system relies upon the measurement of a fluorescent reporter during PCR, in which the amount of emitted fluorescence is directly proportional to the amount of the PCR product in a reaction. Here, we describe the use of TaqMan® probe-based one-tube Real-time reverse transcription quantitative polymerase chain reaction (RT-qPCR) for the detection and quantification of the Chikungunya virus (CHIKV) RNA.

Keywords CHIKV, RNA, Quantification, TaqMan ®, RT-qPCR

1 Introduction

Chikungunya virus (CHIKV) is a small (~70 nm in diameter), enveloped, mosquito-borne Alphavirus within the family *Togaviridae*. The CHIKV genome is approximately 11,800 nucleotides in length, constituting a positive-sence single-stranded RNA. It is primarily transmitted to humans through bites of *Aedes* spp. mosquitoes, causing Chikungunya fever [1, 2]. The disease presents with an abrupt onset of fever, arthralgia (joint pain), rash, and headache. Although most cases are mild and the disease is resolved spontaneously, it leads to chronic arthritis persisting for months or even years in some cases [3]. Since 2004, CHIKV has spread globally, posing a severe public health threat. Thus, rapid and accurate detection of CHIKV is critical for diagnosis, surveillance, and disease outbreak control [4, 5].

Viral RNA quantification is essential for deciphering viral infection dynamics, evaluating antiviral therapeutic efficacy, and investigating viral evolution [6]. In the context of CHIKV infection, RNA quantification enables the quantification of viral load in the blood of patients, facilitating the assessment of disease severity and prediction of clinical outcomes. Furthermore, viral RNA quantification plays an

Claudia Filomatori and Eliana F. Castro (eds.), *Chikungunya Virus*, Methods in Molecular Biology, vol. 3067,
https://doi.org/10.1007/978-1-0716-5496-5_4,

important role in mosquito vector surveillance. It is used to assess infection rates and transmission potential within mosquitoes, thereby informing the development of targeted vector control strategies [7].

Real-time reverse transcription quantitative polymerase chain reaction (RT-qPCR) is currently one of the most widely used methods for viral RNA quantification. RT-qPCR is characterized by high sensitivity, strong specificity, excellent reproducibility, and operational simplicity, enabling rapid determination of viral RNA within a short timeframe [8]. RT-qPCR involves reverse transcribing viral RNA into complementary DNA (cDNA) using reverse transcriptase, followed by PCR amplification with specific primers and fluorescent probes or DNA-binding dyes. The changes in fluorescence signal are monitored in real-time, thereby facilitating the quantification of viral RNA. RT-qPCR has been extensively utilized across multple aspects of CHIKV infection management, including diagnosis, viral load monitoring, antiviral drug screening, and vaccine development/evaluation [9].

This protocol aims to provide researchers with a standardized One-Step RT-qPCR method for the quantitative detection of CHIKV RNA. The protocol includes the following steps: sample collection and processing, RNA extraction, setup of the One Step RT-qPCR reaction mixture, configuration of the PCR amplification program, and data analysis. By following this protocol, researchers can obtain reliable and accurate quantitative data for CHIKV RNA, providing strong support for CHIKV-related research and disease management.

2 Materials

1. Virus RNA extraction
   Nucleic Acid Extraction (DNA/RNA) or Purification Kit (Magnetic beads method) (Cat# S10012E, Sansure Biotech, China)
2. Blood sample or CHIKV-infected cell culture supernatant
3. PCR primer sequences
   CHIKV forward primer (5′CATCTGCACYCAAGTGTAC-CA3′)
   CHIKV reverse primer (5′GCGCATTTTGCCTTCG-TAATG3′)
4. CHIKV-Probe sequence (5′ GCGGTGTACACTGCCTGTGACYGC3′)
   Fluorescence labeling: 5′TAMRA, 3′BHQ2
5. CHIKV ssDNA sequence:
   5′CACAACATCTGCACCCAAGTGTACCACAAAAG-TATCTCCAGGCGGTGTACACTGCCTGTGACCGC-CATTGTGTCATCGTTGCATTACGAAGGCAAAATG-CGCACTAC3′
6. RT-qPCR reagent: One Step Prime Script III RT-qPCR Mix

3 Methods

3.1 Primer and ssDNA Design

1. This quantitative TaqMan probe-based assay uses primers and probes designed against the conserved genomic regions of CHIKV (*nsP2* gene).
2. The full-length sequence of CHIKV is obtained from NCBI, and the conserved regions of the viral sequence are determined through sequence comparison (*see* **Note 1**) (https://www.ncbi.nlm.nih.gov/nucleotide/).
3. Primers are designed using Primer Premier 5.0 software.
4. After synthesis, verification is performed: gradient annealing temperature test (55°C–68°C), single peak analysis of melting curve, and amplification efficiency 90%–110%.
5. Each probe is located immediately downstream of the primer binding region (with a 1–5 bp interval), The probes are 20–30 bp in length, and have a T_m value 5–10 °C higher than that of the corresponding primers [10]. The 5′ end is labeled with a reporter group (TAMRA), and the 3′ end is labeled with a quencher group (BHQ2).

3.2 Sample Preparation

1. Collect blood samples using EDTA-K_2 vacuum blood collection tubes with an inert fel separator for nucleic acid testing.
2. Centrifuge within 4 h of collection at the speed of 3000 ×*g* for 10 min. Store at 4°C if RNA extraction will be performed within 48 h. Otherwise, store plasma at −70°C or below (ultra-low-temperature freezer) until RNA is extracted. Avoid repeated freeze–thaw cycles (≤3 cycles recommended to prevent RNA degradation).
3. For high-throughput sample testing, the blood samples can be pooled into mini pools before the extraction process. Pooled testing reduces costsbut yields inferior analytical sensitivity compared with individual sample testing.

3.3 Virus RNA Extraction

The sample (300 μL plasma or CHIKV-infected cell culture supernatant) containing target nucleic acids undergoes cell lysis using a lysis buffer, followed by specific recognition and high-affinity binding of nucleic acids with magnetic beads. Subsequently, a magnetic separator immobilizes the bead-nucleic acid complexes onto the tube wall. Through sequential washing, elution, and purification, high-purity nucleic acids are obtained. Follow the protocol as described in the kit insert (Cat# S10012, Sansure Biotech, China) (*see* **Note 2**). All solutions and buffers indicated below are included in this kit.

1. Equilibrate all components at room temperature for 30 min, then mix thoroughly before use. If an internal standard is required, add it to Extraction Solution 1.

2. Prepare 1.5 mL microcentrifuge tubes according to the number of samples and label them. Add 300 μL of Extraction Solution 1 (Lysis Buffer) to each tube.
3. Add 300 μL of the test sample to each tube. Vortex for 10 s, then centrifuge briefly.
4. Add 100 μL of Extraction Solution 2 (Magnetic Beads) to each tube. Vortex for 10 s to mix thoroughly, then incubate at room temperature for 30 min.
5. Centrifuge briefly. Place the tubes on the magnetic separator. After 3 min, slowly aspirate and discard the supernatant, taking care not to disturb the magnetic bead pellet.
6. Wash and Purify: Add 600 μL of a mixture of Extraction Solution 3 and Extraction Solution 4 to each tube. Vortex to mix for 5 s, centrifuge briefly, then return the tubes to the magnetic separator.
7. After 3 min, the supernatant will separate into two layers. Insert the pipette tip to the bottom of the tube and slowly aspirate and discard the liquid from the bottom layer. Wait 1 min, then aspirate and discard any remaining liquid at the bottom of the tube completely.
8. Add 30 μL of Elution Buffer to each tube to resuspend the beads from the tube walls. Pipette up and down three to four times to mix. Incubate at room temperature for 10 min. Place the tubes back on the magnetic separator for 3 min. Carefully aspirate the nucleic acid eluate from the bottom of the tube and transfer it to a new microcentrifuge tube.
9. Store extracted RNA at −70°C (*see* **Note 3**).

3.4 Absolute Quantification of Virus RNA

One Step PrimeScript™ III RT-qPCR Mix is a specialized kit designed for one-step Real-Time RT-qPCR using the probe-based method (5′ nuclease assay). The reverse transcription (RT) reaction and quantitative PCR (qPCR) occur sequentially in a single reaction tube, ensuring a streamlined workflow and simplified operation.

1. Perform one-step probe-based real-time RT-PCR by using One Step PrimeScript™ III RT-qPCR Mix.
2. Prepare each PCR reaction in a 20 μL reaction volume containing 2 μL of RNA extraction, 10 μL of 2 × One Step PrimeScript III RT-qPCR Mix, 0.4 μL(10μM) of probe, 0.4 μL(10μM) of each primer (CHIKV forward primer and CHIKV reverse primer), and 6.8 μL ultrapure water. The thermal profile consists of a 5-min reverse transcription step at 52°C, 10 s of polymerase activation at 95°C, followed by 40 cycles of PCR at 95°C of denaturation for 5 s, 60°C of annealing for 30 s. The melt curve analysis is carried out from 65°C to 95°C with an increment

of 0.5°C for 5 s. Include a no-template control (NTC) and a positive control (CHIKV) for each PCR run (*see* **Note 4**).

3. After PCR, generate a standard curve by plotting a threshold cycle (Cq)/crossing points of different standard dilutions against log of ssDNA copy number (see Sect. 3.5).
4. CHIKV RNA quantification of an unknown sample: The RNA load of the unknown sample can be extrapolated from the standard curve by the Cq value (Fig. 1).

3.5 Construction of the Standard Curve

1. Synthesize in vitro the target ssDNA fragment of CHIKV, measure its concentration, and convert it to copy number using the formula: Copies/μL = [DNA concentration (μg/μL) × 6.02 × 10^{17}]/(transcript length × 330). Perform tenfold serial dilutions $10^8 \rightarrow 10^1$ copies/μL (three replicate wells per dilution) (*see* **Note 5**).
2. Establish the linear range by fitting a best-fit line to the data by regression analysis and include the range where the value R^2 for this line is ≥0.99. To establish the 95% LLOD, test on a single run ten replicates of five twofold dilutions extending from, and including, the lower limit of the linear range. Calculate the 95% LLOD using probit analysis.

4 Notes

1. The *nsP2* gene sequence of the Chikungunya virus strain S27-African prototype (GenBank accession no. AF369024.2) is used for primer and probe design. The potential primers are selected by identifying highly conserved regions in nsP2 among 20 different CHIKV genotypes of four genetic lineages of CHIKV using the Primer Premier software. The specificity of selected primers and probes is checked by BLAST analysis against the NCBI nucleotide database of CHIKV and related alphaviruses such as Ross River virus (9057 strain, GenBank accession no. GQ433358.1), Semliki Forest virus (NIV8913590 strain, GenBank accession no. MH426977.1), and Sindbis virus (SE_JA220877_2022 strain, GenBank accession no. PP879160.1). Other than the primers and probe sequences listed here, other published primers can also be used in this PCR assay. Researchers can opt to design their own primer sequences.
2. All reagent components should be equilibrated to room temperature for 30 min and mixed thoroughly before use. RNase-free pipette tips must be used during operations to prevent RNase contamination. Sample loading via elution is recommended. For fluorescence quantification, PCR instruments

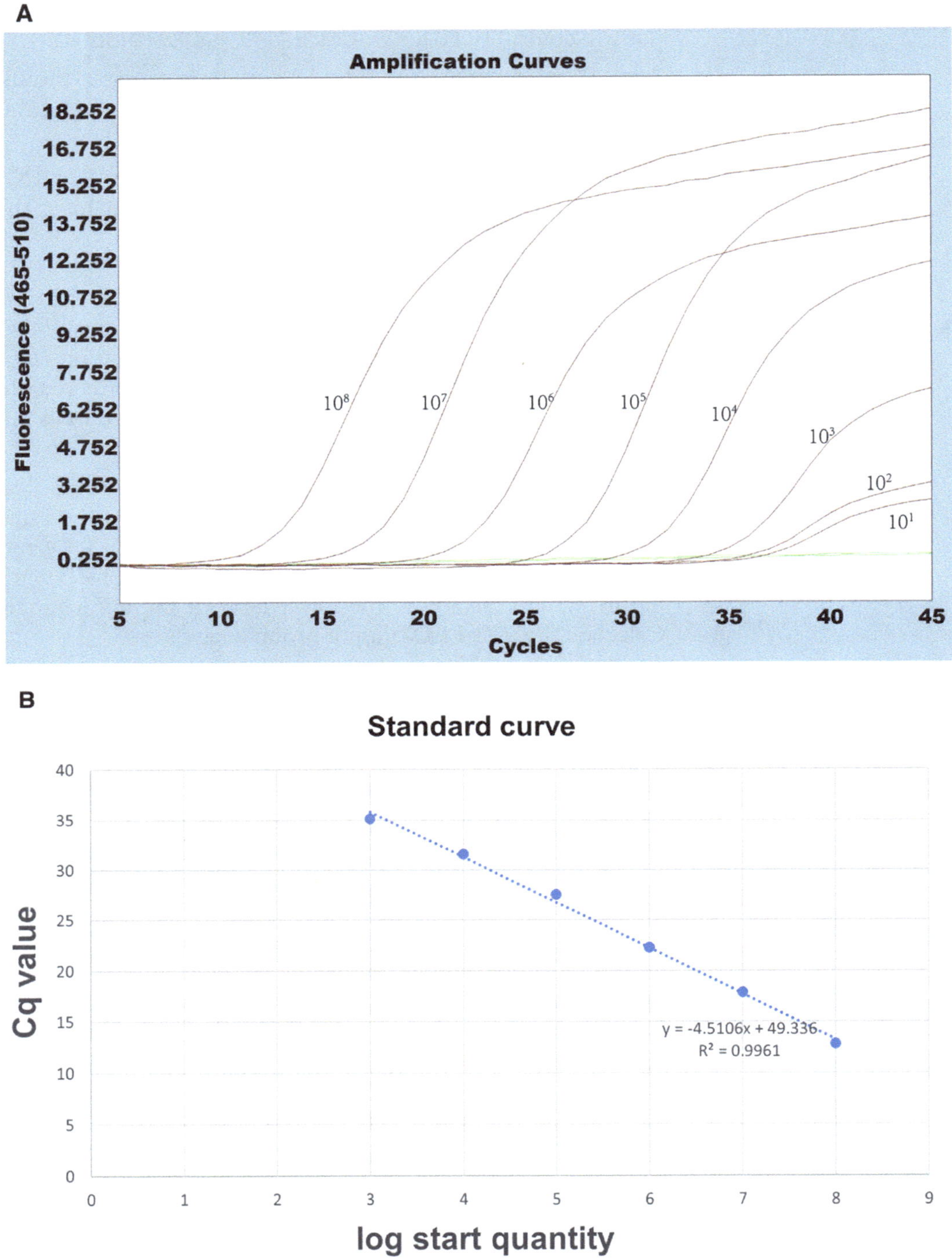

Fig.1 A representative of TaqMan® real-time RT-PCR assay for the detection of the CHIKV *nsP2* gene. (**A**) Sensitivity of assay as shown in the amplification plot from left to right is curves of decreasing concentration of CHIKV ssDNA ranging from 2.0×10^8 to 2.0×10^1 copies/μL in a serial tenfold dilution; (**B**) The standard curve is generated from the Cq values obtained against the known concentration of tenfold serially diluted CHIKV ssDNA. The standard curve shows a correlation coefficient of 0.9961 and a detection limit of 20 ssDNA copies/μL

insensitive to amplification with magnetic beads—such as the ABI7500—bead-containing samples may be directly loaded.

3. Automatic nucleic acid extractor based on the magnetic beads method can be used.
4. Ensure that the correct detection channel is activated according to fluorescence labeling of the TaqMan probe.
5. When performing tenfold dilution, precise pipetting is required. When diluting from high to low concentrations, the pipette tip needs to be changed each time.

References

1. Silva LA, Dermody TS (2017) Chikungunya virus: epidemiology, replication, disease mechanisms, and prospective intervention strategies. J Clin Invest 127(3):737–749. https://doi.org/10.1172/JCI84417
2. de Souza WM, Lecuit M, Weaver SC (2025) Chikungunya virus and other emerging arthritogenic alphaviruses. Nat Rev Microbiol 23(9): 585–601. https://doi.org/10.1038/s41579-025-01177-8
3. Chopra A, Venugopalan A (2025) Chikungunya and other viral arthritis. Best Pract Res Clin Rheumatol 39(2):102068. https://doi.org/10.1016/j.berh.2025.102068
4. Ribeiro Dos Santos G, Jawed F, Mukandavire C, Deol A, Scarponi D, Mboera LEG et al (2025) Global burden of chikungunya virus infections and the potential benefit of vaccination campaigns. Nat Med 31:2342–2349. https://doi.org/10.1038/s41591-025-03703-w
5. Custer B, Grebe E, Buccheri R, Bakkour S, Stone M, Capuani L et al. (2023) Surveillance for Zika, Chikungunya, and dengue virus incidence and RNAemia in blood donors at 4 Brazilian Blood Centers during 2016-2019. J Infect Dis 227(5):696–707. https://doi.org/10.1093/infdis/jiac173
6. Andrew A, Citartan M, Wong KA, Tang TH, Magdline Sia Henry S, Ch'ng ES (2023) Analytical and clinical evaluation of a TaqMan real-time PCR assay for the detection of chikungunya virus. Microbiol Spectr 11(4):e0008823. https://doi.org/10.1128/spectrum.00088-23
7. Yan X, Wang X, Li J, Li B, He B, Huang L et al. (2024) Preliminary survey of three mosquito-borne viruses using a self-established multiplex RT-qPCR assay in Chinese blood donors. J Virol Methods 331:115055. https://doi.org/10.1016/j.jviromet.2024.115055
8. Bong D, Sohn J, Lee S-JV (2024) Brief guide to RT-qPCR. Mol Cells 47(12):100141. https://doi.org/10.1016/j.mocell.2024.100141
9. Bustin SA (2024) Improving the quality of quantitative polymerase chain reaction experiments: 15 years of MIQE. Mol Asp Med 96:101249. https://doi.org/10.1016/j.mam.2024.101249
10. Busi F (2020) Fluorescent oligonucleotide probes for the quantification of RNA by real-time qPCR. Methods Mol Biol 2113:263–280. https://doi.org/10.1007/978-1-0716-0278-2_18

Check for updates

Chapter 5

Determination of Chikungunya Virus Infectious Titer Using Plaque Assays

Sebastiano Volpe and Claudia Filomatori

Abstract

The plaque-forming assay is the gold-standard method for measuring infectious viral particles. In this technique, lytic viruses infect and destroy host cells, but their spread is restricted by a viscous overlay medium. As a result, new viral particles can only infect neighboring cells, leading to localized clear areas known as plaques, which become visible after staining the remaining living cells. The number of plaques reflects the number of infectious viral particles initially present in the sample and is reported as plaque-forming units (PFU) per sample volume. In this protocol, we describe step by step how to perform a plaque-forming assay to determine the concentration of chikungunya virus in a cell culture supernatant.

Keywords Viral plaque assay, Viral titration, Plaque-forming-units

1 Introduction

The plaque assay is a classic virological technique, and arguably, no other method has contributed as profoundly to the advancement of the field. Originally developed to study viruses that infect bacteria, the assay was adapted in 1952 by Renato Dulbecco to quantify animal viruses in cultured cells [1, 2]. This adaptation enabled, for the first time, both quantitative and qualitative measurement of viruses in animal virus research.

Although variations of the assay exist, and even automated counting technologies have been developed [3–5], the general procedure is as follows (Fig. 1a). The viral stock is subjected to serial dilutions and inoculated into monolayers of cultured cells. After a brief incubation to allow viral adsorption and entry, the cells are overlaid with a nutrient semi-solid medium (commonly agar or carboxymethyl cellulose). This overlay prevents newly produced progeny virions from spreading to distal regions of the plate but still allows them to infect immediately adjacent cells (Fig. 1b). Each infectious viral particle then produces a localized zone of cell destruction known

Claudia Filomatori and Eliana F. Castro (eds.), *Chikungunya Virus*, Methods in Molecular Biology, vol. 3067,
https://doi.org/10.1007/978-1-0716-5496-5_5,

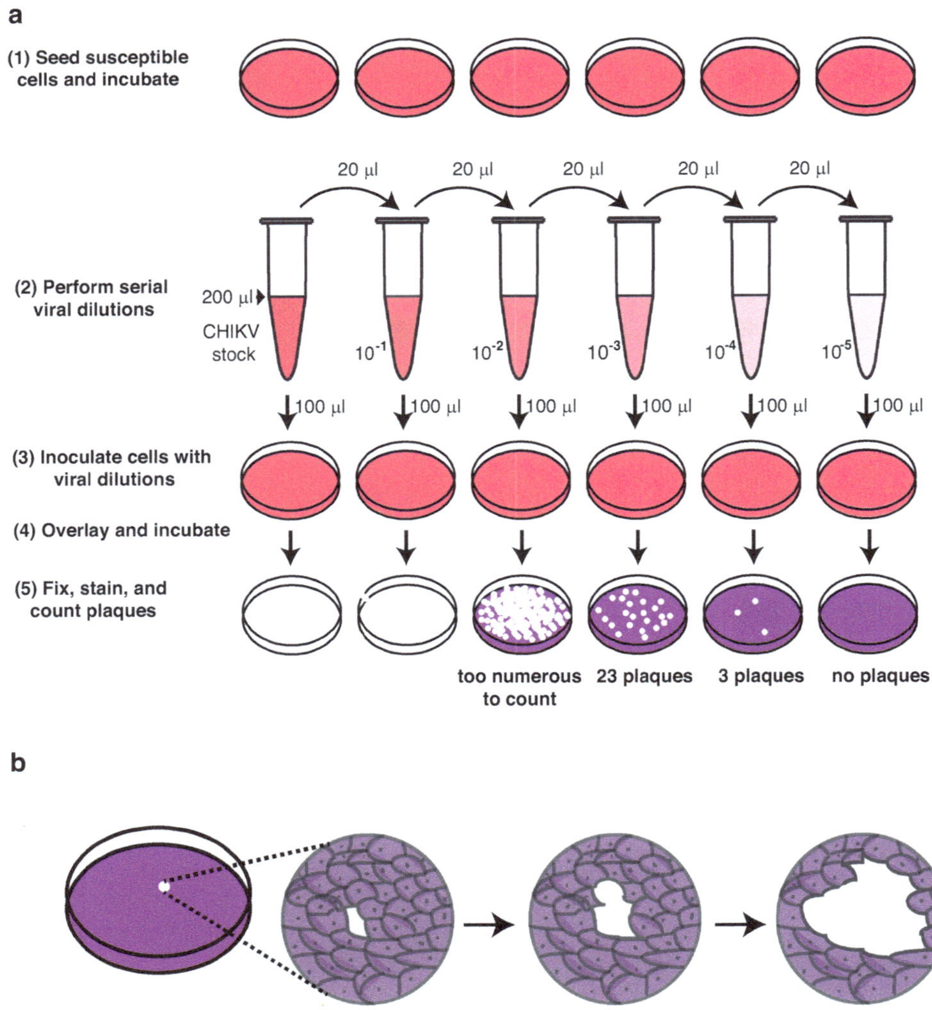

Fig. 1 **(a)** Schematic diagram of the plaque assay protocol. (1) The day before infection, seed healthy Vero cells to obtain a 90%–100% confluent monolayer. (2) Serially dilute viral samples tenfold. (3) Add aliquots of each dilution to the Vero cells. (4) Cover the cells with an overlay medium and incubate. (5) Fix and stain the monolayer. (**b**) Plaque-forming units. Virus produced from an infected cell can infect neighboring cells. When infected cells are lysed, clear regions free of cells—known as plaques—appear in the monolayer

as a plaque, which becomes visible after staining with a vital dye. By counting these plaques, researchers can calculate the concentration of infectious viral particles in the sample, expressed as plaque-forming units (PFU) per sample volume.

The plaque assay was transformative because it provided a quantitative and reproducible method for measuring virus infectivity and helped standardize virus titrations across laboratories. Beyond simple quantification, the assay offers additional applications: it provides insights into viral growth kinetics and virulence factors [6, 7] and is widely used in antiviral drug discovery [8, 9] and vaccine development [10]. To this day, it remains one of the most valuable and indispensable tools for assessing the biological activity of virus preparations.

While alternative methods for viral quantification exist [3, 11], such as quantitative PCR (qPCR), ELISA, or electron microscopy, these typically measure viral genomes, proteins, or physical particles rather than functional infectivity. In contrast, the plaque assay specifically quantifies the number of infectious viral particles capable of initiating a productive replication cycle. This distinction is crucial because the total number of physical particles in a virus preparation often greatly exceeds the number of truly infectious particles [3].

Although the plaque assay can be applied to many animal viruses that induce cell lysis [12–15], the protocol must be optimized for each viral species. In this chapter, we describe in detail the procedure for quantifying chikungunya virus (CHIKV) infectious particles, along with practical tips to improve the accuracy and reliability of the assay.

2 Materials

2.1 Cell Plating

1. Cells: Vero cells (*Cercopithecus aethiops kidney*, ATCC CCL-81) (*see* **Note 1**).
2. Maintenance culture medium: high-glucose Dulbecco's Modified Eagle Medium (DMEM) supplemented with 5% fetal bovine serum (FBS) and 1% P/S (penicillin/streptomycin). Store at 4°C (*see* **Notes 2 and 3**).
3. Cell washing buffer: Dulbecco's phosphate-buffered saline (PBS). Store at 4°C.
4. 2.5% (v/v) trypsin (10×). The 10× trypsin can be stored at 4°C until the expiration date or at −20°C for 1 year. To prepare the 1× working solution, dilute 1 mL of 2.5% (v/v) trypsin in 9 mL of PBS. Store at 4°C for 3 months.
5. Neubauer counting chamber.
6. Inverted microscope.
7. 5% CO_2 incubator at 37°C.
8. Sterile plasticware (1000 and 200 μL pipette tips, 35 mm dishes, and 24-well plates (*see* **Note 4**).
9. Other: pipettes (1000 and 200 μL).

2.2 Viral Dilutions and Infections

1. Virus: CHIKV samples to be analyzed (*see* **Note 5**).
2. DMEM, free of FBS.
3. 1.6% (w/v) methylcellulose. Weigh 1.6 g of methylcellulose and prepare a 100-mL solution. Slowly sprinkle the methylcellulose

powder into cold water while stirring gently. Autoclave 15 min at 121°C to ensure sterility. Once cooled, store at 4°C until use. Before using, check that the solution is homogeneous; if separation occurs, gently mix by inverting (*see* **Note 6**).

4. Overlay medium: 2% (v/v) FBS, 0.8% (v/v) methylcellulose, 1% (v/v) P/S, and DMEM. Prepare just prior to use. In a sterile container, add 1 mL of FBS, 500 μL of P/S solution, 25 mL of 1.6 % (w/v) methylcellulose solution, and 25 mL of DMEM. Mix gently but thoroughly to ensure a homogeneous solution. Avoid creating bubbles.
5. Other: CO_2 (5%) incubator, sterile plasticware, and pipettes.

2.3 Staining

1. Fixing solution: 10% (v/v) formaldehyde in water. Leave at 4°C for current use (*see* **Note 7**).
2. Staining solution: 0.1% (w/v) crystal violet, 20% (v/v) ethanol, in water. Store in an amber bottle or in a dark place to protect from light.
3. Other: Pipettes, Pasteur pipette or dropper.

3 Methods

3.1 Cell Plating

1. The day before, examine the monolayer of cultured Vero cells under the microscope to ensure that they are confluent, healthy, and free of contamination (Fig. 2).

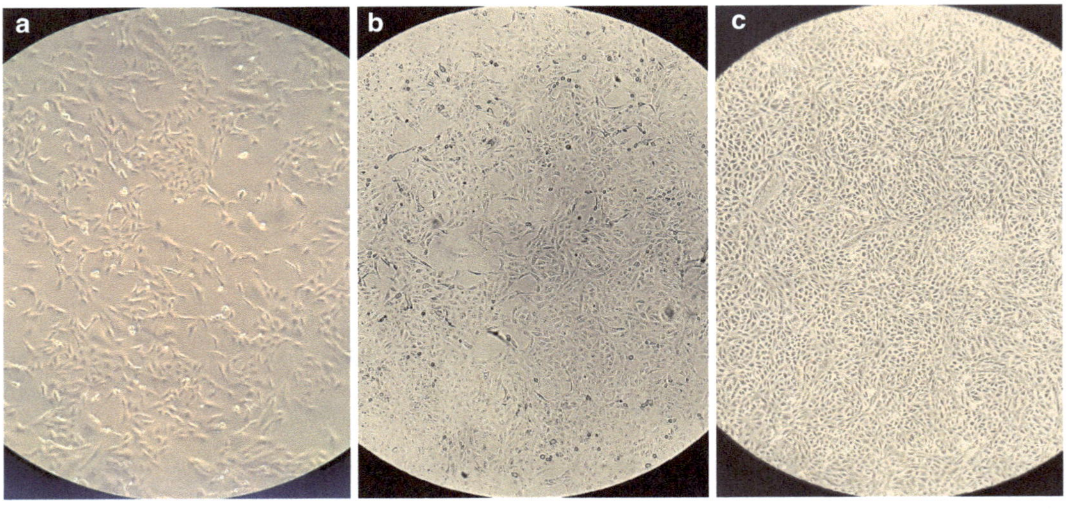

Fig. 2 Vero cells cultured to different confluences. Images show cells at approximately 60% (**a**), 80% (**b**), and 100% (**c**) confluency observed under bright field using an inverted Nikon Eclipse TE300 microscope with a 20× objective lens

2. Remove the maintenance culture medium and gently wash the cells with a washing buffer (300 μL buffer per 35 mm plate).
3. Remove the washing buffer and add 0.25% (v/v) trypsin (300 μL trypsin per 35 mm plate). Ensure the trypsin covers the entire cell layer, then incubate at 37°C for 5 min (*see* **Note 8**).
4. Check under the microscope to ensure the cells are fully detached.
5. Add maintenance culture medium to bring the final volume up to 3 mL to neutralize the trypsin and dilute the cells. Pipette gently to ensure the cells are well separated from each other.
6. Quantify the cells using a Neubauer counting chamber (Fig. 3). First, clean the chamber surface and coverslip with

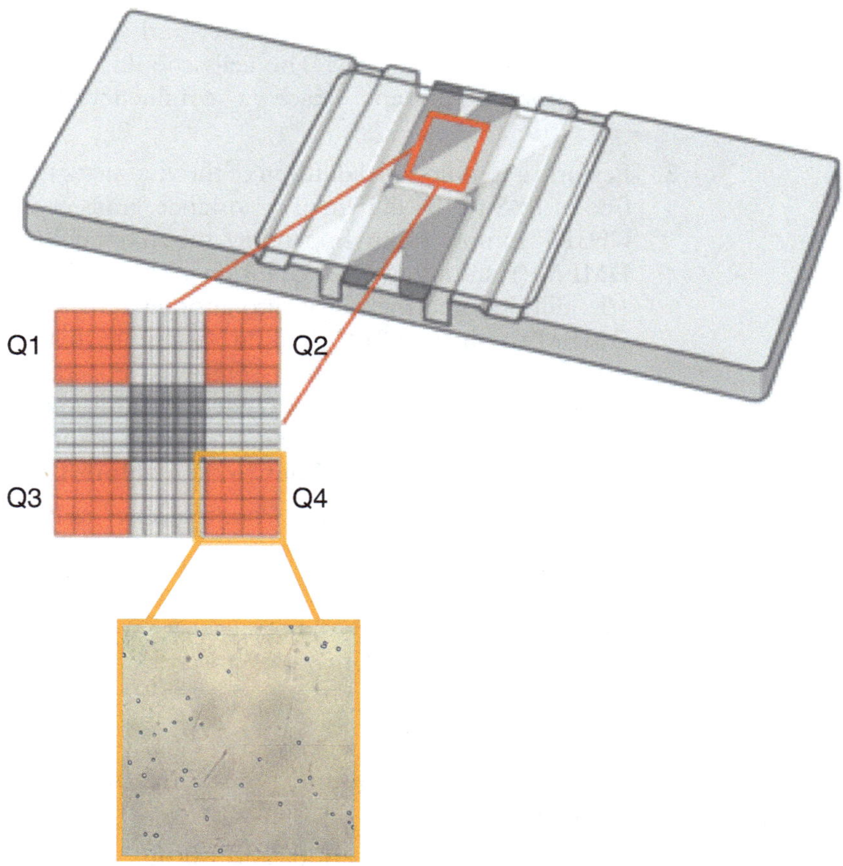

Fig. 3 Cell counting using a Neubauer chamber. Representative image showing the quadrants (Q_1–Q_4) of the chamber. Cells are carefully loaded into the counting chamber, and the number of cells in each quadrant is counted under an inverted microscope at 20× magnification. The average cell count per square is used to estimate the total cell concentration in the sample. The image shows a picture illustrating an optimal number of cells for counting

ethanol 70% (v/v) to remove dust and oil. Place the coverslip, then load 15–20 μL of the cell suspension. Under the microscope, count the cells located in the four large corner quadrants. Calculate cell concentration using Eq. 1:

$$N°\text{cell/mL} = [(Q_1 + Q_2 + Q_3 + Q_4)/4] \times 10^4 \quad (1)$$

where Q_1, Q_2, Q_3, and Q_4 are the numbers of cells counted in each of the four quadrants.

7. After calculating the concentration, plate the volume corresponding to 100,000 cells per well in a 24-multiwell plate (*see* **Note 9**).
8. Add maintenance culture medium to each well to bring the final volume up to 600 μL (*see* **Note 10**).
9. Incubate the cells overnight at 37°C in a CO_2 incubator.

3.2 Viral Dilutions and Infections

1. The next day, check the confluency and viability of the cells before starting the assay. The cells should display their typical morphology and reach a confluency of 90%–100% (*see* **Note 11**).
2. Perform a tenfold serial dilution of the viral stock using DMEM free of FBS as the diluent. For instance, transfer 20 μL of the CHIKV sample to an Eppendorf tube containing 180 μL of DMEM (resulting in a 10^{-1} dilution). Then, take 20 μL of this 10^{-1} dilution and transfer it into another tube with 180 μL of DMEM to obtain a 10^{-2} dilution. Continue this process until reaching the desired dilution (Fig. 1a). For each viral sample, replicate wells must be included. In addition, it is recommended to include a positive control with a known viral titer and a negative control without virus (*see* **Note 12**).
3. Remove the culture medium from the cells (*see* **Note 13**).
4. Add drop by drop, 100 μL of the serially diluted virus samples, along with the positive and negative controls, to the prelabeled wells. Then, add 100 μL of prewarmed DMEM to prevent the cells from drying out (*see* **Notes 14 and 15**).
5. Incubate the plate at 37°C in a 5% CO_2 incubator for 1 h (*see* **Note 16**).
6. During the incubation time, prepare the overlay medium (*see* **Notes 17 and 18**).
7. After the incubation period, add 800 μL of overlay medium directly to each well (*see* **Note 19**).
8. Incubate the plates at 37°C in a 5% CO_2 atmosphere until distinct plaques form. CHIKV plaques typically develop within 3 days of incubation. Plaque formation can be monitored

periodically by examining the wells under a microscope (*see* **Notes 20 and 21**).

3.3 Staining

1. Once plaques are observed, add 1 mL of 10% (v/v) formaldehyde to the well to fix the cells (*see* **Note 22**).
2. Incubate for 1 h at room temperature.
3. Wash the plate by gently immersing it in a tray of water.
4. Repeat the wash step five times.
5. Invert the plate onto tissue paper and allow it to dry.
6. Add staining solution to each well (*see* **Note 23**).
7. Incubate at room temperature until the plaques formed by viral infection become visible; often, 30 min are sufficient.
8. Wash the plate by gently immersing it in a tray of water.
9. Repeat the wash step until the excess stain is removed.
10. Invert the plate onto tissue paper and allow it to dry.

3.4 Determining the Viral Titer

1. Observe the size and morphology of the plaques (Figs. 4 and 5). Use negative and positive controls as references: the negative control should be uniformly stained violet, while the positive control will display plaques consistent with the viral titer (*see* **Note 24**).
2. Select wells containing between 15 and 50 plaques.
3. Count the plaques in each selected well (*see* **Note 25**) and calculate the average from the replicates.

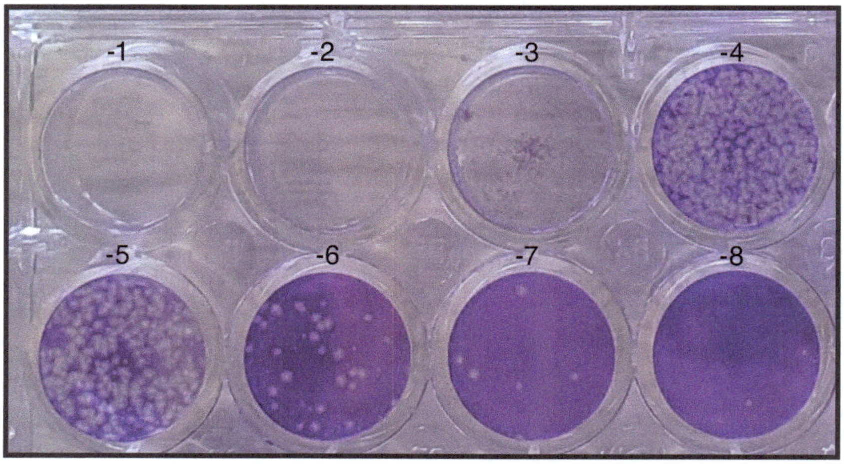

Fig. 4 Plaque assay in Vero cells inoculated with ten-fold serial dilutions of a CHIKV stock. Thirty plaques are observed at the 10^{-6} dilution. Considering the dilution factor and the sample volume (0.1 mL), the estimated viral titer is 3×10^8PFU/mL

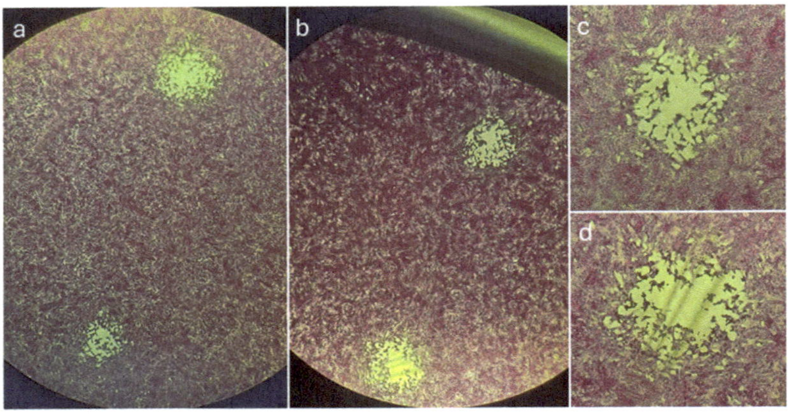

Fig. 5 **Representative** images of CHIKV plaques under an inverted Nikon Eclipse TE300 microscope. The lightly stained regions correspond to plaques after 3 days of incubation. Panels (**a**) and (**b**) are captured with a 10× objective lens, while panels (**c**) and (**d**) are captured with a 20× objective lens

4. Calculate the viral titer by using Eq. 2:

$$\text{Viral titer} = \frac{\#\text{PFU}}{D\ x\ V} \tag{2}$$

where #PFU is the average number of plaque-forming units, D is the dilution factor, and V is the sample volume in mL (*see* **Note 26**).

3.5 Viral Growth Curves

An important application of the plaque assay is the characterization of viral replication dynamics in cell cultures through viral growth curves. Analysis of these curves typically reveals distinct phases of the viral replication cycle: an initial eclipse phase, followed by an exponential growth phase, and finally a plateau where virus production stabilizes. Such studies are critical for comparing the replication kinetics of different viral strains, mutants, or experimental conditions, and they provide valuable insights into viral pathogenesis and transmission potential.

1. Seed susceptible cells in plates and allow them to reach 80%–90% confluency (*see* **Note 27**).
2. The next day, infect cells at a defined multiplicity of infection (MOI), which is the ratio of infectious particles per cell (*see* **Note 28**).
3. Calculate and prepare the appropriate virus dilution in PBS or serum-free medium, according to the viral stock titer and the desired MOI.
4. Remove the culture medium from the cells (*see* **Note 13**).

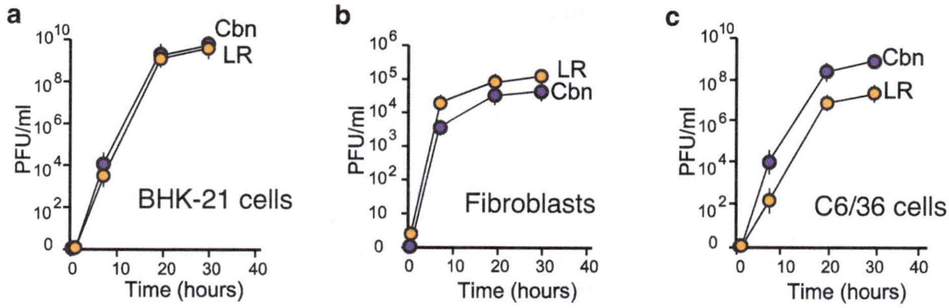

Fig. 6 Comparative growth kinetics of two CHIKV strains: CHIKV-LR (from the 2005 epidemics in La Reunion Island, East-Central-South African lineage, GeneBank# AM258994.1) and CHIKV-Cbn (from the 2013 Caribbean epidemics, Asian lineage, GeneBank# LN898111.1). Infection was performed in three cell lines at an MOI of 0.1. **(a)** In mammalian BHK-21 cells, the growth curves show that both viruses replicate similarly. **(b)** The same occurs in human fibroblasts, although CHIKV-LR exhibits a slight advantage over CHIKV-Cbn. **(c)** In mosquito C6/36 cells CHIKV-Cbn shows enhanced growth kinetics in mosquito cells, reaching viral titers approximately tenfold higher than those of CHIKV-LR. Error bars represent standard deviations of the mean ($n=3$, data analysis by two-way ANOVA with Bonferroni's test)

5. Add the virus dropwise onto the cell monolayer and then add a minimal volume of serum-free medium to cover the cells (*see* **Notes 14 and 15**).
6. Allow the virus to adsorb for 60 min on ice or at 37°C (*see* **Notes 16 and 29**).
7. Remove the inoculum and wash the cells five times with PBS to remove any unbound virus.
8. Add fresh medium supplemented with 2% FBS.
9. Collect supernatants at multiple time points post-infection (*see* **Note 30**). Keep samples frozen at −80°C until titration.
10. Determine the viral titer of each collected sample using a plaque assay.
11. Plot the viral titer (PFU/mL) on the y-axis against the time post-infection on the x-axis to generate the growth curve (*see* **Note 31**). To illustrate, Fig. 6 shows representative viral growth curves comparing the replication kinetics of two CHIKV strains in mammalian and mosquito cells.

4 Notes

1. Different cell lines in which CHIKV induces lysis can be used for plaque assays. Alternatively, BHK-21 cells (*Mesocricetus auratus hamster kidney*, ATCC, CCL-10) may be employed. In our experience, however, plaque-forming units are more easily visualized in Vero cells.

2. If lyophilized powders are purchased, they should be resuspended in pyrogen-free water and sterilized by filtration through cellulose acetate membrane discs (0.22 μm pore size).
3. FBS should be heat-inactivated by incubation at 56°C for 30 min, then aliquoted and stored at −20°C or −80°C, depending on the intended storage duration. During incubation, gently shake or invert the bottle every 5–10 min to prevent temperature gradients.
4. Cell culture dishes and plates should be pretreated to promote cell attachment.
5. Virus stocks may originate from various sources, including the supernatants of CHIKV-infected mammalian or mosquito cell cultures, or patient sera. These stocks are aliquoted and stored at −80°C.
6. Methylcellulose powder may not dissolve completely at first and may form lumps, which is normal. Methylcellulose dissolves better as the temperature drops, and the solution should become viscous and homogeneous. To aid dissolution, stir at low temperature (4°C–10°C) using a magnetic stirrer. If needed, allow it to stir overnight.
7. Formaldehyde is toxic and a known irritant; always handle it in a well-ventilated area or fume hood. Wear gloves and eye protection. Avoid inhalation and skin contact.
8. If the cells do not easily detach from the well, one approach is to prewash them with 1× trypsin before adding the final trypsin solution.
9. Before plating, gently but thoroughly mix the cell suspension with a pipette to ensure the cells are evenly suspended and have not settled.
10. After plating, gently rock the plate in a circular or zigzag motion to help cells spread uniformly across the well surface.
11. The viability of the cells should be checked before starting the experiment. If the cells are not healthy, they may detach from the plate, and plaques will not form properly.
12. Positive and negative controls are essential for troubleshooting.
13. When changing the medium, it should be done carefully and slowly to avoid disturbing the cell monolayer, as this may affect the cells in the area and ultimately impact plaque formation.
14. To optimize infection efficiency, infections should be carried out in the minimal volume necessary, while carefully preventing the cells from drying out. For a well of a 24-multiwell plate, 200 μL is sufficient.

15. It is recommended to use serum-free medium during the infection.
16. Gently rock the plate every 15 min during incubation to ensure even distribution of the virus.
17. Alternatively, sterile 5% SeaPlaque agarose in PBS may be used as an overlay medium.
18. Adjusting the methylcellulose concentration in the overlay medium can significantly influence plaque morphology. Higher concentrations increase the medium's viscosity, which restricts viral diffusion, resulting in smaller and more sharply defined plaques. Conversely, lower concentrations may lead to larger plaques.
19. It is not necessary to remove the inoculum before adding the overlay medium.
20. Once the overlay medium has been added, avoid moving the plate or subjecting it to sudden impacts. Such movements may cause the medium to shift, leading to the spread of virus particles beyond their confined foci. As a result, the plate may develop large, cell-free areas that prevent accurate resolution and counting of individual plaques.
21. Regularly monitor plaque development under the microscope during incubation. Extending infection time too long can cause plaques to merge, while stopping too early may result in plaques too small to detect accurately.
22. Do not remove the overlay medium before fixing the cells.
23. The staining solution can be added using a dropper or a Pasteur pipette. Typically, two drops are enough to cover the entire well.
24. If the background staining is not a uniform violet color, this may indicate uneven cell plating, inadequate washing, insufficient fixation, or cell death during the assay. This can make it difficult to distinguish true plaques from the background staining, which affects the accuracy of the results.
25. For more accurate counting, it is recommended to use a fine-tip permanent marker.
26. For example, if the numbers of the plaques are 20, 30, and 40 in triplicates at the dilution of 10^{-6} and the virus sample volume is 100 μL (0.1 mL):

$$\text{Viral titer} = \frac{30 \quad \text{PFU}}{10^{-6} \quad \times \quad 0.1 \quad \text{mL}}$$

$$\text{Viral titer} = 3 \times 10^{8}\,\text{PFU/mL}$$

27. Mammalian cell lines (e.g., BHK-21, A549, and Huh-7 cells, or human fibroblasts) or mosquito cell lines (e.g., C6/36 cells) can be used.
28. A low MOI (0.01 or 0.1, corresponding to approximately one virus per 100 or 10 cells, respectively) allows multiple rounds of infection. In turn, a high MOI (1 or 10, corresponding to appoximately 1 or 10 viruses per cell, respectively) is used for one-step growth curves to study a single replication cycle.
29. For a one-step growth curve, perform the adsorption step at low temperature (e.g., for 1 h on ice) before shifting the culture to 37°C for incubation, in order to synchronize the viral replication cycle.
30. The CHIKV replication cycle lasts approximately 7–8 h. When performing a one-step growth curve, samples can be collected every 2–3 h (e.g., at 0, 2, 4, 6, 8, 10, and 12 h post infection). If a multistep growth curve is desired, samples can be harvested at longer intervals (e.g., 0, 8, 24, 36, and 48 h post infection) to monitor viral replication over multiple cycles.
31. When plotting growth curves, use a logarithmic scale for viral titer to better visualize the exponential growth phase.

References

1. Dulbecco R (1952) Production of plaques in monolayer tissue cultures by single particles of an animal virus. Proc Natl Acad Sci 38:747–752
2. Cooper PD (1962) The plaque assay of animal viruses, In: Advances in virus research, Elsevier, pp 319–378
3. Payne S (2017) Methods to study viruses, In: Viruses, Elsevier, pp 37–52
4. Baer A and Kehn-Hall K (2014) Viral concentration determination through plaque assays: using traditional and novel overlay systems. J Vis Exp 52065
5. Emi A, Suzuki Y, Odake C et al. (2025) Development of an automated plaque-counting program for the quantification of the Chikungunya virus. Sci Rep 15: 12429
6. Filomatori CV, Bardossy ES, Merwaiss F et al (4) AD RNA recombination at Chikungunya virus 3'UTR as an evolutionary mechanism that provides adaptability. PLoS Pathog 15:e1007706
7. Tsetsarkin KA, Vanlandingham DL, McGee CE et al. (2007) A single mutation in chikungunya virus affects vector specificity and epidemic potential. PLoS Pathog 3:e201
8. Abdelnabi R, Neyts J, Delang L (2015) Towards antivirals against chikungunya virus. Antiviral Res 121:59–68
9. Pohjala L, Utt A, Varjak M et al. (2011) Inhibitors of alphavirus entry and replication identified with a stable chikungunya replicon cell line and virus-based assays. PLoS ONE 6:e28923
10. Akahata W, Yang Z-Y, Andersen H et al. (2010) A virus-like particle vaccine for epidemic Chikungunya virus protects nonhuman primates against infection. Nat Med 16: 334–338
11. MacLachlan NJ, Dubovi EJ (2017) Virus replication. In: MacLachlan NJ, Dubovi EJ (Eds.) Fenner's Veterinary Virology. 5th edn. Academic Press, London, 17–45
12. Neupane B and Bai F (2023) Quantification of West Nile virus by plaque-forming assay, In: Bai F (ed) West Nile virus, Springer US, New York, NY, pp 9–14
13. Juarez D, Long KC, Aguilar P et al. (2013) Assessment of plaque assay methods for alphaviruses. J Virol Methods 187:185–189
14. Shabram P, Vellekamp G, Ruan Q et al (2016) Purification of adenovirus, In: Adenoviral vectors for gene therapy, Elsevier, pp 197–230
15. Hartley JW, Rowe WP (1963) Tissue culture cytopathic and plaque assays for mouse hepatitis viruses. Exp Biol Med 113:403–406

Check for updates

Chapter 6

Automated Counting of Chikungunya Virus–Formed Plaques

Akino Emi and Youichi Suzuki

Abstract

The cytopathic effect (CPE) is a key to understanding the pathogenicity of the chikungunya virus (CHIKV) and is useful for evaluating viral infectivity in vitro. The plaque assay, based on the CPE induced by viral infection, provides an infectious virus titer for test samples. Additionally, the assay is simple, quantitative, and cost-effective. Thus, the plaque assay remains the gold standard for determining the infectivity of CHIKV. However, manually counting of virus-induced plaques with the naked eye or under the microscope is often time-consuming and labor-intensive. Therefore, automated plaque-counting software would improve the consistency and efficiency of plaque counting. We recently developed plaQuest, a stand-alone Windows software that enables rapid, reliable plaque counting for CHIKV. In this chapter, we describe the basic procedure for detecting and counting the CHIKV-formed plaques in a 24-well plate.

Keywords Chikungunya virus, Plaque assay, Plaque counting, Windows software, plaQuest

1 Introduction

The chikungunya virus (CHIKV) is an enveloped RNA virus that belongs to the *Alphavirus* genus of the Togaviridae family, which is a causative agent for chikungunya fever (CHIKF) [1]. CHIKV is transmitted to humans through bites of infected *Aedes aegypti* and *Aedes albopictus* mosquitoes. While CHIKV infection is usually symptomatic, the clinical manifestation of CHIKF is generally a mild self-limiting febrile illness accompanied by myalgia and arthralgia [2, 3]. However, studies have demonstrated that prolonged joint pain persists for months to years in 30%–40% of infected patients, negatively impacting their quality of life (QOL) [1, 4–6]. Thus, appropriate management in the early stages of the disease is critical for improving outcomes in CHIKV-infected patients.

CHIKV has been reported to exhibit a broad spectrum of tissue and cellular tropisms in vitro and in vivo [1, 7–10]. In particular, CHIKV infection induces massive cell death across a variety of cell types, underlying CHIKV pathogenicity [7, 9, 11, 12]. A previous study reported long-term CHIKV infection in the liver in nonhuman

Claudia Filomatori and Eliana F. Castro (eds.), *Chikungunya Virus*, Methods in Molecular Biology, vol. 3067,
https://doi.org/10.1007/978-1-0716-5496-5_6,

primates with abnormally high levels of hepatocyte death, implicating the persistent symptoms observed in humans [2, 13, 14]. Cytopathogenicity, generally referred to as the cytopathic effect (CPE), is an important indicator of CHIKV infectivity. The plaque assay is a CPE-based method for titrating viruses, in which a circular zone of infected cells, or plaques, is produced by infectious virus particles [15, 16]. If the virus infection causes cell death, as is the case with CHIKV, the circular zone can be visualized by staining with a vital dye. Since plaques are considered to result from single virus infections, the plaque assay can provide a measure of viable virions (i.e., the infectious titer) in test samples. Additionally, the plaque assay is simple and cost-effective. A variation of the CPE-based plaque assay uses a virus-specific antibody combined with an indicator molecule to visualize the circular area of virus-infected cells (immunostain-based plaque assay), which is especially useful for titrating non- or less-cytopathic viruses [17].

For the aforementioned reasons, the plaque assay is the gold standard technique for measuring infectious titers in CHIKV research. However, the typical procedure for plaque assay requires researchers to count virus plaques that are large enough to be seen with the naked eye. This manual counting process is labor-intensive and time-consuming for researchers, especially in drug development studies where many samples are examined. To reduce this workload, automated plaque-counting programs have been developed for certain RNA viruses, including influenza virus, dengue virus (DENV), and hepatitis C virus (HCV) [18–24]. We have recently designed a standalone software program running on Windows named plaQuest that uses open-source algorithms from the OpenCV library (https://opencv.org) to enable rapid and reliable counting of CHIKV plaques using a scanned image of a plaque assay plate [25]. This chapter outlines the standard procedure for automatically counting CHIKV plaques in the CPE-based plaque assay using plaQuest. The counting procedure consists of four steps: (i) image acquisition of a plaque assay plate using a scanner, (ii) import of the plate data to plaQuest, (iii) setting the parameters for adequate recognition of the plate's wells and the plaques within them, and (iv) counting the plaques in all wells in the plate with the adjusted parameters. Optional functions include measuring the pixel size of the plaque area and outputting a plate image showing the plaque count for each well that can be output. In our experimental settings, plaQuest was applicable to plaque counting for DENV and severe acute respiratory syndrome coronavirus 2 (SARS-CoV-2), as well as an immunostain-based plaque assay [25]. Nevertheless, to obtain reliable results with plaQuest, it is critical to optimize the conditions of the plaque assay (e.g., cell type, cell density, virus dilution, cultivation period, and staining method) to ensure that a reasonable number of clear, well-separated plaques is produced. If the plaque assay plates are sufficiently optimized for plaque detection in plaQuest, however, this automatic counting software offers substantial advantages over manual plaque counting for researchers working on virus titration, thereby increasing assay throughput in CHIKV study.

2 Materials

1. Scanner (e.g., EPSON GT-X980 flatbed scanner).
2. Windows PC (specifications of our PC: Intel i7-1135G7 processor, 8 GB of RAM, a 512 GB solid-state drive (SSD), and a 64-bit Windows 11 Home operating system (OS), monitor with a resolution of 1920×1080)
3. plaQuest software application that runs on a Windows PC. The installation package is available upon request from our laboratory (contact: youichi.suzuki@ompu.ac.jp).

3 Methods

The detailed protocol for the CPE-based plaque assay to measure the titer of CHIKV on 24-well plates using Vero (green monkey kidney) cells (*see* **Note 1**) and crystal violet staining [12] is described in Chap. 6 of this book.

3.1 Installation of plaQuest

1. Download and unzip the "plaQuest package.zip" file to the desired location on a Windows PC (*see* **Note 2**).
2. The package consists of the plaQuest folder, which contains executable (.exe) and dynamic link library (.dll) files (*see* **Note 3**), as well as two plaque assay demonstration images for CHIKV (CHIKV demo plate.jpg) and Zika virus (ZIKV demo plate.jpg) files.

3.2 Scanning of the Plaque Assay Plate

1. Using a flatbed scanner, scan the plaque assay plate as a JPEG image at a resolution of 1200 dpi and a size of 127×83 mm, covering exactly the perimeter of the plate (*see* **Note 4**).

3.3 Importing the Plate Image and Parameter Settings for Culture Well Recognition

1. Launch the plaQuest software (Fig. 1a) by double-clicking on "plaQuest.exe" in the plaQuest folder.
2. Import the scanned JPEG file by clicking "Import image" (Fig. 1b, yellow frame). The plate image will appear in the Whole View panel.
3. Select the appropriate staining type ("CPE"-based or "Immunostain"-based) and plate format (6-, 12-, 24-, 48-, or 96-well plate) for the plaque assay to be analyzed using "Mode" and "Plate type" (Fig. 2a, *see* **Note 5**).

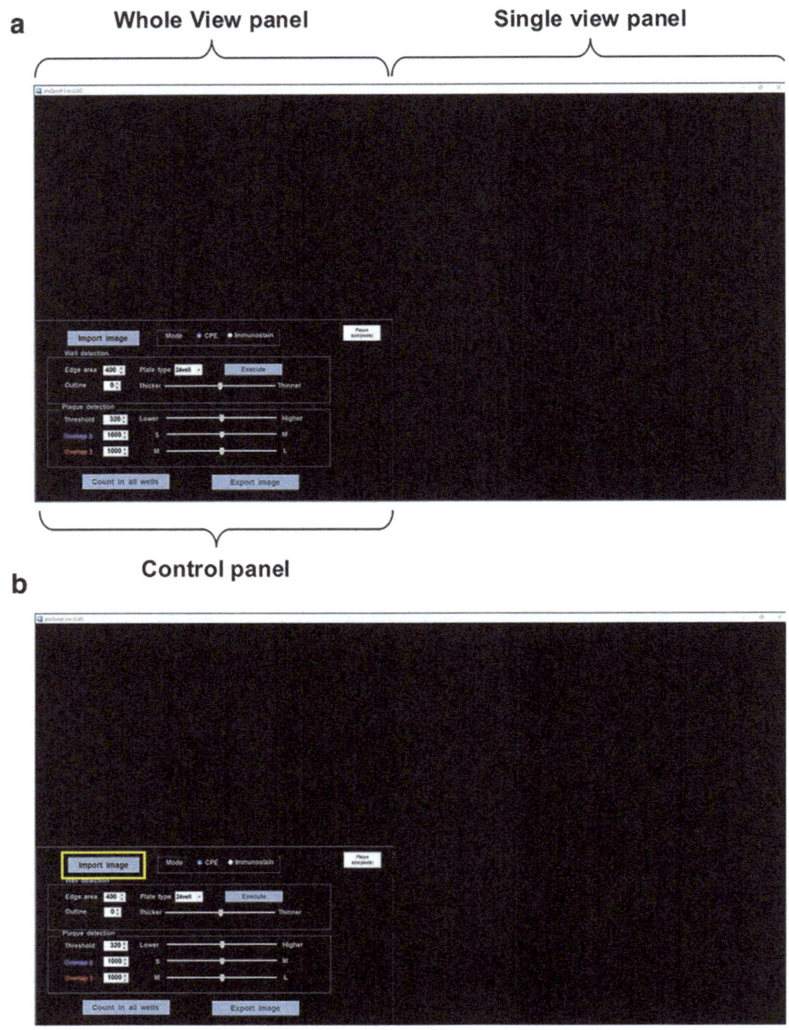

Fig. 1 The plaQuest automated plaque-counting software. (**a**) The plaQuest user interface consists of three areas. The imported plate image appears in the Whole View panel (upper left), and an enlarged image of the selected well appears in the Single View panel (right). The control panel for adjusting well- and plaque-detection parameters is located in the bottom left corner. (**b**) Import of scanned plate image. The following procedures illustrate how to analyze the "CHIKV demo plate.jpg" file in the plaQuest package

4. Click "Execute" to recognize the outline of each culture well (Fig. 2b, yellow frame).
5. If the wells are not outlined accurately, adjust the "Edge area" parameter to between 100 and 1200. Then click "Execute" to check whether all wells are recognized correctly (Fig. 2c).
6. Double-click on any well in the Whole View panel to display an enlarged image of that well in the Single View panel (Fig. 3a).

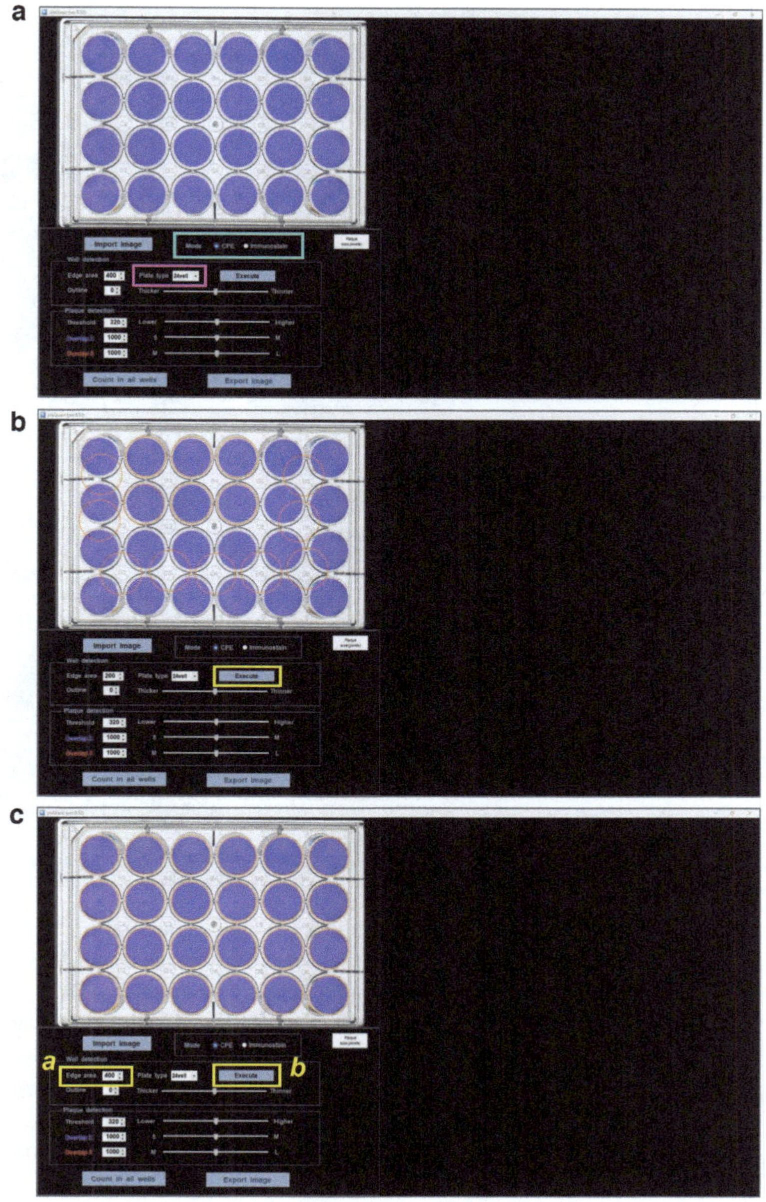

Fig. 2 Well-recognition procedure in the Whole View panel. (**a**) Selection of the type of plaque assay. The staining mode (CPE- or immunostain-based plaque assay, light blue frame) and plate type (6-, 12-, 24-, 48-, or 96-well plate, magenta frame) must be selected before adjusting the detection parameter settings. In this example (using CHIKV demo plate.jpg), "CPE" and "24-well" are chosen. (**b**) Recognition of well outlines in the imported plate image. The orange circles in the Whole View panel indicate the locations that plaQuest has recognized as wells. (**c**) Adjustment of the well-recognition parameter. If the wells are not outlined accurately, as shown in Fig. 6, adjust the "Edge area" parameter to between 100 and 1200 (yellow frame *a*), and click "Execute" (yellow frame *b*) to apply the new setting

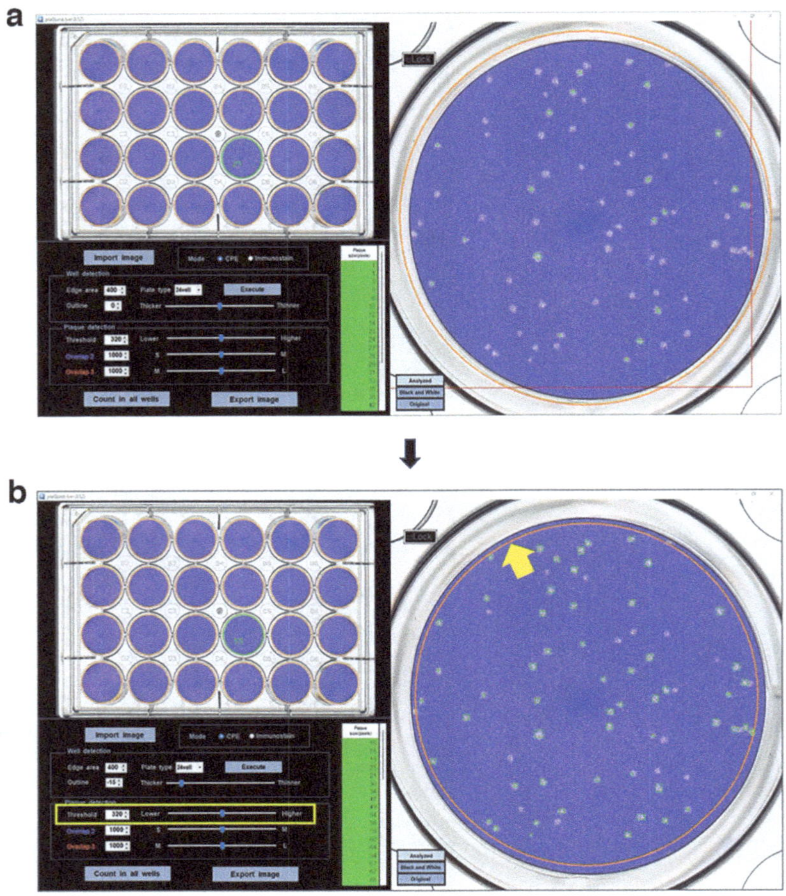

Fig. 3 Adjustment of well outline in the Single View panel. (**a**) Enlarged image of a plaque assay well in the Single View panel. By double-clicking a well on the plate image in the Whole View panel (the C4 well is selected in this example), an enlarged image of the selected well is displayed in the Single View panel. This operation also displays the plaque count of the selected well in the Whole View panel. In this example, the number "27" is shown in green on the C4 well. It also displays the plaque-detection status with green quadrilaterals in the Single View panel. (**b**) Adjusting the area for plaque counting. Adjust the orange-circled area where plaque counting should be performed (yellow arrow in the Single View panel) using the "Outline" parameter (yellow frame)

7. The orange circle in the Single View panel shows the area in which plaques are counted. By default, the plaque-counting area is set outside the well. Use the "Outline" parameter to properly set the area inside the well (Fig. 3b).
8. If the well is still not recognized correctly, even after adjusting the "Edge area" parameter, left-click on an orange circle while holding down the Ctrl key (the circle will turn green). Then move the circle to the desired location in the Whole View panel (Fig. 4).

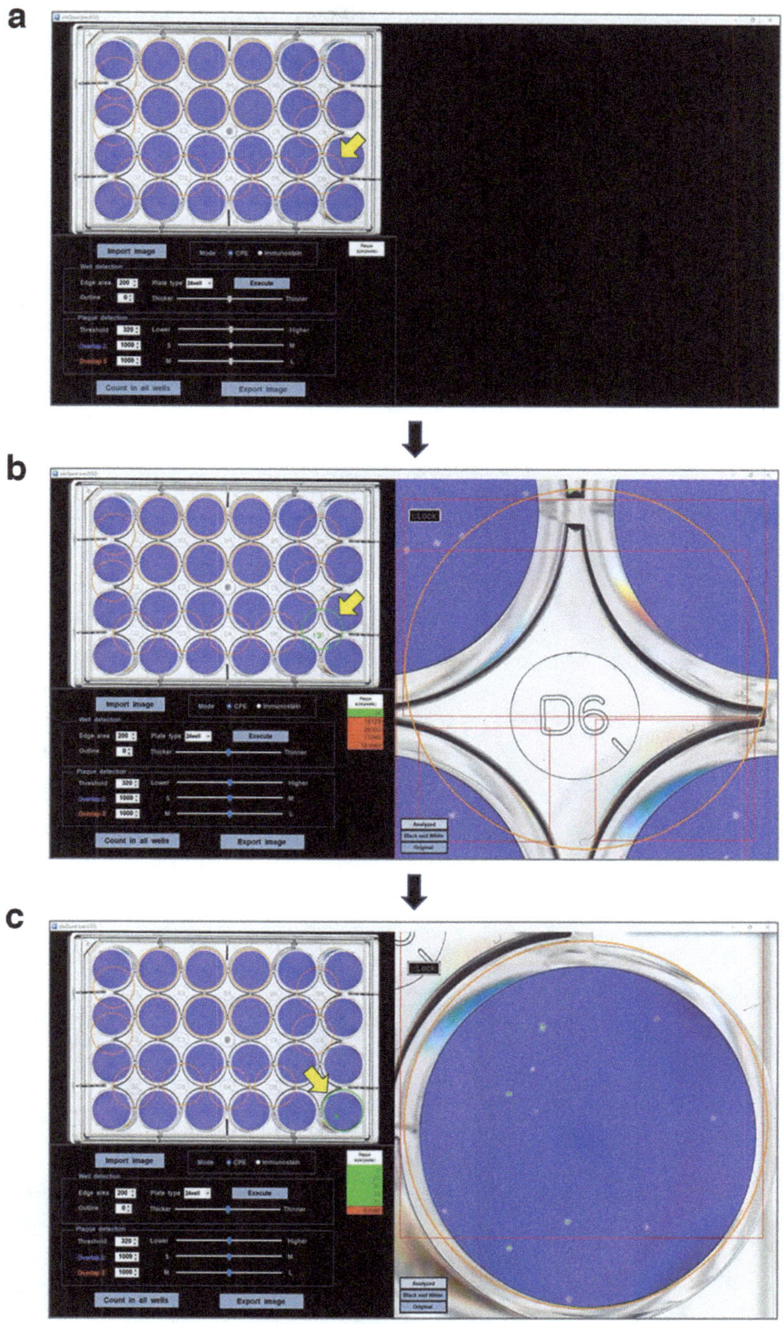

Fig. 4 Manual adjustment of the plaque-counting area. If the "Edge area" parameter fails to recognize the wells correctly (**a**, yellow arrow), select the orange circle to be adjusted by left-clicking while holding down the Ctrl key (**b**), and then drag it directly to the desired well (**c**) in the Whole View panel

3.4 Plaque Counting

1. After setting the parameters for well recognition, double-clicking on a single well in the Whole View panel displays the initial

plaque-detection status in the well. These detected plaques are indicated by green quadrilaterals in the Single View panel. The initial plaque count for the selected well is also displayed in green numbers in the Whole View panel (Fig. 3).

2. The default plaque-detection settings will likely not detect all plaques. Therefore, adjust the plaque-detection sensitivity using the "Threshold" parameter (Fig. 5).
3. You can check the plaque-detection status of other wells using the adjusted parameter settings by double-clicking the target well (*see* **Note 6**).
4. If multiple plaques that are close together or overlapping are counted as a single plaque (e.g., light blue and magenta arrows in Fig. 6a), it can be defined as two plaques or three plaques by changing the "Overlap 2" (light blue frame in Fig. 6b) and "Overlap 3" (magenta frame in Fig. 6b) parameters, respectively (*see* **Note 7**).
5. In a green-lined quadrilateral area where the "Overlap 2" or "Overlap 3" parameter does not accurately define the plaque count (e.g., yellow arrow in Fig. 7a), you can manually assign a plaque number by left-clicking near the quadrilateral to open a small dialog box (the quadrilateral line will turn yellow; Fig. 7b). Then, change the number (Fig. 7c). Be sure to click "Lock" to fix the modified plaque number (Fig. 7d, e).
6. After adjusting the necessary parameter settings, click "Count in all wells" to count the number of plaques in all wells. The

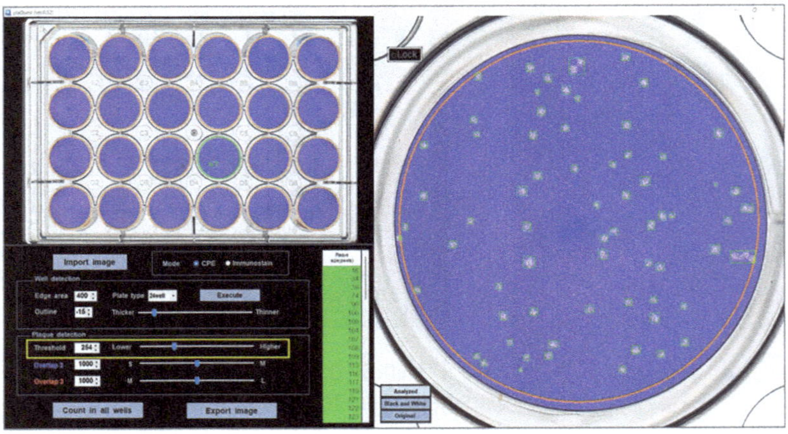

Fig. 5 Adjustment of plaque detection sensitivity. Small or faint plaques are generally not detected with the initial configurations in the Single View panel (see Fig. 3b). In this case, adjust the plaque-detection sensitivity using the "Threshold" parameter (yellow frame). Double-clicking on a well in the Whole View panel updates the plaque count

plaque numbers will then be displayed in the individual wells of the Whole View panel (yellow frame *a* in Fig. 8, *see* **Note 8**).

3.5 Optional Functions of plaQuest

1. Export the plate image showing the number of plaques in each well as a JPEG file by clicking "Export image" (yellow frame *b* in Fig. 8).
2. The "Plaque size" panel (Fig. 9) lists the pixel sizes of the quadrilaterals in a well from smallest to largest. Click on a column to display the corresponding plaque quadrilateral with a yellow line in the Single View panel (e.g., *b* in Fig. 9). Please note that this function indicates the pixel size of the quadrilateral area surrounding the plaque(s), not the plaque size.

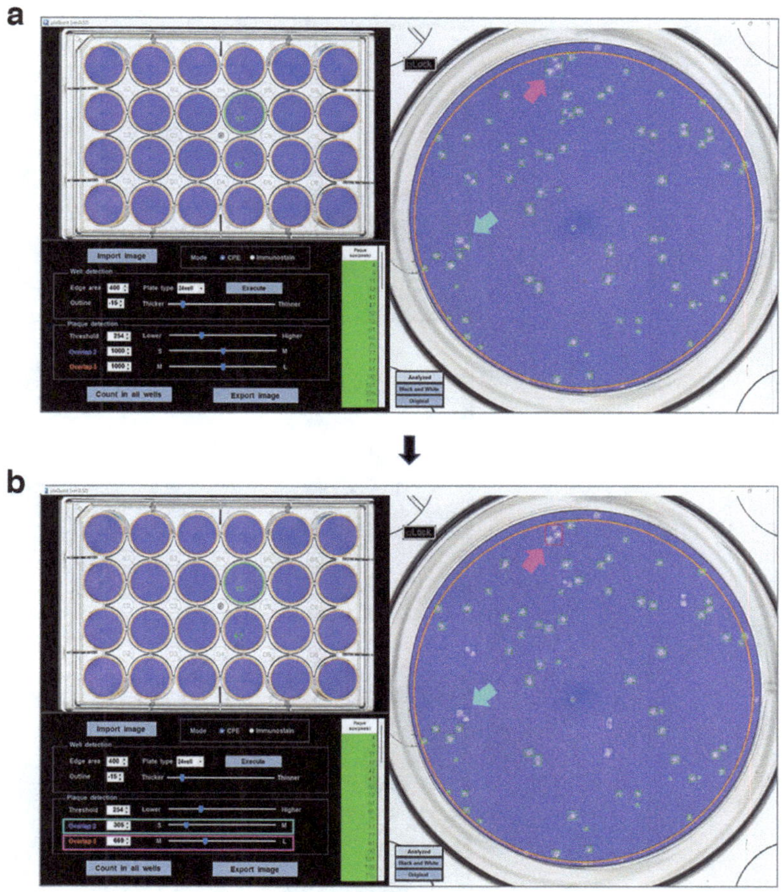

Fig. 6 Definitions of two- and three-plaque clusters. Clusters of two (light blue arrow) or three (magenta arrow) plaques that are detected as a single plaque due to their proximity or overlap (**a**) can be defined as two or three separate plaques, respectively, using the "Overlap 2" (light blue frame) or "Overlap 3" (magenta frame) functions (**b**)

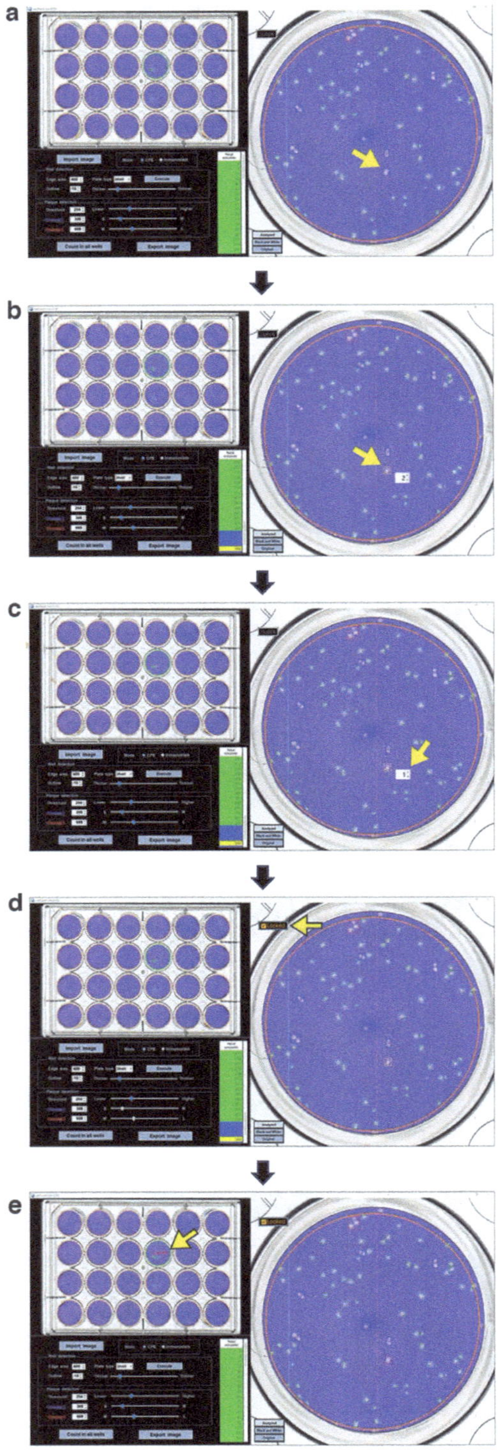

Fig. 7 Manual assignment of a plaque count. The number of plaques enclosed in the green quadrilateral can be changed manually as desired. Select the quadrilateral for which you would like to change the plaque count by left-clicking near it (**a**, yellow arrow in this example). This operation changes the quadrilateral to yellow and opens a small dialog box displaying the initial plaque number (**b**; the plaque number is "2" in this example). Then, change the number as desired (**c**; in this example, the number is changed to "1"), and click "Lock" (**d**; the yellow arrow indicates the "Lock" button) to fix the count number. Double-clicking the target well in the Whole View panel displays the new plaque count on the well (**e**, yellow arrow)

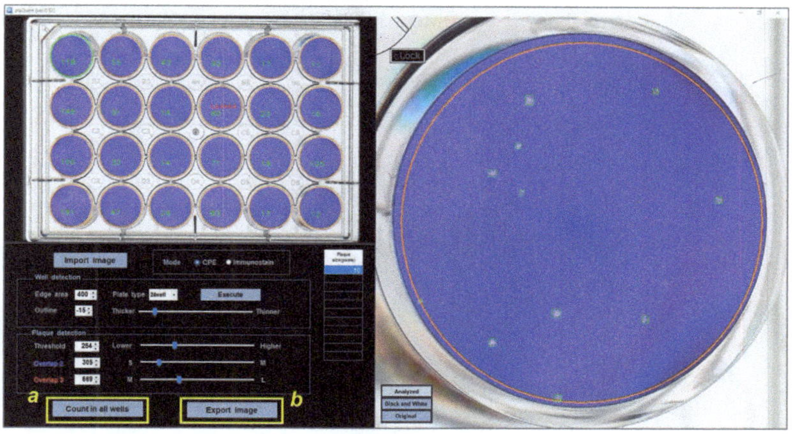

Fig. 8 Operation to obtain the plaque count result. After setting all parameters, click "Count in all wells" (yellow frame *a*) to process the plaque counting for the entire plate image. During counting, the display of the "Count in all wells" button changes to "Processing." Clicking "Export image" (yellow frame *b*) saves the plate image with the plaque counts as a JPEG file in any directory on the PC

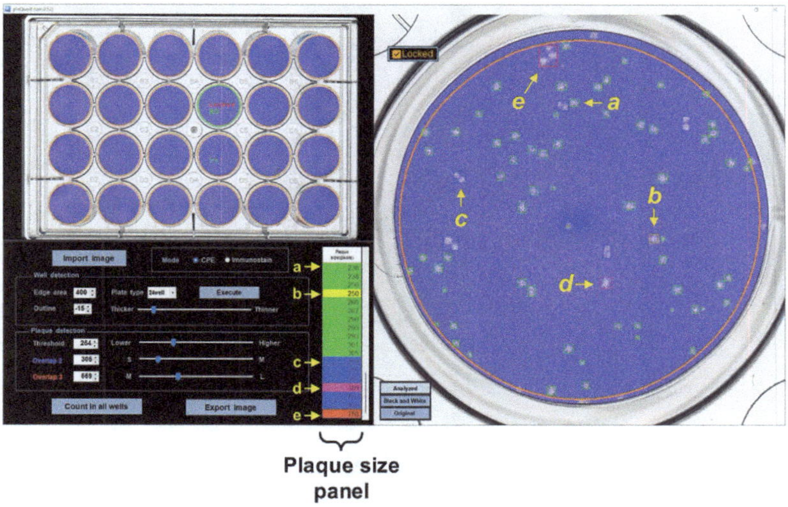

Fig. 9 Plaque size panel. This panel lists the pixel sizes of the detected quadrilaterals. (a, green column) A quadrilateral detected as a single plaque. (b, yellow column) A selected quadrilateral. (c, blue column) A quadrilateral defined as two plaques by the "Overlap 2" parameter (see Fig. 6). (d, magenta column) A quadrilateral whose plaque count is manually assigned (see Fig. 7). (e, red column) A quadrilateral defined as three plaques by the "Overlap 3" parameter (see Fig. 6).

3. The Single View panel offers three view modes: (i) "Analyzed" image view, which displays plaque-detection results (Fig. 10a); (ii) "Black and White" image view, which shows a binarized image (Fig. 10b). This is a processed data image required for

plaque detection and counting [25]; and (iii) “Original” well image view (Fig. 10c).

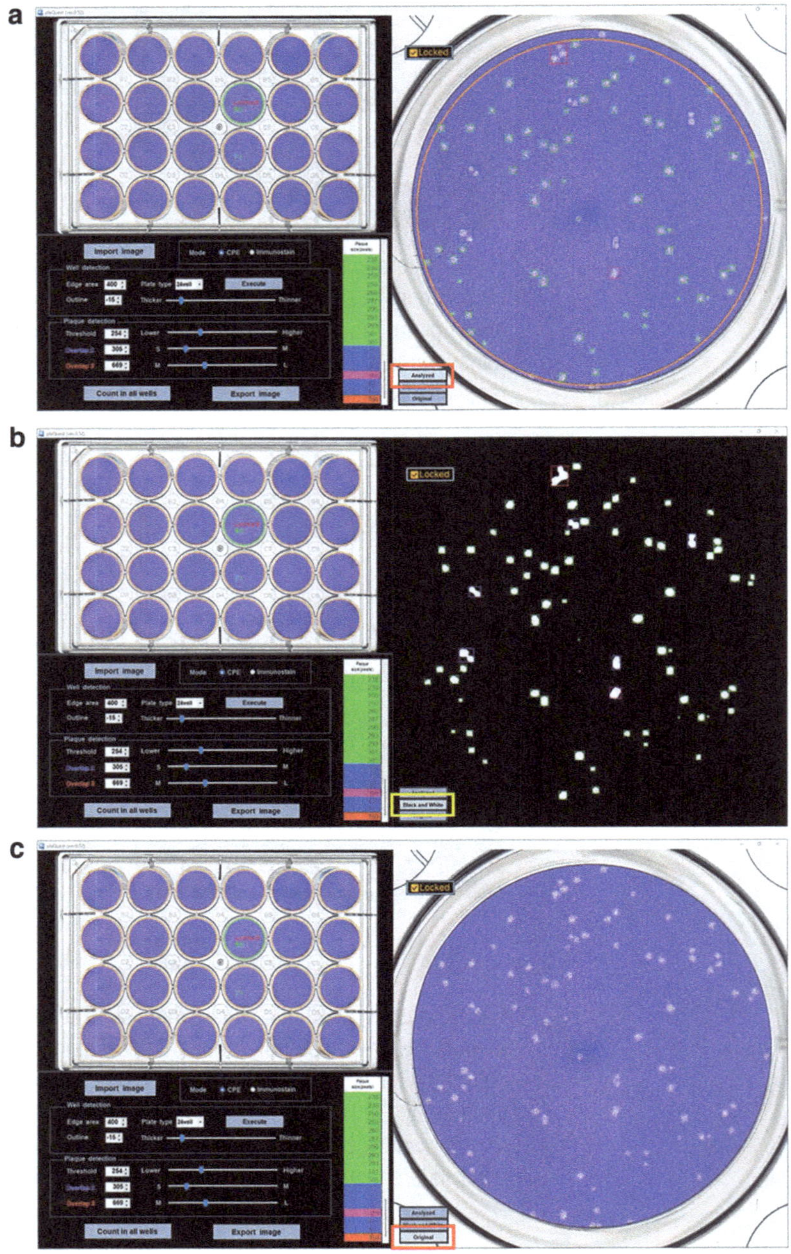

Fig. 10 Three different views of the Single View panel. (**a**) Analyzed view (by clicking the “Analyzed” button [red frame]). (**b**) Black and white view (by clicking the “Black and White” button [yellow frame]). (**c**) Original view (by clicking the “Original” button [red frame])

4 Notes

1. plaQuest is primarily designed and optimized for counting plaques in 24-well plates. However, it has been shown that plaQuest can also analyze other plate formats—such as 6-, 12-, 48-, and 96-well plates—if clear, separated plaques are produced in the plaque assay [25].
2. The plaQuest folder can be placed in any directory on a Windows PC, enabling the plaQuest software to be run from the desktop or a USB flash drive.
3. The plaQuest folder contains three dynamic link library (.dll) files ("OpenCvSharp.dll," "OpenCvSharp.Extensions.dll," and "OpenCvSharpExtern.dll") that are necessary to run the program.
4. When scanning a plaque assay plate, you can choose from grayscale (8- or 16-bit) or color (24- or 48-bit) options. We typically select the 48-bit color mode. However, the optical resolution must be set to 1200 dpi. While the capture size does not need to be exactly 127 × 83 mm, it should be kept as close as possible to the size to ensure subsequent detection of culture wells in the plaque assay plate.
5. As mentioned in Sect. 1, plaQuest has been shown to be applicable to immunostain-based plaque assays for viruses that exhibit weaker cytotoxicity in cultured cells. Examples include the SARS-CoV-2 Omicron variant and DENV [25]. Additionally, it has been demonstrated that plate types other than the 24-well plate—such as 6-, 12-, 48-, and 96-well plates—can be analyzed using plaQuest. However, regardless of the staining type or plate format used, it is essential to optimize the plaque assay conditions to obtain reliable results with plaQuest. Sufficient contrast that clearly distinguishes infected (i.e., plaque) areas from uninfected (i.e., background) areas is a key factor for plaque detection in plaQuest. Based on our experience, it is desirable for the gray-value contrast, as measured by the Plot Profile tool in ImageJ (https://imagej.net/ij/), between the plaque and background (a gray value of plaque area divided by the gray value of background area) to be 1.4 or higher. This contrast appears to be significantly influenced by the staining method, cell type, cell number, and culture duration [25].
6. To update the total number of plaques, double-click a well in the Whole View panel again. The new plaque count, as reflected in the new parameter settings, is displayed on the well.
7. The "Overlap 2" and "Overlap 3" parameters define the number of plaques contained in the recognized area based on their

pixel sizes. Therefore, note that other areas may be counted as two or three plaques if the pixel size is the same, even if they do not actually contain two or three plaques.

8. This process takes between 30 s and 1 min, but it may take longer depending on the processing power of the Windows PC or the total number of plaques to be counted. Please be patient.

Acknowledgments

This work was supported by the Research Program on Emerging and Re-emerging Infectious Diseases of the Japan Agency for Medical Research and Development (Grant No. JP24fk0108656).

References

1. Schwartz O, Albert ML (2010) Biology and pathogenesis of chikungunya virus. Nat Rev Microbiol 8:491–500
2. Mourad O, Makhani L, Chen LH (2022) Chikungunya: an emerging public health concern. Curr Infect Dis Rep 24:217–228
3. Paul BJ, Sadanand S (2018) Chikungunya infection: a re-emerging epidemic. Rheum Ther 5:317–326
4. Staikowsky F, Talarmin F, Grivard P et al (2009) Prospective study of chikungunya virus acute infection in the Island of La Réunion during the 2005–2006 outbreak. PLoS One 4:e7603
5. Soumahoro M-K, Gérardin P, Boëlle P-Y et al (2009) Impact of chikungunya virus infection on health status and quality of life: a retrospective cohort study. PLoS One 4:e7800
6. Kumar R, Ahmed S, Parray HA et al (2021) Chikungunya and arthritis: an overview. Travel Med Infect Dis 44:102168
7. Wikan N, Sakoonwatanyoo P, Ubol S et al (2012) Chikungunya virus infection of cell lines: analysis of the East, Central and South African lineage. PLoS One 7:e31102
8. Roberts GC, Zothner C, Remenyi R et al (2017) Evaluation of a range of mammalian and mosquito cell lines for use in chikungunya virus research. Sci Rep 7:14641
9. Sourisseau M, Schilte C, Casartelli N et al (2007) Characterization of reemerging chikungunya virus. PLoS Pathog 3:e89
10. Her Z, Malleret B, Chan M et al (2010) Active infection of human blood monocytes by chikungunya virus triggers an innate immune response. J Immunol 184:5903–5913
11. Salvador B, Zhou Y, Michault A et al (2009) Characterization of chikungunya pseudotyped viruses: identification of refractory cell lines and demonstration of cellular tropism differences mediated by mutations in E1 glycoprotein. Virology 393:33–41
12. Suzuki Y, Tanaka A, Maeda Y et al (2021) Construction and characterization of an infectious clone generated from chikungunya virus SL11131 strain. Virology 552:52–62
13. Labadie K, Larcher T, Joubert C et al (2010) Chikungunya disease in nonhuman primates involves long-term viral persistence in macrophages. J Clin Investig 120:894–906
14. Couderc T, Lecuit M (2015) Chikungunya virus pathogenesis: from bedside to bench. Antivir Res 121:120–131
15. Dulbecco R (1952) Production of plaques in monolayer tissue cultures by single particles of an animal virus. Proc Nat Acad Sci U S A 38:747–752
16. Condit RC (2007) Principles of virology. In: Knipe DM, Howley PM (eds) Fields virology, 5th edn. Lippincott-Williams & Wilkins, Philadelphia, pp 25–57
17. Flint SJ, Enquist LW, Racaniello VR, Skalka AM (2000) Virus cultivation, detection, and genetics. In: Principles of virology: molecular biology, pathogenesis, and control. ASM, pp 25–56
18. Sullivan K, Kloess J, Qian C et al (2012) High throughput virus plaque quantitation using a flatbed scanner. J Virol Methods 179:81–89
19. Kang W, Shin E-C (2012) Colorimetric focus-forming assay with automated focus

counting by image analysis for quantification of infectious hepatitis C virions. PLoS One 7:e43960
20. Boonyasuppayakorn S, Suroengrit A, Srivarangkul P et al (2016) Simplified dengue virus microwell plaque assay using an automated quantification program. J Virol Methods 237:25–31
21. Culley S, Towers GJ, Selwood DL et al (2016) Infection counter: automated quantification of in vitro virus replication by fluorescence microscopy. Viruses 8:201
22. Katzelnick LC, Escoto AC, McElvany BD et al (2018) Viridot: an automated virus plaque (immunofocus) counter for the measurement of serological neutralizing responses with application to dengue virus. PLoS Negl Trop Dis 12:e0006862
23. Cacciabue M, Currá A, Gismondi MI (2019) ViralPlaque: a Fiji macro for automated assessment of viral plaque statistics. PeerJ 7:e7729
24. Phanomchoeng G, Kukiattikoon C, Plengkham S et al (2022) Machine-learning-based automated quantification machine for virus plaque assay counting. PeerJ Comput Sci 8:e878
25. Emi A, Suzuki Y, Odake C et al (2025) Development of an automated plaque-counting program for the quantification of the chikungunya virus. Sci Rep 15:12429

Chapter 7

Production of Viral Particles from a Chikungunya Virus Infectious Clone

Eugenia S. Bardossy and Claudia Filomatori

Abstract

Chikungunya virus (CHIKV) is a positive-sense single-stranded RNA virus, which poses challenges for its study and genetic manipulation. Because direct mutagenesis of viral RNA genomes is technically impractical, reverse genetics systems are essential tools for investigating viral biology. To enable such approaches, infectious clones containing a full-length cDNA copy of the viral genome are constructed. The cDNA is positioned under the control of a bacteriophage RNA polymerase promoter, allowing commercial RNA polymerases to use the linearized plasmid as a template for the in vitro transcription of full-length viral genomic RNA (gRNA). Importantly, positive-sense viral genomes serve as mRNAs for the translation of viral proteins in a cellular environment, meaning that these transcripts contain all the information required to initiate viral replication. Following transfection into permissive cultured cells, viral proteins are expressed, enabling genome replication and, ultimately, the recovery of infectious particles from the cell supernatant. Here, we describe a detailed procedure for generating CHIKV particles through plasmid linearization, in vitro transcription, and subsequent RNA transfection.

Keywords Infectious clone, In-vitro transcription, Virus rescue

1 Introduction

RNA viruses cannot be directly cloned or genetically manipulated [1]. This is largely due to the chemical instability of RNA and the fact that it cannot be replicated or handled using conventional molecular biology tools. Therefore, the gRNA must first be converted into a complementary DNA (cDNA) form, which allows stable propagation and genetic modification in bacteria. This principle underlies the development of infectious clones—plasmids containing a full-length, replication-competent cDNA copy of a viral genome (Fig. 1).

The development of the first infectious cDNA clones of positive-strand RNA viruses became feasible following the discovery of reverse transcriptase in the early 1970s, which enabled the conversion of viral RNA into complementary DNA and its stable propagation within a plasmid backbone [2, 3]. A major improvement was later

Claudia Filomatori and Eliana F. Castro (eds.), *Chikungunya Virus*, Methods in Molecular Biology, vol. 3067,
https://doi.org/10.1007/978-1-0716-5496-5_7,

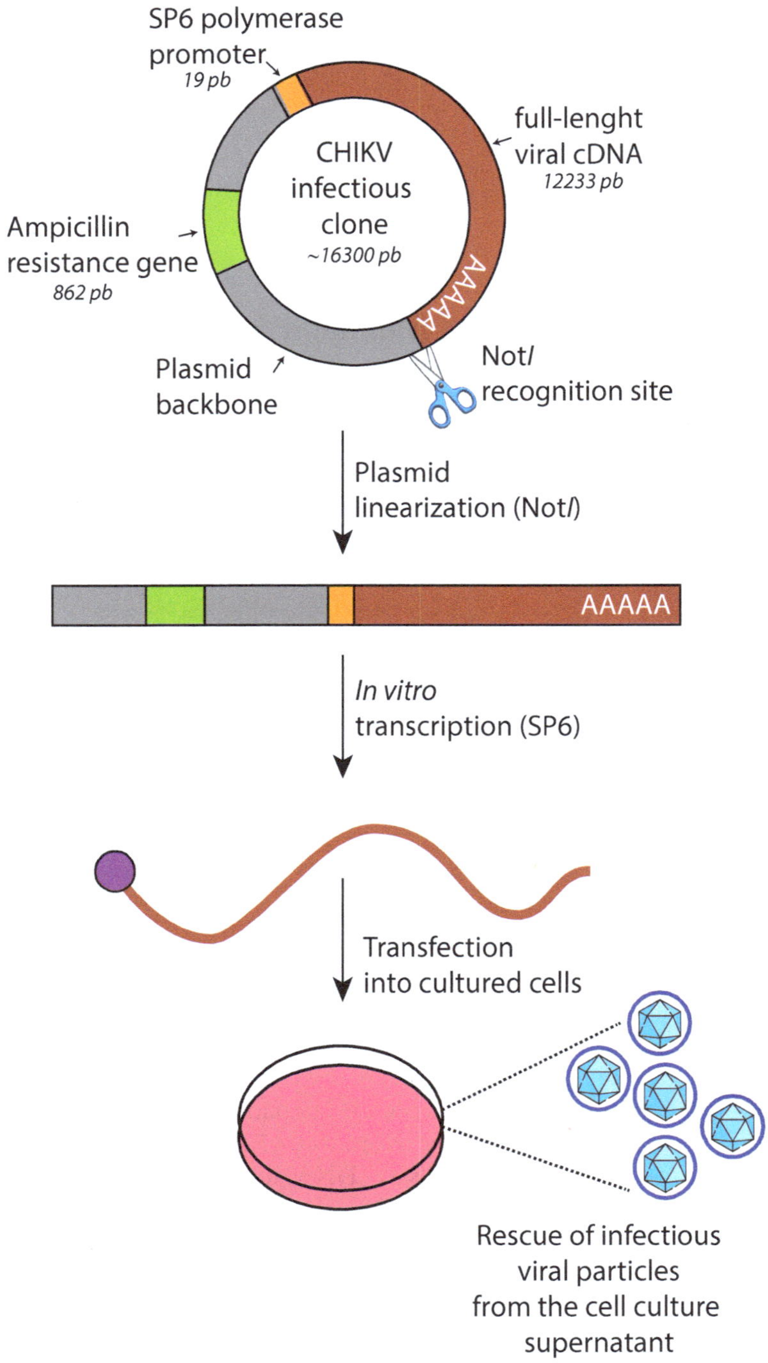

Fig. 1 Schematic representation of the workflow for generating infectious CHIKV from a cDNA infectious clone. The plasmid containing the full-length viral genome under the control of an SP6 promoter is linearized at the *NotI* restriction site to allow efficient in vitro transcription. The resulting capped RNA transcript, bearing a poly(A) tail, is subsequently transfected into permissive cultured cells, where viral replication is initiated. Infectious viral particles are rescued from the cell culture supernatant and can be further amplified or used for downstream applications

achieved by flanking viral cDNAs with a well-defined bacterial promoter at the 5′ end and a unique restriction site at the 3′ end, enabling the in vitro production of viral RNAs with precisely defined termini [4, 5]. The earlier use of the multi-component *E. coli* RNA polymerase for transcription was soon replaced by single-subunit bacteriophage RNA polymerases (SP6, T3, and T7), which became commercially available in recombinant form and can reliably generate long (>1 kb) RNA transcripts from compact promoter elements (~20 bp) [6, 7].

Importantly, the relatively short genomes of many RNA viruses make them particularly suitable for de novo chemical synthesis based on published sequence information, enabling the direct assembly of full-length viral cDNA without requiring infectious material [8]. In addition, numerous infectious cDNA clones are now readily accessible through scientific collaborations or available from biological repositories such as Addgene and BEI Resources.

Because the viral genome is maintained as cDNA within a plasmid, infectious clones can be stably propagated in bacteria under antibiotic selection. In addition, infectious clones serve as templates for the in vitro synthesis of full-length viral gRNA. To achieve this, the plasmid is first linearized using a unique restriction enzyme—located, in the case of chikungunya virus (CHIKV), immediately downstream of the viral poly(A) tail [9, 10]. The linearized cDNA is then used for in vitro transcription by commercially available bacteriophage RNA polymerases, generating capped gRNA molecules that can initiate infection when delivered into permissive cells. Upon entering cells, positive-strand RNA virus genomes are directly translated by cellular ribosomes [11] to produce the viral proteins needed for viral replication. In fact, the CHIKV genome encodes two open reading frames (ORFs) (Fig. 2a) [12]. The first ORF is translated into a polyprotein that is processed into the nonstructural proteins responsible for RNA replication (Fig. 2b) [13]. The replication complex synthesizes a negative-strand RNA intermediate, which is subsequently used to produce gRNA and subgenomic RNA (sgRNA) [14]. The sgRNA encodes the structural proteins, which drive virion assembly, genome encapsidation, and release of infectious particles into the culture supernatant.

Infectious clones have broad and powerful applications in virology. One of their primary uses is the generation of viruses with a defined and genetically homogeneous background, avoiding the diversity that arises during repeated viral passages. This genetic stability is particularly advantageous for maintaining live attenuated vaccine strains, which may otherwise revert to virulence [15]. In addition, infectious clones enable precise genetic manipulation using standard molecular biology techniques, facilitating targeted nucleotide substitutions, insertions, or deletions to investigate viral replication, host interactions, and mechanisms of pathogenesis [10, 16]. Infectious clones are also used to engineer reporter viruses and replicons that express easily quantifiable proteins—such as luciferase

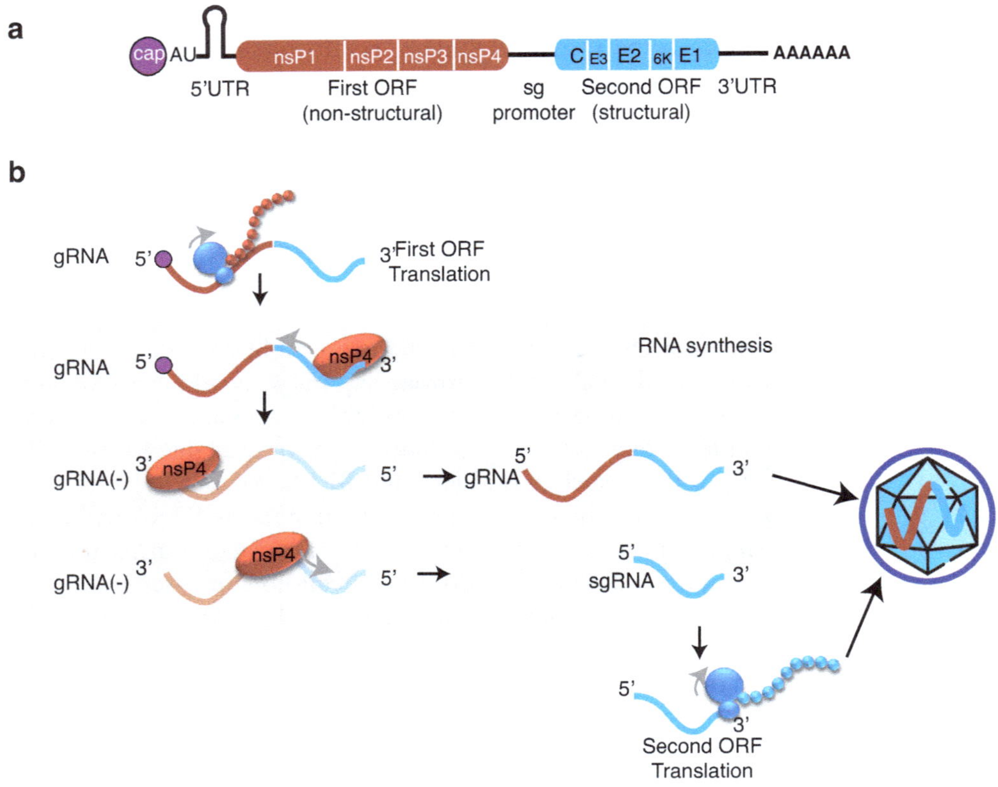

Fig. 2 Genome organization and replication strategy of CHIKV. (**a**) Schematic representation of the positive-sense RNA genome, showing the 5′ cap, the nonstructural polyprotein encoded by the first ORF (nsP1–nsP4), the subgenomic promoter, and the structural polyprotein encoded by the second ORF (C, E3, E2, 6K, and E1), followed by the poly(A) tail at the 3′ end. (**b**) Overview of the viral replication cycle. The first ORF is translated from the gRNA, producing the nonstructural proteins that assemble into the replication complex, in which nsP4 functions as the RNA-dependent RNA polymerase. The complex first synthesizes the negative-strand intermediate (gRNA(−)), which is used as a template to produce both gRNA and sgRNA molecules. The sgRNA encodes the second ORF, which is translated into the structural proteins that encapsidate the gRNA to assemble new infectious particles

or fluorescent markers—in a replication-dependent manner [17–19]. Moreover, infectious clone platforms can be adapted for the delivery of heterologous antigens, supporting the development of novel viral vector vaccines [20].

In summary, infectious clone systems provide a robust, genetically tractable, and versatile platform for reverse genetics, enabling applications ranging from basic research in viral biology to antiviral drug screening, pathogenesis studies, and vaccine design. In this chapter, we describe detailed laboratory procedures for generating infectious CHIKV particles from an infectious clone through plasmid linearization, in vitro transcription of genomic RNA, and subsequent RNA transfection into cultured cells (Fig. 1).

2 Materials

2.1 Plasmid Linearization

1. CHIKV infectious clone (*see* **Note 1**).
2. *Not*I restriction enzyme.
3. 10× Buffer O.
4. Nuclease-free water.
5. 37°C incubator.
6. Micropipettes (P2, P10).
7. RNase- and DNase-free sterile plasticware (10 μL pipette tips, Eppendorf tubes).

2.2 In Vitro Transcription and RNA Quality Assessment

1. In vitro transcription kit (e.g., mMESSAGE mMACHINE™ Transcription Kit, ThermoFisher), including SP6 enzyme mix, 2× NTP/Cap mixture: 10 mM ATP, 10 mM CTP, 10 mM UTP, 2 mM GTP, 8 mM $m^7G(5')ppp(5')G$ cap analog, 20 mM GTP, 10× buffer, 2 U/μL TURBO DNase, and lithium chloride (LiCl) precipitation solution (7.5 M LiCl, 50 mM EDTA). Store at −20°C.
2. 70% (v/v) ethanol.
3. 1% agarose in 1× TAE buffer.
4. 10× TAE buffer: 0.4 M Tris-HCl (pH 8.0), 0.4 M acetate, 10 mM EDTA.
5. DNA-staining dye: ethidium bromide, GelRed, or SYBR Safe.
6. DNA loading dye (often included with molecular weight standards).
7. 1 kb DNA ladder.
8. Micropipettes (P2, P10).
9. RNase- and DNase-free sterile plasticware (10 μL pipette tips, Eppendorf tubes).
10. 37°C incubator.
11. Microcentrifuge.
12. Microwave.
13. Gel documentation system.
14. Ice bucket.
15. Gel casting tray.
16. Electrophoresis chamber with corresponding power supply.
17. Nanodrop or Qubit spectrophotometer.

2.3 Cell Culture

1. Cells: Mammalian BHK-21 cells (*Mesocricetus auratus* hamster kidney, ATCC, CCL-10), Huh-7 cells (human hepatocyte cell line, ATCC, CVCL_0336), and C6/36 cells (*Aedes albopictus*, ATCC, CRL-1660) (*see* **Note 2**).
2. Maintenance culture medium: For Huh-7 cells, DMEM high-glucose medium supplemented with 10% FBS (fetal bovine serum) and 1% penicillin/streptomycin (P/S). For BHK-21 cells, MEM alpha medium supplemented with 10% FBS and 1% P/S. For C6/36 cells, Leibovitz L-15 medium supplemented with 10% FBS, 10% tryptose phosphate, 1% P/S, and amphotericin B. Store at 4°C.
3. Cell washing buffer: Dulbecco's phosphate-buffered saline (PBS). Store at 4°C.
4. 2.5% (v/v) trypsin (10×). The 10× trypsin can be stored at 4°C until the expiration date or at −20°C for 1 year. To prepare the 1× working solution, dilute 1 mL of 2.5% (v/v) trypsin in 9 mL of PBS. Store at 4°C for 3 months.
5. Neubauer counting chamber.
6. Inverted microscope.
7. 5% CO_2 incubator at 37°C.
8. Sterile plasticware (1000 and 200 μL pipette tips, 35 mm dishes, and 24-well plates).
9. Micropipettes (P2, P10, P200, P1000).
10. Electrophoresis chamber with corresponding power supply.
11. Nanodrop or Qubit spectrophotometer.

2.4 gRNA Transfection into Culture Cells

1. Lipofectamine® 2000 Reagent.
2. Opti-MEM® Medium: reduced-serum medium.
3. Infection culture medium: For Huh-7 cells, DMEM high-glucose medium supplemented with 2% FBS (fetal bovine serum) and 1% P/S. For BHK-21 cells, MEM alpha medium supplemented with 2% FBS and 1% P/S. For C6/36 cells, Leibovitz L-15 medium supplemented with 2% FBS, 10% tryptose phosphate, 1% P/S, and amphotericin B. Store at 4°C.
4. Cell washing buffer: Dulbecco's phosphate-buffered saline (PBS). Store at 4°C.
5. Micropipettes (P2, P10, P200, P1000).
6. 5% CO_2 incubator at 37°C.

3 Methods

3.1 Plasmid Linearization

To ensure high-quality RNA synthesis and robust viral titers, it is recommended to begin with a pure, concentrated stock of the infectious clone (Fig. 3a). The integrity of the plasmid can be verified by running an aliquot on an agarose gel; a predominant supercoiled band typically indicates a high-quality template, whereas smearing may suggest degradation that could lead to truncated transcripts. Additionally, if the plasmid stock is aged or was obtained from an external source, it may be beneficial to perform full-plasmid sequencing. This

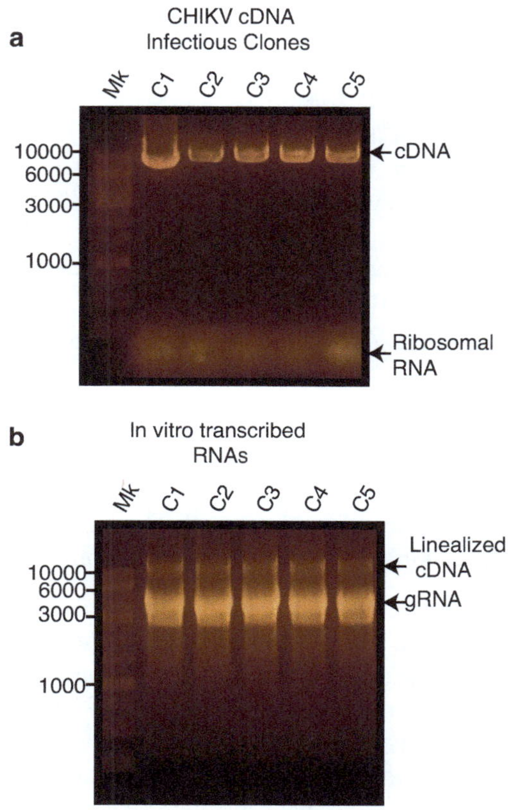

Fig. 3 Quality control of CHIKV infectious clones and in vitro–transcribed CHIKV gRNAs. (**a**) Agarose gel electrophoresis of CHIKV infectious clone plasmids (C1–C5) derived from the Indian Ocean Lineage strain after in-house miniprep isolation. Diffuse low-molecular-weight bands in the front correspond to bacterial ribosomal RNA. Mk indicates the 1 kb DNA molecular size ladder, and size markers are shown on the left in base pairs (bp). Following purification through spin columns and linearization with *NotI* restriction enzyme, the cDNA was used as a template for in vitro transcription using SP6 polymerase. (**b**) Agarose gel electrophoresis of in vitro–transcribed RNAs derived from each infectious clone (C1–C5), showing intact full-length gRNA transcripts. Note that gRNAs migrate faster than the corresponding linearized cDNA templates. The absence of RNA smearing indicates high-quality transcription products. Notably, gRNAs derived from certain CHIKV strains, such as the Caribbean strain, appear as two distinct bands corresponding to different RNA conformations

helps confirm that no spontaneous mutations or deletions—which can occur during bacterial amplification—are present to compromise the functionality of the infectious clone.

After quality confirmation, the circular DNA plasmid must be linearized to serve as a suitable template for in vitro transcription. Here, we describe the procedure for the infectious clones corresponding to the Indian Ocean lineage (IOL) CHIKV [21] and the Caribbean CHIKV lineage [22]. Both clones contain a unique *Not*I recognition site located immediately after the poly(A) sequence in the plasmid, downstream of the viral 3′ UTR.

1. Assemble the restriction enzyme reaction in a sterile 1.5 mL microcentrifuge tube.

Components	Volume	Final conc.
10× Buffer O	3 μL	1×
*Not*I (10 U/μL)	1 μL	10 U
Purified plasmid DNA (*see* **Note 3**)	13 μL	~500 ng/μL
Milli-Q® water	13 μL	
Total volume	30 μL	

2. Incubate at 37°C for 2 h.
3. Run 1 μL of the DNA on a 1% agarose gel to check linearization.

3.2 In Vitro Transcription and RNA Quality Assessment

Perform in vitro transcription using the kits available in your laboratory, following the manufacturer's instructions. Here, in vitro transcription using the mMESSAGE mMACHINE™ Kit from Thermo Fisher Scientific is described. The basic steps are outlined below, but the full instructions are available from the manufacturer.

1. Work in a clean environment and use RNase-free pipettes, tips, and tubes, and with fresh gloves. Use recombinant linearized DNA constructs as templates for transcription with SP6 polymerase in the presence of $m^7G(5')ppp(5')G$ cap structure using the mMESSAGE mMACHINE™ Transcription Kit (Thermo Fisher). Thaw the 2× NTP/Cap mixture on ice and the 10× buffer at room temperature. Briefly centrifuge all reagents before opening the tubes. Assemble the reaction at room temperature (*see* **Note 4**) and mix the reaction by gently pipetting up and down; do not vortex:

Components	Volume	Final conc.
Milli-Q®	0.5 μL	
2× NTP/Cap mixture	2.5 μL	5 mM ATP, 5 mM CTP, 5 mM UTP, 1 mM GTP, 4 mM Cap

Components	Volume	Final conc.
10× buffer	0.5 μL	1×
GTP (*see* **Note 5**)	0.25 μL	1 mM
Enzyme mix	0.5 μL	
Linearized plasmid	1 μL	100 ng/μL
Total	5 μL	

2. Incubate the reaction at 37°C for 2 h.
3. (optional) Add 0.25 μL 2 U/μL TURBO DNase, mix well, and incubate 15 min at 37°C (*see* **Note 6**).
4. (optional) Precipitate RNA with lithium chloride to remove unincorporated nucleotides and most proteins (*see* **Note 7**). Stop the reaction and precipitate the RNA by adding 15 μL nuclease-free water and 15 μL LiCl precipitation solution. Mix thoroughly. Chill for ≥30 min at −80°C. Centrifuge at 4°C for 30 min at maximum speed to pellet the RNA. Carefully remove the supernatant. Wash the pellet once with ~1 mL 70% (v/v) ethanol and re-centrifuge to maximize removal of unincorporated nucleotides. Carefully remove the 70% ethanol and resuspend the RNA in 20 μL nuclease-free water. Determine the RNA concentration and store frozen at −70°C.
5. Quantify the RNA by UV light absorbance reading the A260 of 1:10 dilution of the reaction. For single-stranded RNA, 1 A260 unit corresponds to 40 μg/mL, so the RNA yield can be calculated as follows: A260 × dilution factor ×40 = μg/mL RNA (*see* **Note 8**).
6. Run 1 μL of the RNA product on a 1% agarose gel to check integrity (*see* **Note 9**) (Fig. 3b). For some CHIKV strains (such as the Caribbean strain RNA), it is usual to see two bands, which correspond to different RNA conformations.
7. Store the remaining RNA at −80°C until use.

3.3 Cell Plating

1. Grow BHK-21 or Huh-7 cells at 37°C in a 5% CO_2 atmosphere, and C6/36 cells at 28°C without CO_2 control, each cell line in its corresponding maintenance medium.
2. The day before the RNA transfection, examine the monolayer of cultured cells under the microscope.
3. Remove the maintenance culture medium and gently wash the cells with the washing buffer (300 μL buffer per 35 mm plate).

4. Remove the washing buffer and add 0.25% (v/v) trypsin (300 μL trypsin per 35 mm plate). Ensure the trypsin covers the entire cell layer, then incubate at 37°C for 5 min (*see* **Note 10**).
5. Check under the microscope to ensure the cells are fully detached.
6. Add maintenance culture medium to bring the final volume up to 3 mL to neutralize the trypsin and dilute the cells. Pipette gently to ensure the cells are well separated from each other.
7. Quantify the cells using a Neubauer counting chamber. First, clean the chamber surface and coverslip with 70% (v/v) ethanol to remove dust and oil. Place the coverslip, then load 15–20 μL of the cell suspension. Under the microscope, count the cells located in the four large corner quadrants. Calculate cell concentration using Eq. (1):

$$N^{\circ}\text{cell/mL} = [(Q_1 + Q_2 + Q_3 + Q_4)/4] \times 10^4 \qquad (1)$$

where Q_1, Q_2, Q_3, and Q_4 are the numbers of cells counted in each of the four quadrants.

8. After calculating the concentration, plate the volume corresponding to 50,000 BHK-21 or Huh-7 cells, or 10,000 C6/36 cells per well in a 24-multiwell plate (*see* **Notes 11 and 12**).
9. Add maintenance culture medium to each well to bring the final volume up to 600 μL (*see* **Note 13**).
10. Incubate the cells overnight at 37°C in a CO_2 incubator (BHK-21 and Huh-7) or at 28°C without CO_2 (C6/36).

3.4 gRNA Transfection into Culture Cells

1. The next day, assess cell confluency and viability before starting the assay. The cells should exhibit their typical morphology and reach 70%–90% confluency at the time of transfection (*see* **Note 14**).
2. Dilute 1 μL Lipofectamine® 2000 Reagent in 50 μL Opti-MEM® Medium per well of a 24-well plate (*see* **Notes 15 and 16**).
3. Dilute 500 ng gRNA in 50 μL Opti-MEM® Medium per well of a 24-well plate (*see* **Note 17**).
4. Add diluted gRNA to diluted Lipofectamine® 2000 Reagent (1:1 ratio).
5. Incubate 20 min at room temperature.
6. Remove the culture medium from the cells.
7. Add 100 μL Opti-MEM® Medium to the cells (*see* **Note 18**).
8. Add drop by drop, 1:1 RNA–lipid complex to cells.

9. Incubate BHK-21 or Huh-7 cells at 37°C, and C6/36 cells at 28°C, for 2–3 h.
10. Remove the transfection mix, wash the cells with the washing medium, and add 600 µL of fresh infection medium to each well (*see* **Note 19**).
11. Incubate at 37°C or 28°C depending on the cell line.
12. Examine the monolayer of infected cells daily under the microscope to monitor the onset and progression of cytopathic effect (only visible in mammalian cell lines; *see* **Notes 20 and 21**). Once cytopathic effect becomes evident (typically after 2–3 days), collect the cell-culture supernatants to rescue viral particles.
13. If desired, clarify by centrifuging at 3000 × *g* for 5 min at 4°C. This step removes cell debris and dead cells. Transfer the supernatant into a new tube.
14. Collect a 10 µL aliquot of the supernatant to perform viral particle titration following the protocol described in Chap. 5 (*see* **Note 22**).
15. Store the virus stocks at −80°C until further use (*see* **Note 23**).

4 Notes

1. Different CHIKV infectious clones have been generated from viral isolates obtained from patient serum during various outbreaks. For instance, an infectious clone representing the Indian Ocean lineage (IOL) was derived from viruses circulating during the 2005–2006 epidemics in La Réunion Island [21]. Similarly, an infectious clone corresponding to the Caribbean lineage was generated from viruses isolated during the 2013 outbreak in the Americas [22].
2. Any other permissive cell line can be used for RNA transfection. Some cell lines, such as mammalian BHK-21 or mosquito C6/36 cells, are deficient in antiviral defense mechanisms and therefore support higher viral yields. In contrast, other cell lines, including human Huh-7 and mosquito U4.4 cells, are immunocompetent and restrict viral replication. Additionally, transfection efficiency varies depending on the cell type. For example, human HEK cells are easily transfected, whereas mosquito cell lines are more difficult to transfect, resulting in a lower proportion of transfected cells.
3. DNA templates must be free of contaminating proteins and RNA, as impurities can interfere with the efficiency of in vitro

transcription reactions. Most commercially available spin-column plasmid purification kits produce DNA of sufficient quality for this purpose. Alternatively, template DNA can be further purified using phenol–chloroform extraction.

4. Keep the 10× Reaction Buffer at room temperature while assembling the reaction. The spermidine in the 10× Reaction Buffer can coprecipitate the template DNA if the reaction is assembled on ice.
5. For long templates (>5 kb), such as CHIKV cDNA, supplementing the reaction with additional GTP is recommended.
6. This step is particularly important when downstream analyses, such as sequencing of progeny viral genomes, are planned.
7. This step is required when spectrophotometric quantification is intended, as any unincorporated nucleotides or residual template DNA in the reaction mixture would artificially contribute to the absorbance reading.
8. The amount of RNA synthesized should be about 1 μg/μL. While UV spectrophotometry is common, using fluorometric-based quantification (e.g., Qubit) is recommended for higher precision. These reagents are more accurate than UV methods as they specifically measure RNA, minimizing interference from residual DNA or contaminants and ensuring a more reliable quantification for downstream transfection.
9. To remove RNases from the electrophoresis chamber, add six to ten pellets of sodium or potassium hydroxide to the chamber, then fill with distilled or Milli-Q water. Incubate for 15 min at room temperature. Afterward, thoroughly rinse the chamber several times with distilled water to ensure complete removal of the hydroxide.
10. If the cells do not easily detach from the well, one approach is to prewash them with 1× trypsin before adding the final trypsin solution.
11. Before plating, gently but thoroughly mix the cell suspension with a pipette to ensure the cells are evenly suspended and have not settled.
12. Transfections can be scaled up or down in format, as needed.
13. After plating, gently rock the plate in a circular or zigzag motion to help cells spread uniformly across the well surface.
14. If the cells are overly confluent, transfection efficiency decreases, whereas if they are too sparsely confluent, they may be more sensitive to Lipofectamine, resulting in a marked reduction in cell viability.

15. Transfection works better in a reduced-serum medium such as Opti-MEM; otherwise, a non-FBS-supplemented maintenance medium may be used.
16. Mosquito C6/36 cells are more difficult to transfect than mammalian BHK-21 cells and may require a higher amount of Lipofectamine® Reagent (1.5–2.0 μL per well). Also, specialized versions such as Lipofectamine 3000, LTX, or Effectine may be preferred to transfect other target cell lines.
17. A master mix can be prepared to accommodate all wells to be transfected.
18. The addition of extra Opti-MEM to the wells prevents the cells from drying out.
19. Lipofectamine can be toxic to cells; therefore, prolonged incubation with DNA–Lipofectamine complexes is not recommended. However, toxicity may vary depending on the cell type and the confluency at the time of transfection, so incubation time can be extended under certain conditions. Typically, 2–3 h is sufficient to allow complex entry while minimizing cytotoxic effects.
20. In mammalian cells, CHIKV infection typically induces a cytopathic effect characterized by cell rounding, shrinkage, and detachment from the monolayer due to virus-mediated shut-off of host functions and cell death. The optimal time to collect CHIKV stocks is when cytopathic effect reaches approximately 70% of the mammalian cell monolayer, before extensive cell lysis occurs.
21. Mosquito cell lines typically do not exhibit a noticeable cytopathic effect upon CHIKV infection, even when viral replication is efficient, and they can sustain persistent infections.
22. In general, viral titers obtained after transfection are not very high (typically 10^4–10^5PFU/mL). Titer can be increased (up to ~ 10^9 PFU/mL) by using the transfection supernatant to infect fresh cells. However, the disadvantage of performing successive passages is that RNA virus genomes accumulate adaptive mutations during replication in cell culture.
23. Avoid repeated freezing and thawing of viral stocks, as the titer decreases with each freeze–thaw cycle.

References

1. (2022) Reinventing positive-strand RNA virus reverse genetics, In: Advances in virus research. Elsevier, pp 1–29
2. Taniguchi T, Palmieri M, Weissmann C (1978) Qβ DNA-containing hybrid plasmids giving rise to Qβ phage formation in the bacterial host. Nature 274:223–228
3. Racaniello VR, Baltimore D (1981) Cloned poliovirus complementary DNA is infectious in mammalian cells. Science 214:916–919

4. Ahlquist P, Janda M (1984) cDNA cloning and in vitro transcription of the complete brome mosaic virus genome. Mol Cell Biol 4:2876–2882
5. Ahlquist P, French R, Janda M et al. (1984) Multicomponent RNA plant virus infection derived from cloned viral cDNA. Proc Natl Acad Sci U S A 81:7066–7070
6. Janda M, French R, Ahlquist P (1987) High efficiency T7 polymerase synthesis of infectious RNA from cloned brome mosaic virus cdna and effects of 5′ extensions on transcript infectivity. Virology 158:259–262
7. Van Der Werf S, Bradley J, Wimmer E et al. (1986) Synthesis of infectious poliovirus RNA by purified T7 RNA polymerase. Proc Natl Acad Sci U S A 83:2330–2334
8. Mueller S, Coleman JR, Wimmer E (2009) Putting synthesis into biology: a viral view of genetic engineering through de novo gene and genome synthesis. Chem Biol 16:337–347
9. Filomatori CV, Bardossy ES, Merwaiss F et al (2019) RNA recombination at chikungunya virus 3'UTR as an evolutionary mechanism that provides adaptability. PLoS Pathog 15(4):e1007706
10. Bardossy ES, Volpe S, Alvarez DE et al. (2023) A conserved Y-shaped RNA structure in the 3'UTR of chikungunya virus genome as a host-specialized element that modulates viral replication and evolution. PLoS Pathog 19:e1011352
11. Baltimore D (1971) Expression of animal virus genomes. Bacteriol Rev 35:235–241
12. Rupp JC, Sokoloski KJ, Gebhart NN et al. (2015) Alphavirus RNA synthesis and non-structural protein functions. J Gen Virol 96:2483–2500
13. Tan YB, Chmielewski D, Law MCY et al. (2022) Molecular architecture of the Chikungunya virus replication complex. Sci Adv 8:eadd2536
14. Lemm JA, Rice CM (1993) Roles of nonstructural polyproteins and cleavage products in regulating Sindbis virus RNA replication and transcription. J Virol 67:1916–1926
15. Kohara M, Abe S, Kuge S et al. (1986) An infectious cDNA clone of the poliovirus sabin strain could be used as a stable repository and inoculum for the oral polio live vaccine. Virology 151:21–30
16. Bardossy ES, Volpe S, Suzuki Y et al. (2024) Molecular basis of RNA recombination in the 3′UTR of chikungunya virus genome. Nucleic Acids Res 52:9727–9744
17. Filomatori CV, Lodeiro MF, Alvarez DE et al. (2006) A 5' RNA element promotes dengue virus RNA synthesis on a circular genome. Genes Dev 20:2238–49
18. Villordo SM, Filomatori CV, Sánchez-Vargas I et al. (2015) Dengue virus RNA structure specialization facilitates host adaptation. PLoS Pathog 11:e1004604
19. Villordo SM, Carballeda JM, Filomatori CV et al. (2016) RNA structure duplications and flavivirus host adaptation. Trends Microbiol 24:270–283
20. Quintana V, Caillava J, Byk LA et al. (2025) Improvement in the potency of a N1-methylpseudouridine-modified self-amplifying RNA through mutations in the RNA-dependent RNA polymerase. J Biol Chem 301:110487
21. Coffey LL, Vignuzzi M (2011) Host alternation of chikungunya virus increases fitness while restricting population diversity and adaptability to novel selective pressures. J Virol 85:1025–1035
22. Stapleford KA, Moratorio G, Henningsson R et al. (2016) Whole-genome sequencing analysis from the chikungunya virus Caribbean outbreak reveals novel evolutionary genomic elements. PLoS Negl Trop Dis 10:e0004402

Chapter 8

Molecular Cloning and Reverse Genetics

Eugenia S. Bardossy

Abstract

This chapter describes a detailed molecular biology protocol for introducing specific point mutations into the chikungunya virus (CHIKV) genome using a reverse genetics strategy. The method utilizes an overlapping PCR-based approach to generate a mutated DNA fragment, which is then cloned into a pre-engineered CHIKV infectious clone plasmid. The protocol covers all major steps, from the initial PCRs to create the mutated insert to its digestion and ligation into the vector. It also includes procedures for bacterial transformation, colony screening via PCR and Sanger sequencing to confirm the mutation, and plasmid purification via miniprep. The document is structured with a clear introduction, a list of all required reagents and equipment, and a step-by-step methods section.

Keywords Reverse genetics, Overlapping PCR, Chikungunya virus, Molecular cloning, Site-directed mutagenesis

1 Introduction

Introducing targeted mutations into the genomes of RNA viruses is challenging due to the inherent instability of RNA molecules and the limited tools available for direct RNA manipulation. This led to the development of reverse genetics, a powerful molecular biology technique that allows us to manipulate the viral genome of RNA viruses in a more stable DNA format. Reverse genetics involves converting the viral RNA genome into a complementary DNA (cDNA) molecule, which is then cloned into a plasmid to create an infectious clone [1]. A key advantage of this approach is the ability to introduce specific mutations using site-directed mutagenesis techniques [2].

A common method to introduce mutations into DNA samples is overlapping PCR-based mutagenesis (Fig. 1) [3]. This technique uses complementary oligonucleotide (oligo) primers and PCR to generate two DNA fragments with overlapping ends. These fragments are combined in a subsequent "fusion" reaction where the overlapping ends anneal, allowing each strand to serve as a primer for the extension of the complementary strand. The resulting full-length product

Claudia Filomatori and Eliana F. Castro (eds.), *Chikungunya Virus*, Methods in Molecular Biology, vol. 3067,
https://doi.org/10.1007/978-1-0716-5496-5_8,

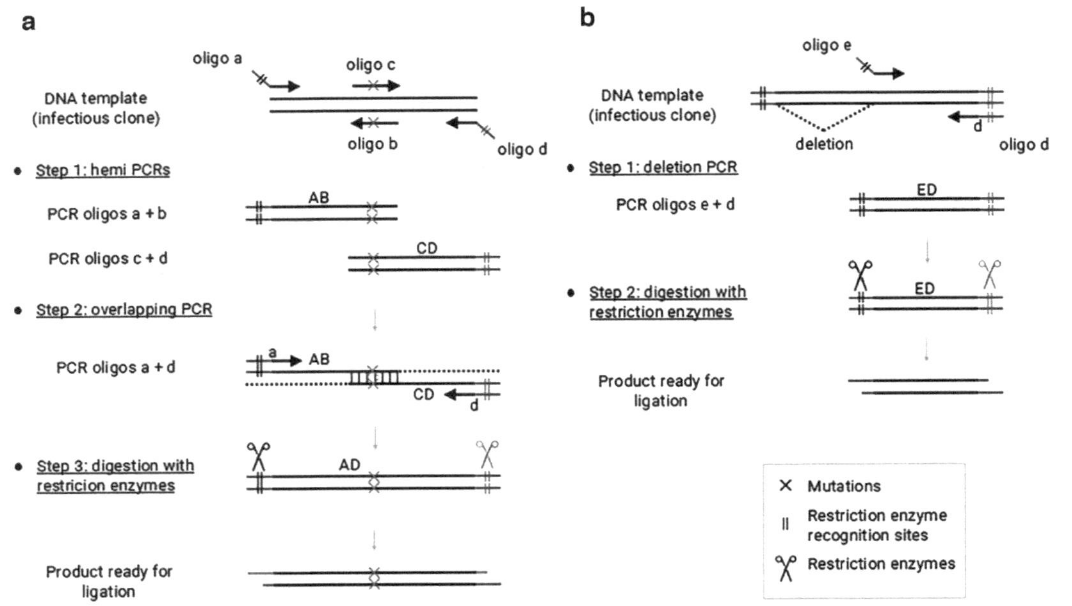

Fig. 1 Overlapping PCR-based mutagenesis. (**a**) Strategy to introduce point mutations. Two initial hemi-PCRs are performed using primers that introduce the desired mutation and overlap in the mutated region. Next, a second PCR is performed using the products of the first two reactions as templates, along with the external primers. The final product is then digested with restriction enzymes, preparing it for ligation into the vector. (**b**) Alternative strategy for introducing deletions. Alternatively, to the method shown in (**a**), deletions can be introduced by using a primer that binds after the desired deletion site. This primer must contain a restriction site. The resulting PCR product is then digested with the appropriate restriction enzyme, making it ready for ligation

is then amplified by PCR. Specific mutations can be introduced by incorporating nucleotide changes into the overlapping oligo primers. At the same time, specific restriction enzyme recognition sites can be introduced in the external oligos to facilitate the subsequent insertion of the PCR product into the infectious clone plasmid. This strategy allows for the precise introduction of single-nucleotide substitutions, deletions, or insertions, providing a flexible and robust platform for studying the function of specific viral genes or sequences.

This chapter describes a protocol for introducing mutations into the chikungunya virus (CHIKV) genome. The method uses a CHIKV Caribbean (CHIKV-Cbn) infectious clone [4], which contains the full-length cDNA genome preceded by an SP6 promoter sequence (*see* **Note 1**). To facilitate the exchange between the mutated and wild-type DNA fragments, two unique restriction enzyme sites flanking the mutation site are required: one upstream (enzyme a) and one downstream (enzyme d). This arrangement establishes a cloning cassette that enables the straightforward exchange of the wild-type sequence with a mutated one [5]. The workflow is as follows:

1. A mutated 3′UTR fragment is created using overlapping PCR.

2. The PCR product and the infectious clone are digested with enzyme a and enzyme d.
3. The digested PCR product is ligated into the digested vector, replacing the wild-type DNA with the mutated one.
4. The ligation product is used to transform competent bacterial cells.
5. Individual colonies are screened by PCR to identify those containing the desired mutation.
6. Finally, positive clones are grown in bacterial broth, and the plasmid is purified by plasmid minipreparation (miniprep).

2 Materials

2.1 PCR Reactions

1. Taq DNA Polymerase (5 U/μL) (*see* **Note 2**).
2. AccuPrime™ Taq DNA Polymerase (Invitrogen®) or similar.
3. 10× PCR buffer without Mg^{2+}.
4. 10× AccuPrime™ PCR Buffer (Invitrogen®) or similar.
5. 10 mM dNTP mixture.
6. 25 mM $MgCl_2$.
7. 10 μM forward and reverse oligos (oligos a, b, c, d) (*see* **Note 3**).
8. CHIKV-Cbn plasmid DNA (diluted to 10 ng/μL).
9. Milli-Q® water (molecular biology grade).

2.2 Agarose Gel Electrophoresis

1. 1% agarose in 1× TAE buffer.
2. 10× TAE buffer: 0.4 M Tris-HCl pH 8.0, 0.4 M acetate, 10 mM EDTA.
3. DNA-staining dye: ethidium bromide, GelRed, or SYBR Safe.
4. DNA loading dye (often included with molecular weight standards).
5. 1 kb DNA ladder.
6. 100 bp DNA ladder.

2.3 DNA Purification

1. Silica PCR purification kit (e.g., from Qiagen®).
2. Gel Extraction PCR kit (e.g., from Qiagen®).

2.4 Digestion and Ligation

1. Restriction enzyme a (10 U/μL).
2. Restriction enzyme d (10 U/μL).
3. 10× corresponding buffer for the restriction enzymes chosen.

4. T4 DNA Ligase (5 Weiss U/μL).
5. CHIKV-Cbn plasmid DNA (stock ~500 ng/μL).
6. Competent *E. coli* DH5α cells.
7. Luria-Bertani (LB) broth.
8. Luria-Bertani (LB) agar plates.
9. Ampicillin.

2.5 Colony Screening

1. Exonuclease I (*Exo* I).
2. Shrimp Alkaline Phosphatase (SAP).
3. Sanger sequencing primers (annealing in the plasmid, outside the 3′UTR).

2.6 Plasmid Miniprep

1. Miniprep kit.

2.7 Equipment

1. Thermal cycler (PCR machine).
2. Microcentrifuge.
3. Vortex mixer.
4. Micropipettes (P2, P10, P200, P1000).
5. Ice bucket.
6. Gel casting tray.
7. Electrophoresis cubet with corresponding power supply.
8. Gel documentation system (Gel Doc): An imager with a UV transilluminator and a camera to visualize and capture images of the ethidium bromide-stained DNA bands.
9. Microwave.
10. Scalpel or razor blade (to excise gel bands).
11. NanoDrop or Qubit spectrophotometer.
12. Heat block or water bath.
13. Shaking incubator.
14. 37°C incubator.

3 Methods

This section outlines the steps for introducing point mutations into the CHIKV genome using the strategy shown in Fig. 1a. For instructions on introducing deletions, *see* **Note 4**.

3.1 Hemi-PCRs

To generate the mutated fragment, perform two separate hemi-PCR reactions. The PCR 1 uses forward oligo “a” and reverse oligo “b,” while the PCR 2 uses forward oligo “c” and reverse oligo “d.” The template for both reactions is the CHIKV-Cbn plasmid, diluted to a concentration of 10 ng/μL.

To prevent contamination, assemble the PCR reactions in a clean, DNA-free environment. The use of dedicated micropipettes and aerosol-resistant barrier tips is highly recommended.

1. Add the components in a sterile 0.2 mL microcentrifuge tube on ice (*see* **Note 5**):

Components	Volume	Final conc.
10× PCR buffer without Mg^{2+}	3 μL	1×
10 mM dNTP mixture	0.6 μL	0.2 mM each
25 mM $MgCl_2$	3.6 μL	3 mM
10 μM Forward Primer	1.5 μL	0.5 μM
10 μM Reverse Primer	1.5 μL	0.5 μM
10 ng/μl Template DNA	1 μL	0.3 ng/μL
Taq DNA Polymerase (5 U/μL) (*see* **Note 2**)	0.2 μL	1 unit
Milli-Q® water	18.6 μL	
Total volume:	30 μL	

2. Cap the tubes, briefly vortex to mix the contents, and centrifuge to collect the liquid at the bottom of the tube.
3. Perform the PCR cycles in a thermocycler (*see* **Note 6**):

Step	Temperature	Time	Number of cycles
Initial denaturation	95°C	5 min	1×
Denaturation	95°C	30 s	
Annealing	58°C	30 s	35×
Extension	72°C	1 min (1 min per 1 kb)	
Final extension	72°C	5 min	1×
Final hold	18°C	∞	

4. Analyze the amplification products by running 1 μL of each PCR product on a 1.5% agarose gel. Visualize the DNA bands with a DNA-staining dye (e.g., ethidium bromide) and use a molecular weight ladder to confirm the size of the product.

   If a single band of the expected size is observed, purify the entire PCR product using a silica PCR purification kit. If multiple bands are present, run the entire PCR product in a gel, excise the specific band from the gel, and purify it using a gel extraction kit (*see* **Note 7**).

5. After purification, run a small aliquot of the sample on an agarose gel to verify the product's purity and integrity.
6. Quantify the purified PCR product using a NanoDrop spectrophotometer or a similar method. A concentration of at least 10 ng/μL is required for the next step. Store the samples at −20°C until needed.

3.2 Overlapping PCR

This section details the protocol for performing overlapping PCR using the purified hemi-PCR products (Template DNA I and II) as templates and external primers "a" and "d." This method uses a high-fidelity polymerase, such as the AccuPrime enzyme, to minimize errors during amplification (*see* **Note 8**).

1. Add these components to a sterile, thin-walled 0.2-mL PCR tube on ice:

Components	Volume	Final conc.
10× AccuPrime™ PCR Buffer	5 μL	1×
10 μM Forward Primer a	1.5 μL	0.3 μM
10 μM Reverse Primer d	1.5 μL	0.3 μM
Template DNA I (AB)	1 μL	10–20 ng
Template DNA II (CD)	1 μL	10–20 ng
AccuPrime™ Taq DNA Polymerase	0.5 μL	0.02 units
Milli-Q® water	to 50 μL	
Total volume:	50 μL	

2. Cap the tubes, briefly vortex to mix the contents, and centrifuge to collect the liquid at the bottom.
3. Perform the PCR cycles in a thermocycler (*see* **Note 9**):

Step	Temperature	Time	No. of cycles
Initial denaturation	95°C	5 min	1×
Denaturation	95°C	30 s	
Annealing	58°C	30 s	35×
Extension	68°C	1 min (1 min per 1 kb)	
Final extension	68°C	5 min	1×
Final hold	18°C	∞	

4. Analyze the amplification product by running a small aliquot on an agarose gel. Visualize the DNA band with a DNA-staining

dye and an appropriate molecular weight ladder to confirm the specific size of the full-length product.

If a single band of the expected size is observed, purify the entire PCR product using a silica PCR purification kit.

If multiple bands are present, run the entire PCR product in a gel, excise the specific band from the gel, and purify it using a gel extraction kit.

5. After purification, run a small aliquot of the sample on an agarose gel to verify its purity and integrity.
6. Quantify the purified PCR product using a NanoDrop spectrophotometer or a similar method. Store the samples at −20°C until needed.

3.3 Digestion of Insert and Vector with Restriction Enzymes

The next step is to digest both the vector (infectious clone) and the overlapping PCR product with restriction enzymes a and d, which flank the site of mutation within the CHIKV genome. For the vector, perform the digestion and then purify the largest fragment whose ends will be ready for cloning. The digestion of the overlapping PCR product eliminates the blunt ends, making it ready for ligation. Please refer to the manufacturer's instructions to find the optimal buffers and incubation conditions for dual-enzyme digestions.

1. Prepare the two digestion reactions in sterile 1.5 mL microcentrifuge tubes.

Vector digestion	**Volume**
10× Buffer	3 μL
Enzyme a (10 U/μL) Enzyme d (10 U/μL)	1 μL 1 μL
DNA CHIKV-Cbn plasmid (~500 ng/μL)	13 μL
Milli-Q® water	Up to 30 μL
Total volume:	30 μL
Insert digestion	**Volume**
10× Buffer	3 μL
Enzyme a (10 U/μL) Enzyme d (10 U/μL)	1 μL 1 μL
Overlapping PCR (minimum 50 ng/μL)	18 μL
Milli-Q® water	up to 30 μL
Total volume:	30 μL

Incubate the tubes at the temperature and time indicated by the manufacturer.

2. Run the digested products on an agarose gel.

To purify the vector, use a 1% agarose gel with a 1 kb ladder to distinguish the vector band, which should be larger than 12 kb. Excise the corresponding band from the gel and purify it using a gel extraction kit.

To purify the insert, use a 1.5% agarose gel with a 100 bp ladder and purify the correct band using a gel extraction kit.

3. Run 1 μL of each purified product (vector and insert) on a new agarose gel to check their quality and quantity. Use a Nanorop spectrophotometer to quantify the DNA concentration of each sample.

3.4 Ligation

The vector and insert from the previous step will be ligated using T4 DNA ligase. The ligation product will then be transformed into competent *E. coli* DH5α cells.

1. Assemble the following ligation reaction in a sterile 1.5 mL microcentrifuge tube (*see* **Note 10**).

Components	Volume
10× T4 DNA Ligase Buffer	1 μL
T4 DNA Ligase (5 Weiss U/μL)	0.5 μL
Vector	1 μL
Insert	4 μL
Milli-Q® water	3.5 μL
Total volume:	10 μL

Incubate the tube in a water bath at room temperature overnight.

3.5 Bacterial Transformation

The following day, proceed with the bacterial transformation (*see* **Note 11**).

1. Place the ligation tube on ice for 5 min.
2. Add 100 μL of chemically competent DH5α *E. coli* suspension to the tube. Gently mix by tapping the tube; do not pipette up and down.
3. Incubate the mixture on ice for 30 min.
4. Perform a heat shock by placing the tube in a heat block or water bath at 42°C for exactly 90 s.
5. Immediately place the tube back on ice for 2 min to cool it down.
6. Add 900 μL of LB broth to the tube. Incubate at 37°C with agitation (250 rpm) for 45 min to allow the bacteria to recover.
7. Centrifuge the tube at 2500× *g* for 4 min to pellet the bacteria.

8. Carefully decant the supernatant, leaving approximately 100 μ L of media. Resuspend the bacterial pellet in the remaining media.
9. Plate the entire bacterial suspension onto an LB agar plate containing ampicillin (at a final concentration of 100 μg/mL). Use a sterile spreader or glass beads to evenly distribute the bacteria.
10. Incubate the plate at 37°C overnight (for 18–20 h).

3.6 Colony Screening

This protocol describes colony PCR using external primers, followed by SAP/Exo treatment of the amplicons before Sanger sequencing. If the sequencing results are positive, the corresponding bacterial clone can be grown in a large volume of LB broth with ampicillin for subsequent plasmid extraction via miniprep.

1. Visually inspect the LB agar plate and select at least ten colonies to screen. Prepare a master mix for the colony PCR using the following recipe.

Components	**Volume (1×)**	**Volume (10×)**
10× PCR buffer without Mg^{2+}	1 μL	10 μL
10 mM dNTP mixture	0.2 μL	2 μL
25 mM $MgCl_2$	1.8 μL	18 μL
10 μM Forward Primer	0.5 μL	5 μL
10 μM Reverse Primer	0.5 μL	5 μL
Taq DNA Polymerase (5 U/μL)	0.1 μL	1 μL
Milli-Q® water	5.9 μL	59 μL
Total volume:	10 μL	100 μL

Dispense 10 μL of the PCR master mix into individual 0.2 mL PCR tubes. Use a sterile pipette tip to pick a single bacterial colony and transfer it directly into a PCR tube. Then, using the same tip, inoculate 5 mL of LB broth with ampicillin in a separate culture tube. Repeat this process for each colony you are screening. Ensure all tubes are correctly labeled to match each PCR sample with its corresponding LB broth culture.

2. Perform the PCR in a thermocycler:

Step	**Temperature**	**Time**	**No. of cycles**
Initial denaturation	95°C	5 min	1×
Denaturation	95°C	30 s	
Annealing	58°C	30 s	35×

Step	Temperature	Time	No. of cycles
Extension	68°C	1 min (1 min per 1 kb)	
Final extension	68°C	5 min	1×
Final hold	18°C	∞	

In the meantime, incubate the LB broth cultures at 37°C overnight with shaking (250 rpm). The next day, store the grown bacterial cultures at 4°C while you wait for the sequencing results.

3. Analyze the PCR products by running 1 μL of each sample on a 1.5% agarose gel. Visualize the DNA bands with a DNA-staining dye and use a molecular weight ladder to confirm the size of the amplified product.
4. For samples with the correct band size, proceed with ExoI-SAP treatment to remove residual primers and dNTPs before sequencing. To treat 7 μL of the PCR product in a total 10 μ L volume, add 0.15 μL of Exonuclease I (20 U/μL), 0.9 μL of SAP (1 U/μL), and 1.95 μL of water. Incubate the samples in a thermocycler at 37°C for 20 min, followed by 85°C for 15 min.
5. Send the treated samples to a sequencing facility for Sanger sequencing. Samples can be stored at −20°C until they are sent.
6. Check the Sanger sequencing chromatograms to identify clones with the desired mutations. Align the wild-type 3′UTR sequence with your Sanger sequences using appropriate software (e.g., SnapGene, Benchling). Select the positive clones for further use.

3.7 Miniprep of Positive Clones

For the clones that showed positive sequencing results, take the corresponding LB broth culture that was stored at 4°C and use a Miniprep kit to purify the plasmid DNA.

1. Purify the plasmid DNA from the bacterial culture following the manufacturer's instructions for the specific Miniprep kit being used.
2. Check the integrity of the purified plasmid by running 1 μL on a 1% agarose gel.
3. Quantify the DNA concentration using a Nanodrop or Qubit spectrophotometer. This value is your plasmid stock concentration for the mutant infectious clone.
4. Label the tube clearly and store the plasmid stock at −20°C for future use.

4 Notes

1. In this protocol, the infectious clone corresponding to the Caribbean strain of CHIKV is used. However, other CHIKV infectious clones, representing lineages such as the Indian Ocean Lineage (IOL) and African lineages, are available in the scientific community.
2. Other DNA polymerases can be used instead of Taq. Enzymes, such as Q5 High-Fidelity DNA Polymerase or Pfu DNA Polymerase, are generally preferred for cloning applications because they possess proofreading activity, which results in higher fidelity (fewer errors) and increased specificity compared to Taq polymerase.
3. Follow the general rules for designing good primers. The main guidelines are:
   Length: Aim for a length of 18–25 nucleotides (though 18–20 is often sufficient).
   Secondary structure: Check for the presence of internal hairpins and primer-dimer-forming structures such as palindromes or self-complementarity.
   Melting temperature (Tm): The forward and reverse primers should have a similar Tm (within 5°C).
   Target Tm: Aim for a Tm in the range of 55–65°C.
   3′end: It is preferable to have a G or C in the last one or two positions at the 3′end of the primers (known as a "GC clamp") to promote stable binding.
   Several software tools exist to help you design and analyze your primers; many of these are free and available online.
4. Introducing deletions into an infectious clone is much easier than introducing point mutations if the desired deletion is located immediately next to a restriction enzyme site. This approach does not require overlapping PCR, but rather only one simple PCR, as illustrated in Fig. 1b. The key step is to design a primer that binds precisely after the region you intend to delete. You must also include the restriction site in the 5′end of this primer (the same restriction site that is present in the template at the beginning of the desired deletion). After the PCR is performed, the product is ready for digestion and ligation. Upon ligation, the final plasmid will have successfully deleted the region that was excluded when you designed your primer.
5. Taq polymerase kits vary in format. Some provide a pre-mixed master mix that already contains the enzyme; others supply the master mix and the enzyme separately. Always consult the manufacturer's instructions for reaction assembly. Regardless

of the kit format, the primers and template DNA must always be added to initiate the PCR.

6. These are standard starting conditions. You may need to adjust the PCR program—specifically the annealing temperature and extension time—based on the specific primers (oligos), the expected product size, and the polymerase being used (e.g., check the extension temperature if using an enzyme other than Taq).
7. A wide range of commercial kits exists for the purification of PCR amplicons, whether directly from the reaction mix or after extraction from an agarose gel slice. You should first utilize the kits currently available in the lab. However, be aware that these kits are often optimized for specific amplicon sizes. If you experience significant mass loss (low yield) during purification, consider switching to an alternative kit brand, as different manufacturers use varied chemistries that may provide superior recovery for your specific DNA fragments.
8. Other high-fidelity polymerases, such as Q5 High-Fidelity DNA Polymerase, may be used as alternatives to AccuPrime. If a substitution is made, you must adjust the corresponding extension temperature.
9. For any PCR reaction performed using this protocol, the following parameters must be optimized by the user:
   Annealing temperature (Ta): Must be calculated and adjusted based on the melting temperature (Tm) of the specific oligonucleotide primers used.
   Extension time: Must be set based on the length of the expected amplicon (typically 30 s to 1 min/kb of DNA).
   If no amplicon is detected in the initial attempt, you may try lowering the annealing temperature (Ta). This change increases the probability of primer binding to the template DNA. Conversely, if you observe multiple nonspecific bands, you should increase the annealing temperature (Ta) to enhance primer specificity.
10. The ideal molar ratio for the insert to the vector in a ligation reaction is typically 3:1 (insert:vector). This means you must adjust the masses of the DNA fragments based on their size and concentration to achieve this optimal ratio for the ligation step. If the initial attempt does not yield positive clones, you should try testing alternative ratios (e.g., 1:1, 5:1) as part of your troubleshooting.
11. You may use in-house prepared chemically competent *E. coli* or commercially available competent cells. If using in-house bacteria, first verify the transformation efficiency is equal to or higher than 1×10^7 colony-forming units (CFU) per μg of DNA when transformed with a standard plasmid such as pUC19.

References

1. Racaniello VR, Baltimore D (1981) Cloned poliovirus complementary DNA is infectious in mammalian cells. Science 214:916–919
2. Hutchison CA, Phillips S, Edgell MH et al (1978) Mutagenesis at a specific position in a DNA sequence. J Biol Chem 253:6551–6560
3. Ho SN, Hunt HD, Horton RM et al (1989) Site-directed mutagenesis by overlap extension using the polymerase chain reaction. Gene 77:51–59
4. Stapleford KA, Moratorio G, Henningsson R et al (2016) Whole-genome sequencing analysis from the chikungunya virus Caribbean outbreak reveals novel evolutionary genomic elements. PLoS Negl Trop Dis 10:e0004402
5. Filomatori CV, Bardossy ES, Merwaiss F et al (2019) RNA recombination at chikungunya virus 3'UTR as an evolutionary mechanism that provides adaptability. PLoS Pathog 15:e1007706

Chapter 9

A Restriction-Free Cloning Approach for Molecular Engineering of Plasmids

Nicole T. Cerf and Claudia Filomatori

Abstract

Molecular cloning by PCR amplification using a highly processive, high-fidelity DNA polymerase represents a robust and versatile technique for the precise manipulation of nucleic acid sequences. This approach enables the insertion, replacement, or modification of specific DNA fragments within a cloning vector, thereby generating an accurate copy of a gene or viral segment for downstream applications, such as protein expression, site-directed mutagenesis, and structural or functional analyses. The use of processive, high-fidelity polymerases significantly reduces the occurrence of base substitution errors, ensuring sequence integrity throughout the amplification process. Traditionally, restriction enzymes have been employed to facilitate directional cloning; however, alternative methods allow for mutagenesis without the need for unique and specific restriction sites and can be applied to virtually any cloning or seamless DNA assembly strategy. In this chapter, we describe a restriction enzyme–free and ligation-free PCR-based protocol widely applicable to any circular vector. This method enables targeted mutagenesis of the chikungunya virus (CHIKV) genome, offering a fast, efficient, and reliable strategy for generating mutant constructs suitable for virological and molecular studies.

Keywords Molecular engineering, Restriction-free cloning

1 Introduction

The polymerase chain reaction (PCR) has become an indispensable tool for molecular cloning, as it enables the rapid and efficient amplification of specific DNA fragments in quantities suitable for downstream applications. Beyond simple amplification, PCR allows the deliberate incorporation of additional sequence features—such as recombination tags or site-specific modifications—through the design of custom primers. These properties make PCR a flexible platform for DNA engineering, particularly in cases where restriction sites are not conveniently located.

Among the available DNA polymerases, high-fidelity DNA polymerases from *Pyrococcus furiosus* are particularly well-suited for cloning, mutagenesis, and sequencing applications that require

Claudia Filomatori and Eliana F. Castro (eds.), *Chikungunya Virus*, Methods in Molecular Biology, vol. 3067,
https://doi.org/10.1007/978-1-0716-5496-5_9,

maximal accuracy [1–4]. Its engineered proofreading (3′→5′ exonuclease) activity provides an error rate over 100 times lower than that of standard *Taq* polymerase, minimizing unwanted mutations during amplification [5]. In addition, high-fidelity DNA polymerases typically generate blunt-ended PCR products and have progressively improved in their ability to amplify longer plasmids, from initial ranges of 3–10 kb to as much as 20–30 kb, depending on template complexity and enzyme properties [6, 7]. Phusion™ High-Fidelity DNA Polymerase, which combines a *Pyrococcus*-like enzyme with a processivity-enhancing domain, exemplifies these advances, enabling efficient amplification of long and complex targets with superior fidelity [8].

The development of highly processive, high-fidelity DNA polymerases has enabled powerful variants of restriction-free cloning approaches, such as the *megaprimer*-based method. This approach consists of two rounds of PCR (Figs. 1 and 2). The first reaction generates a short fragment—the *megaprimer*—that incorporates the desired mutation deletion or insertion through primers carrying the modified sequence. Importantly, when introducing point mutations or deletions, a single PCR can be performed using the target plasmid as a template and a forward primer carrying the desired nucleotide changes together with its reverse complementary counterpart. However, it is generally preferable to generate a *megaprimer*, which provides greater complementarity to the target plasmid and thereby increases the efficiency of the cloning process (Fig. 1). Moreover, if different substitutions at the same site are desired, distinct mutagenic forward primers can be used in combination with a common reverse primer.

The *megaprimer*-based method also offers a highly adaptable strategy for gene assembly. In this case, the *megaprimer* consists of the full-length gene to be inserted, flanked by regions homologous to the insertion site of the target plasmid (Fig. 2).

Once synthesized, the *megaprimer* and the target plasmid are used in a second PCR reaction to extend and amplify the entire plasmid of interest, yielding a full-length product that contains the desired modification.

Following the second round of PCR, the plasmid used as the template is digested with *Dpn*I to degrade the methylated parental DNA, and the newly synthesized full-length linear plasmid can be directly transformed into competent bacterial cells, where it subsequently circularizes. Finally, plasmid DNA is isolated, and the success of the cloning is verified by restriction enzyme digestion and confirmed by Sanger sequencing.

Overall, this technique provides a fast, cost-effective, and enzyme-free alternative to traditional restriction-based cloning, and it has proven particularly useful in applications requiring precise sequence manipulation, such as viral mutagenesis or protein engineering.

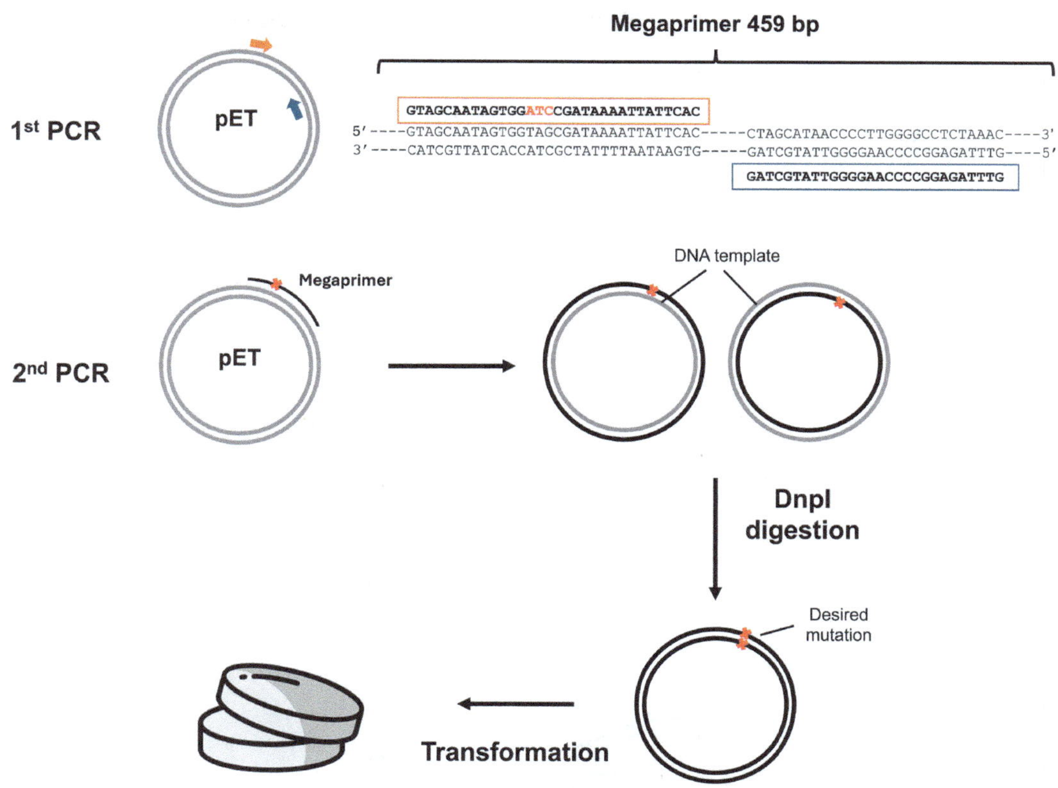

Fig. 1 Schematic representation of the restriction-independent cloning protocol for chikungunya virus mutant generation. In the first PCR, a plasmid (pET) containing chikungunya virus sequences is used as a template. The orange and blue arrows represent the forward and reverse primers, respectively, which delimit the region to be amplified to generate the *megaprimer*. The forward primer contains the desired mutation, which will be incorporated into the *megaprimer* sequence. The second PCR uses the *megaprimer* obtained from the first reaction. *Dpn*I digestion removes the methylated parental plasmid DNA, and the resulting mutated plasmid is subsequently transformed into competent bacteria

2 Materials

2.1 PCR

1. Template: genomic DNA, plasmid, cDNA (*see* **Note 1**).
2. Forward and reverse primers, 10 nM each.
3. dNTP mix: 10 mM dATP, 10 mM dCTP, 10 mM dTTP, and 10 mM dGTP.
4. HF Buffer (5×).
5. 2 U/μL Phusion DNA polymerase.
6. Agarose gel.
7. Ethidium bromide (*see* **Note 2**).
8. Loading buffer.
9. GeneRuler 1 kb DNA Ladder.

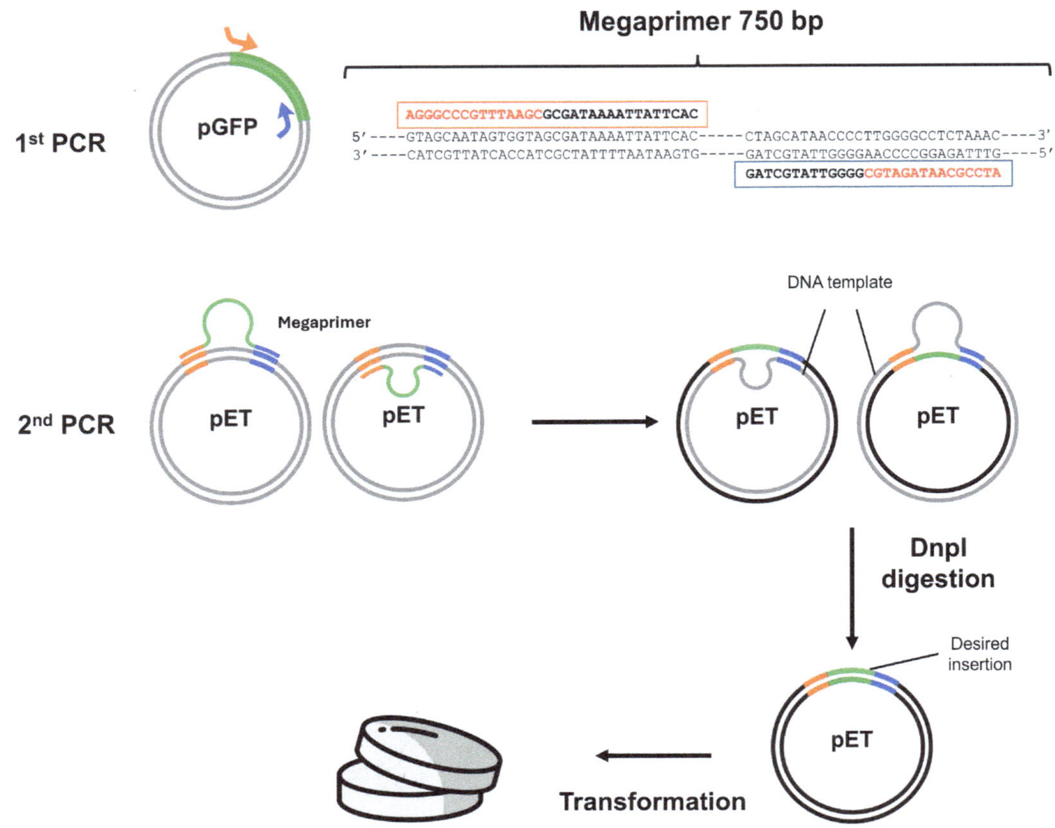

Fig. 2 Schematic representation of the restriction-independent cloning protocol for the insertion of a reporter gene into a plasmid (pET) containing chikungunya virus sequences. In the first PCR, primers are designed to hybridize with the pGFP plasmid at the regions corresponding to the beginning and end of the GFP gene. Each primer also contains an additional non-complementary sequence, which serves to direct the insertion into the pET plasmid. As a result, the first PCR generates a *megaprimer* consisting of the GFP sequence. In the second PCR, this *megaprimer* hybridizes with the target insertion site in the pET plasmid, allowing the incorporation of the GFP gene. The *Dpn*I digestion step selectively degrades the methylated template plasmid, allowing the newly synthesized mutant plasmid to be transformed into competent bacteria

10. 20 U/μL *Dpn*I enzyme.
11. Nuclease-free water.

2.2 DNA Purification

1. Scalpel.
2. QG Buffer: 5.5 M guanidine thiocyanate, 20 mM Tris-HCl buffer (pH 6.6)
3. Silica (*see* **Note 3**).
4. 3 M sodium acetate (pH 5.0).
5. PE Buffer: 80% (v/v) ethanol, 10 mM Tris-HCl (pH 7.5).
6. EB Buffer: 10 mM Tris-HCl (pH 8.5).

2.3 Mini Plasmid DNA Preparation

1. SN1: 50 mM Tris-HCl (pH 8), 10 mM EDTA.
2. SN2: 0.2 M NaOH, 1% SDS.
3. SN3: 3 M sodium acetate.
4. Isopropanol.
5. 70% (v/v) ethanol.

2.4 Equipment

1. Agarose gel electrophoresis chamber.
2. Long-wave UV transilluminator.
3. Thermal cycler.
4. Microcentrifuge tubes.
5. Heat block or water bath.
6. Vortex.
7. Refrigerated microcentrifuge.

Other: pipettes (2, 10, 20 μL), 0.2- or 0.5-mL nuclease-free microcentrifuge tubes.

3 Methods

3.1 First PCR: Megaprimer Synthesis

1. Maintain all reaction components on ice.
2. Gently mix all the components of the reaction before use.
3. In a sterile 0.2-mL microcentrifuge tube, mix the reagents for the first round of PCR (*see* **Notes 4 and 5**):

Reaction mix	Volume (μL)	Final
Water	9.65	
HF Buffer (5×)	3	1×
10 mM dNTPs	0.3	200 μM
15 ng/μL plasmid	1	1 ng/μL
10 nM Forward primer	0.45	0.3 μM
10 nM Reverse primer	0.45	0.3 μM
2 U/μL Phusion polymerase	0.15	0.02 U/μL

4. Quickly transfer the reactions to a thermocycler preheated to the denaturation temperature (98°C) (*see* **Note 6**).
5. Begin thermocycling (*see* **Notes 7–12**):

Step	Temperature (°C)	Time (s)	Cycles (n°)
Initial denaturation	98	30	1
Denaturation	98	10	35

Step	Temperature (°C)	Time (s)	Cycles (n°)
Annealing	65	30	35
Extension	72	15–30 s/kb	35
Final extension	72	600	1
Hold	4	∞	1

6. When thermocycling is complete, the integrity and molecular weight of the fragments should be checked on a 1% agarose gel (*see* **Note 13**).

3.2 Purification of the Megaprimer

1. Visualize the bands under a long-wave UV transilluminator.
2. Cut out the desired DNA fragment with a clean scalpel (*see* **Note 14**).
3. Place the gel slice into a pre-weighed 1.5 mL tube and record the mass.
4. Add three volumes of QG Buffer for every 100 mg of gel slice when purifying DNA fragments ranging from 100 bp to 4 kb (*see* **Note 15**).
5. Add 15 μL of silica.
6. Add 10 μL of 3 M sodium acetate (pH 5.0).
7. Mix by vortexing.
8. Incubate at 50–60°C for 10–15 min, vortexing every 2–3 min, until the gel is completely dissolved.
9. Centrifuge for 1 min at 3200 × *g*.
10. Discard the supernatant.
11. Add 500 μL of QG Buffer and vortex to resuspend.
12. Centrifuge for 1 min at 3200 × *g*.
13. Discard the supernatant.
14. Add 500 μL of PE Buffer and vortex to resuspend.
15. Centrifuge for 1 min at 3200 × *g*.
16. Discard the supernatant.
17. Repeat Steps 14–16.
18. Air-dry the pellet for 15 min.
19. Resuspend in 20 μL of water or EB Buffer and vortex to mix.
20. Incubate at 50–60°C for 10–15 min.
21. Centrifuge for 2 min at 13,000 × *g*.
22. Transfer the supernatant to a clean tube and discard the pellet.

3.3 Second PCR

1. Maintain all reaction components on ice.
2. Gently mix all the components of the reaction before use.
   In a sterile 0.2-mL microcentrifuge tube, mix the reagents for the second round of PCR (*see* **Note 16**):

Reaction mix	Volume (μL)	Final
Water	9.55	
HF Buffer (5×)	3	1×
10 mM dNTPs	0.3	200 μM
15 ng/μL plasmid	1	1 ng/μL
75 ng/μL *megaprimer*	1	5 ng/μL
2 U/μL Phusion polymerase	0.15	0.02 U/μL

3. As PCR conditions for the amplification of the final mutated product, use the following (*see* **Note 17**):

Step	Temperature (°C)	Time (s)	Cycles (n°)
Initial denaturation	98	30	1
Denaturation	98	30	1
Annealing	65	90	1
Extension	72	15 s/kb	1
Denaturation	98	30	1
Annealing	63	90	1
Extension	72	15 s/kb	1
Denaturation	98	30	1
Annealing	61	90	1
Extension	72	15 s/kb	1
Denaturation	98	30	1
Annealing	59	90	1
Extension	72	15 s/kb	1
Denaturation	98	30	14
Annealing	65	90	14
Extension	72	15 s/kb	14
Final extension	72	300	1
Hold	4	∞	1

4. Analyze the PCR product on a 1% agarose gel.
5. Purify the DNA fragment following the procedure described in Sect. 3.2.

3.4 *Dpn*I Digestion and Transformation

1. Transfer the purified product of the second PCR into a 1.5 mL Eppendorf tube.
2. Transfer a negative control for digestion into another 1.5 mL Eppendorf tube (HF Buffer, plasmid vector, and water).
3. To each tube, add 0.3 μL of *Dpn*I (recommended: 1 μL per 50 μL of PCR product or 50 μL of negative control).
4. Incubate the reaction at 37°C for 1.5–2 h in a heat block or water bath.
5. Inactivate the enzyme by incubating for 20 min at 80°C in a heat block or water bath (optional).
6. Transform *E. coli* competent cells (e.g., DH5α or XL1-Blue strains) with 2–5 μL of the *Dpn*I-treated PCR product or the negative control using a standard heat-shock or electroporation protocol.
7. Spread 50–200 μL of the transformation mixture on LB agar plates containing the appropriate antibiotic for plasmid selection. Incubate overnight at 37°C (*see* **Note 18**).

3.5 Homemade Miniprep

1. Inoculate colonies transformed with the plasmid into 3–5 mL of selective LB medium and incubate overnight at 37°C with shaking (180–220 rpm) (*see* **Note 19**).
2. Transfer 1.5 mL of the overnight culture to a microcentrifuge tube and centrifuge at 13,000× *g* for 1 min to pellet the cells. Discard the supernatant. Refill the same tube with another 1.5 mL of culture, centrifuge again, and repeat until the entire culture has been pelleted in the same tube.
3. Discard the supernatant and keep the pellet.
4. Add 200 μL of Solution 1 (SN1) to the pellet and vortex until fully resuspended.
5. Add 200 μL of Solution 2 (SN2).
6. Gently invert the tube six times to mix.
7. Incubate for 5 min at room temperature (*see* **Note 20**).
8. Add 200 μL of Solution 3 (SN3).
9. Incubate on ice for 10 min.
10. Centrifuge at 16,000× *g* for 10 min at 4°C.
11. Transfer the supernatant to a new tube containing 600 μL of isopropanol.
12. Invert the tubes four times, or until the solution is completely homogenized.
13. Incubate on ice for 20 min.
14. Centrifuge at 16,000× *g* for 20 min at 4°C.
15. Carefully discard the supernatant.

16. Wash the pellet with 500 μL of 70% (v/v) ethanol (do not resuspend).
17. Centrifuge for 5 min at 4°C at 16,000 × *g*.
18. Discard the supernatant completely.
19. Allow the pellet to air-dry.
20. Resuspend the DNA pellet in 20–50 μL of sterile water.
21. Confirm the integrity and sequence of the modified region by Sanger sequencing using suitable primers.

4 Notes

1. Employing purified, high-quality DNA templates is essential for optimal PCR results.
2. As a safer alternative to ethidium bromide, DNA can be visualized in agarose gels using SYBR Safe or GelRed, which offer high sensitivity and reduced mutagenic risk, and can be detected with either UV or blue-light transilluminators.
3. To prepare the silica, weigh 1 g of silica, add 10 mL of PBS, mix, and let it settle for 2 h. Repeat this procedure three to four times and then let it settle overnight. The following day, resuspend the silica in 10 mL of QG Buffer.
4. Phusion DNA Polymerase should be added last to prevent the 3′→5′ exonuclease activity.
5. If the template DNA is obtained from a cDNA synthesis reaction, the volume added should be less than 10% of the total reaction volume.
6. If needed, perform a quick spin to bring all liquid to the bottom of the tube.
7. Due to the nature of the Phusion™ High-Fidelity DNA Polymerase, the optimal reaction conditions may differ from PCR protocols for standard DNA polymerases. For instance, due to the high salt concentration in the reaction buffer, the Phusion™ High-Fidelity DNA Polymerase does not follow the common annealing rules and tends to work better at elevated denaturation and annealing temperatures.
8. Use 15–30 s/kb for extension. Do not exceed 1 min/kb.
9. It is critical that the Phusion™ High-Fidelity DNA Polymerase be added as the final component in the PCR mixture, since its 3′→5′ exonuclease activity can degrade primers in the absence of dNTPs.
10. Other high-fidelity polymerases may be used for *megaprimer* synthesis. It is essential that the chosen polymerase generates

blunt-ended PCR products and does not add non-templated adenine overhangs.

11. Two buffers are provided with the enzyme: 5× Phusion™ HF Buffer and 5× Phusion™ GC Buffer. The error rate of the polymerase in the HF Buffer is lower than in the GC Buffer. Therefore, the HF Buffer should be used as the default buffer for high-fidelity amplification. However, GC Buffer can improve the performance of Phusion™ High-Fidelity DNA Polymerase on some difficult or long templates, such as GC-rich templates or those with complex secondary structures.
12. In some cases, low PCR product yields may be obtained, requiring optimization of reaction conditions. Among the parameters to adjust, increasing the $MgCl_2$ concentration and/or adding DMSO (recommended for GC-rich amplicons) can improve amplification efficiency.
13. To prevent cross-contamination, analyze PCR products in a workspace separate from where PCR reactions are assembled.
14. The scalpel can be cleaned with 70% (v/v) ethanol to prevent contamination of the DNA sample.
15. As the density of agarose gel is approximately 1 g/mL, you can assume that 100 mg ≈ 100 μL of gel. For DNA fragments smaller than 100 bp, add six volumes of buffer. For fragments larger than 4 kb, add three volumes of QG Buffer and two volumes of water. When working with a 2% (w/v) agarose gel, use six volumes of QG Buffer.
16. An optimal *megaprimer*-to-primer ratio of 5:1 is recommended.
17. Alternatively, 25–35 PCR cycles with annealing performed at a single temperature may also be used. However, better results are often obtained by starting with initial PCR cycles in which the annealing temperature decreases by 2°C per cycle (from 65°C to 59°C), followed by 14 additional cycles with a constant annealing temperature of 65°C.
18. To confirm the presence of the expected modification before plasmid isolation, it is recommended to pick several colonies and perform a colony PCR using specific primers flanking the mutated or inserted region. Then, perform the miniprep using the positive colonies.
19. Ideally, 5–20 colonies should be picked, inoculated, and screened, depending on the relative number of colonies obtained from the PCR transformation compared to the negative control.
20. This incubation time is critical and must be strictly followed. Exceeding the recommended time may lead to plasmid denaturation or contamination with chromosomal DNA.

References

1. Zeng F, Hao Z, Li P et al (2017) A restriction-free method for gene reconstitution using two single-primer PCRs in parallel to generate compatible cohesive ends. BMC Biotechnol 17:32
2. Bhat S, Bialy D, Sealy JE et al (2020) A ligation and restriction enzyme independent cloning technique: an alternative to conventional methods for cloning hard-to-clone gene segments in the influenza reverse genetics system. Virol J 17:82
3. Lundberg KS, Shoemaker DD, Adams MWW et al (1991) High-fidelity amplification using a thermostable DNA polymerase isolated from Pyrococcus furiosus. Gene 108:1–6
4. Wang Y (2004) A novel strategy to engineer DNA polymerases for enhanced processivity and improved performance in vitro. Nucleic Acids Res 32:1197–1207
5. Dolgova AS, Stukolova OA (2017) High-fidelity PCR enzyme with DNA-binding domain facilitates de novo gene synthesis. 3 Biotech 7:128
6. Van Den Ent F, Löwe J (2006) RF cloning: a restriction-free method for inserting target genes into plasmids. J Biochem Biophys Methods 67:67–74
7. Hemsley A, Arnheim N, Toney MD et al (1989) A simple method for site-directed mutagenesis using the polymerase chain reaction. Nucleic Acids Res 17:6545–6551
8. Bond SR, Naus CC (2012) RF-Cloning.org: an online tool for the design of restriction-free cloning projects. Nucleic Acids Res 40: W209–W213

Chapter 10

Expression and Purification of nsP2 Protease of Chikungunya Virus

Jesús J. Barraza Sánchez and Santiago E. Faraj

Abstract

The nonstructural protein 2 (nsP2) of chikungunya virus (CHIKV) is an essential cysteine protease involved in viral polyprotein processing and replication complex regulation. Biochemical and structural studies of nsP2 require the availability of highly pure and enzymatically active protein. In this chapter, we describe a straightforward protocol for the heterologous expression and purification of CHIKV nsP2 protease domain (nsP2pro) in *Escherichia coli*. The protein is expressed at low temperature as a His-tagged construct and purified under native conditions by immobilized metal ion affinity chromatography, followed by dialysis for buffer exchange. The procedure typically yields milligram quantities of highly pure (>95%) and active nsP2pro, and has been successfully applied to wild-type and mutant variants for enzymatic assays and inhibitor screening.

Keywords Chikungunya, nsP2 protease, Expression, Purification

1 Introduction

The nonstructural protein 2 (nsP2) of chikungunya virus (CHIKV) is a multifunctional enzyme that plays a central role in the viral replication cycle. In addition to its N-terminal RNA helicase domain, nsP2 harbors a C-terminal cysteine protease domain responsible for the cleavage of the viral precursor polyprotein nsP1234 into the four nonstructural proteins (nsP1, nsP2, nsP3, and nsP4). Through its proteolytic activity, nsP2 coordinates the temporal assembly of the replication complex and thus orchestrates the switch from negative- to positive-strand RNA synthesis [1–4]. Beyond its enzymatic functions, nsP2 also contributes to host shutoff and immune evasion [5–7].

Biochemical and structural characterization of nsP2 requires a pure and enzymatically active recombinant protein. In this chapter, we describe a protocol for the heterologous expression and purification of the CHIKV nsP2 protease domain (nsP2pro) in *Escherichia coli*. The protein is expressed from a pET28 construct encoding

Claudia Filomatori and Eliana F. Castro (eds.), *Chikungunya Virus*, Methods in Molecular Biology, vol. 3067,
https://doi.org/10.1007/978-1-0716-5496-5_10,

nsP2pro fused to an N-terminal histidine tag and purified under native conditions by immobilized metal ion affinity chromatography (IMAC). Purification involves two main steps: (i) immobilized Ni^{2+} affinity chromatography under native conditions and (ii) dialysis for buffer exchange [8].

This procedure typically yields >95% pure nsP2pro, which remains active for several months. The protocol provides an accessible and reproducible method for obtaining active recombinant nsP2pro suitable for steady-state kinetic assays, structural analyses, and inhibitor screening. The same approach has also been successfully applied to purify nsP2pro variants carrying point mutations.

2 Materials

2.1 Plasmid and Bacteria

1. Expression plasmid bearing the nsP2pro gene, e.g., pET28, which provides resistance to kanamycin.
2. Competent BL21(DE3) cells (*see* **Note 1**).

2.2 Expression

1. LB liquid medium: 1% tryptone, 0.5% yeast extract, 1% NaCl. Sterilized by autoclaving for 20 min at 15 psi on liquid cycle.
2. LB solid medium: 1% tryptone, 0.5% yeast extract, 1% NaCl, 1.5% Bacto agar. Sterilized by autoclaving for 20 min at 15 psi on liquid cycle.
3. Kanamycin solution: 50 mg/mL kanamycin in water, filter-sterilized. Store at −20°C.
4. Chloramphenicol solution: 25 mg/mL chloramphenicol in ethanol. Store at −20°C.
5. Isopropyl-β-D-1-thiogalactopyranoside (IPTG): 1.0 M solution in water, filter-sterilized. Store at −20°C. Alternatively, solid lactose may be used.
6. Thermostatized orbital shaker capable of maintaining 16°C.
7. Photometer or spectrophotometer capable of measuring at 600 nm (OD_{600}).

2.3 Purification

1. Nickel IMAC resin for polyhistidine-tagged protein purification (Ni-NTA).
2. Chromatography columns and tubing.
3. Tris buffer: 1.0 M Tris–HCl, pH 7.5 at 4°C (*see* **Note 2**).
4. Imidazole buffer: 1.0 M imidazole–HCl, pH 7.5 at 4°C (*see* **Note 2**).
5. NaCl solution: 5.0 M NaCl in water.
6. Glycerol.

7. Dithiothreitol (DTT): 1.0 M DTT solution in water. Store at −20°C.
8. Tween-20.
9. Ethylenediaminetetraacetic acid (EDTA): 0.5 M EDTA solution in water, pH 7.5, adjusted with KOH.
10. 5× IMAC buffer: 250 mM Tris–HCl, pH 7.5 at 4°C (1/4 dilution of Tris buffer), 2.5 M NaCl, and 25% glycerol.
11. Lysis buffer: 1/5 dilution of the 5× IMAC buffer, supplemented with 1.0 mM DTT, 1% Tween-20, and 20 mM imidazole–HCl (1/50 dilution of the Imidazole buffer). Freshly prepared.
12. Binding buffer: 1/5 dilution of the 5× IMAC buffer, supplemented with 20 mM imidazole–HCl (1/50 dilution of Imidazole buffer).
13. Washing buffer: 1/5 dilution of the 5× IMAC buffer, supplemented with 100 mM imidazole–HCl (1/10 dilution of Imidazole buffer).
14. Elution buffer: 1/5 dilution of the 5× IMAC buffer, supplemented with 250 mM imidazole–HCl (1/4 dilution of Imidazole buffer).
15. Dialysis buffer: 50 mM Tris–HCl, pH 7.5 at 4°C (1/20 dilution of Tris buffer), 50 mM NaCl, and 20% glycerol.

3 Methods

3.1 Protein Expression

1. Bacteria transformation. Use your preferred *E. coli* transformation procedure to transform the expression plasmid bearing the nsP2pro gene fused to an N-terminal histidine tag into competent BL21(DE3) bacteria (*see* **Note 1**).
2. Plate bacteria on antibiotic selection plates. Use plates with *LB* solid medium supplemented with 50 μg/mL kanamycin (1/1000 dilution of the Kanamycin solution) and 25 μg/mL chloramphenicol (1/1000 dilution of the Chloramphenicol solution), or with the appropriate antibiotics considering the plasmid-encoded resistance and any host-specific antibiotic. Incubate overnight at 37°C.
3. Starter culture. Use a single colony from those selected in the previous step to inoculate 10 mL of LB liquid medium supplemented with 50 μg/mL kanamycin and 25 μg/mL chloramphenicol. Grow overnight in an orbital shaker (180–200 rpm) at 37°C.

4. Expression culture. Inoculate starter culture at a 1/100 dilution into LB liquid medium supplemented with 50 μg/mL kanamycin and 25 μg/mL chloramphenicol, and grow in an orbital shaker (180–200 rpm) at 37°C until OD_{600} reaches 0.4–0.6. Remove the culture from the shaker and place flasks on ice for 15 min.
5. Induce expression by adding IPTG to the culture to a final concentration of 0.5 mM (or 0.5% (w/v) lactose), and incubate the culture overnight (~16 h) in an orbital shaker (120–160 rpm) at 16°C (*see* **Note 3**).
6. Remove the culture from the shaker and immediately cool it on ice.
7. Harvest bacteria by centrifuging the culture at 4000 × *g* for 15 min at 4°C. Carefully discard the supernatant. It is recommended to pre-weigh the vessel prior to the addition of the cell suspension to know the mass of bacteria obtained. Pellets may be stored at −80°C or used immediately for protein purification.

3.2 Purification

All purification steps must be carried out at 4°C, and the protein sample should be kept on ice throughout.

1. IMAC setup. Gently shake the Ni-NTA resin container to completely resuspend the slurry and transfer an appropriate amount of nickel chelating resin (~1–3 mL) to a suitable column (e.g., 1.0 × 10 cm) (*see* **Note 4**). Allow the resin to settle. Allow the storage solution (typically 20% ethanol) to flow through and wash with ten volumes of distilled water. Discard the flow-through.
2. Equilibrate the resin with five bed volumes of Binding buffer (Lysis buffer may be used alternatively). Discard the flow-through. Work in a refrigerated room or keep the column refrigerated otherwise.
3. Resuspend bacteria pellets in 20–40 mL of freshly prepared ice-cold Lysis buffer using a disposable Pasteur pipette or otherwise. This volume is based on the expectation of about 4–5 g of cells/L, i.e., 5 mL for every gram of cells (wet weight).
4. Cell disruption. Sonicate the cell suspension on ice using five to eight pulses of 15 s each, allowing 30–60 s cooling intervals between pulses to prevent overheating. Sonicate until the lysate becomes visibly clarified. Alternatively, other standard bacterial lysis methods, such as the French press, may be used.
5. Centrifuge the cell lysate at ≥10,000 × *g* for 30 min at 4°C to separate the soluble (supernatant) and insoluble (cellular debris) fractions. Transfer the clarified supernatant to a clean tube. Optionally, the supernatant may be filtered through a

0.22 µm syringe filter to remove residual particulates. Retain aliquots of both the clarified supernatant and the insoluble fraction, resuspended in an equivalent volume of buffer supplemented with 2% SDS, for subsequent SDS-PAGE analysis.

6. Load the supernatant at a flow rate of 0.5–1.0 mL/min (*see* **Note 5**). Collect the sample by gravity flow. Save the flow-through for SDS-PAGE analysis.
7. Wash the column with at least ten bed volumes of Washing buffer at a rate of 0.5–1.0 mL/min to remove unbound proteins (*see* **Note 6**).
8. Elute nsP2pro with ten volumes of Elution buffer, collecting the flow-through in five to ten fractions of 1 mL each. Immediately subnith 1 mM EDTA (*see* **Note 7**). The protein usually starts to elute within the first three fractions and can be readily detected using the Bradford protein assay.
9. Run aliquots of the insoluble fraction, clarified soluble lysate, non-retained fraction, wash fractions, and eluate fractions on a 10% SDS-PAGE (*see* Chap. 11) for instructions on hand-cast gel preparation and gel electrophoresis) (*see* **Note 8**). Figure 1 shows a representative SDS-PAGE profile of the different stages of the purification.

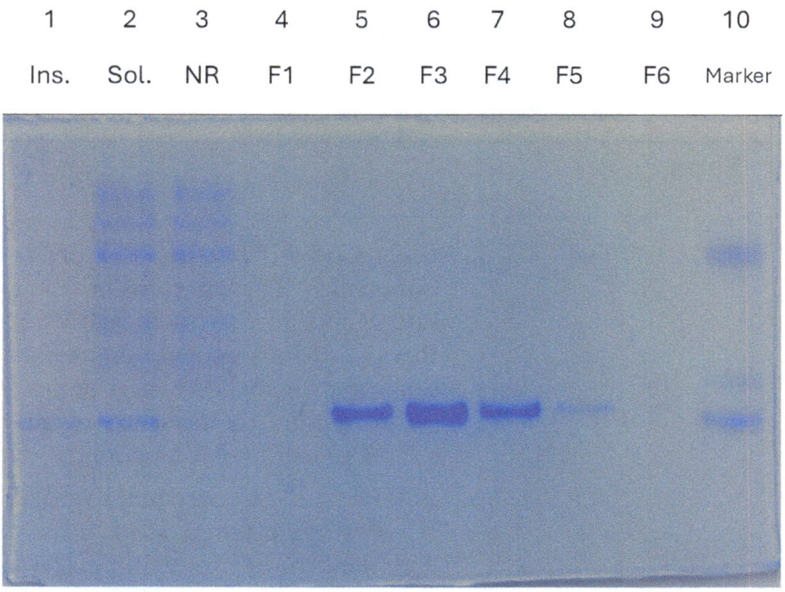

Fig. 1 Representative SDS-PAGE profile of different stages of nsP2pro purification. Lanes 1 and 2 correspond to the insoluble and soluble fractions of the initial cell lysate, respectively. Lane 3 shows the non-retained (flow-through) fraction obtained after loading the clarified lysate onto the IMAC column, and lanes 4–9 show fractions eluted from the IMAC column. The nsP2pro band is observed at approximately 40 kDa

10. Pool the fractions exhibiting the highest purity and protein concentration (*see* **Note 9**).
11. Dialyze the pooled protein fractions against Dialysis buffer at 4°C to remove imidazole using dialysis tubing with an MWCO of 10–20 kDa (*see* **Note 10**). Replace the dialysis buffer after 2–4 h and continue dialysis overnight. Further purification steps, such as His-tag removal, may be considered at this stage (*see* **Note 11**).
12. Centrifuge the sample at ≥10,000 × *g* for 30 min at 4°C to remove insoluble material and aggregates. Transfer the supernatant to a clean tube, aliquot into convenient volumes, and flash-freeze in liquid nitrogen. Protein may be stored in liquid nitrogen or at −80°C (*see* **Note 12**).
13. Immediately before use, centrifuge the enzyme preparation at ≥10,000 × *g* for 30 min at 4°C to remove any aggregates that may have formed during storage. Transfer the supernatant to a clean tube. Determine protein concentration by measuring absorbance at 280 nm and using an extinction coefficient of 51,255 M^{-1} cm^{-1}.

This procedure yields approximately 25 mg of >95% pure nsP2pro per liter of bacterial culture, with a typical k_{cat} value of 2 min^{-1} when measured using a protein-based substrate (see Chap. 11), using enzyme aliquots that have undergone a single freeze–thaw cycle. For quality control, the identity of the purified protein may be confirmed by mass spectrometry.

4 Notes

1. BL21 CodonPlus(DE3)-RIL or Rosetta(DE3)pLysS are recommended. These bacteria bear an extra plasmid that supplies rare tRNAs needed to efficiently translate heterologous genes that use codons uncommon in *E. coli*, improving the expression of proteins with non-optimized codon usage. These strains require the presence of chloramphenicol to maintain selection for the tRNA plasmid.
2. Buffers should be prepared so that the desired pH is achieved at the working temperature. Because buffer p*Ka* values vary with temperature, to obtain buffers at pH 7.5 at 4°C, the pH should be adjusted at 25°C to 6.91 for Tris (pKa at 25°C: 8.072; ΔpKa/°C: −0.028) and to 7.06 for imidazole (pKa at 25°C: 6.993; ΔpKa/°C: −0.021). pH values should be verified prior to use, after cooling to 4°C.
3. Expression at low temperature improves solubility. Under our working conditions, we typically obtain about 50% of

the recombinant protein in soluble state (vs. 100% insoluble when expression was performed at 37°C).

4. The amount of resin required depends on the amount of protein produced. Considering Ni-NTA agarose has a binding capacity of up to 50 mg/mL, a bed volume of 1–3 mL is sufficient for most applications.
5. Alternatively, an FPLC/ÄKTA system with a packed column might be used to obtain higher resolution and more reproducible elution profiles.
6. Note that nsP2pro binds very tightly to the resin. Therefore, extensive washing of the column can be safely performed even at the relatively stringent composition of the Washing buffer (100 mM imidazole–HCl).
7. Collection tubes can be preloaded with a concentrated EDTA solution prior to eluate collection. EDTA inhibits proteases that require metal cofactors, protects the reactive cysteine residue from metal-induced oxidation, and chelates Ni^{2+} ions that may co-elute with the protein and promote aggregation through metal-mediated cross-linking of protein molecules.
8. In order to compare the amount of protein obtained in each fraction (to analyze loss and enrichment), it is advisable to maintain proportionality when loading the soluble, insoluble, and wash fractions for SDS-PAGE.
9. When selecting fractions for pooling, both protein purity and concentration should be considered. In our experience, nsP2pro exhibits a strong tendency to aggregate during concentration steps. As a consequence, concentrating pooled fractions often results in significant protein loss and is therefore not recommended for this protein.
10. The Dialysis buffer should be pre-equilibrated to 4°C prior to use. For sample volumes of 5–10 mL, approximately 250 mL of Dialysis buffer per dialysis step is recommended. After completion of dialysis, retain a portion of the final dialysis buffer to be used as a blank for subsequent protein concentration measurements.
11. If required, the His-tag may be removed at this stage using a suitable site-specific protease, provided that a cleavage site was included in the expression construct. Additional polishing steps (e.g., further chromatography) can also be performed. However, any additional manipulation of the enzyme may lead to substantially reduced yields due to protein aggregation and a pronounced loss of enzymatic activity.
12. Samples should be aliquoted in volumes appropriate for single use and subsequent quantification after thawing. In our experience, nsP2pro can be stored either at −80°C or

in liquid nitrogen with comparable stability. The addition of cryoprotectants such as glycerol (up to 40%) or sucrose (up to 250 mM) provides little, if any, improvement in activity preservation. Importantly, nsP2pro is extremely sensitive to freeze–thaw cycles, with approximately 50% loss of activity per cycle under our working conditions.

Acknowledgments

We are grateful to Prof. Shailly Tomar (Indian Institute of Technology Roorkee) for kindly providing the construct for the expression of the enzyme. This work was supported by Agencia Nacional de Promoción Científica y Tecnológica PICT 598/2020 (2022–2025) and Universidad de Buenos Aires UBACyT 20020220400192BA (2023–2026).

References

1. Rupp JC, Sokoloski KJ, Gebhart NN, Hardy RW (2015) Alphavirus RNA synthesis and non-structural protein functions. J Gen Virol 96:2483–2500. https://doi.org/10.1099/jgv.0.000249
2. Shin G, Yost SA, Miller MT et al (2012) Structural and functional insights into alphavirus polyprotein processing and pathogenesis. Proc Natl Acad Sci U S A 109:16534–16539. https://doi.org/10.1073/pnas.1210418109
3. Rausalu K, Utt A, Quirin T et al (2016) Chikungunya virus infectivity, RNA replication and non-structural polyprotein processing depend on the nsP2 protease's active site cysteine residue. Sci Rep 6:37124. https://doi.org/10.1038/srep37124
4. Lulla V, Karo-Astover L, Rausalu K et al (2018) Timeliness of proteolytic events is prerequisite for efficient functioning of the alphaviral replicase. J Virol 92:e00151-18. https://doi.org/10.1128/JVI.00151-18
5. Akhrymuk I, Kulemzin SV, Frolova EI (2012) Evasion of the innate immune response: the Old World alphavirus nsP2 Protein induces rapid degradation of Rpb1, a catalytic subunit of RNA polymerase II. J Virol 86:7180–7191. https://doi.org/10.1128/JVI.00541-12
6. Fros JJ, Van Der Maten E, Vlak JM, Pijlman GP (2013) The C-terminal domain of chikungunya virus nsP2 independently governs viral RNA replication, cytopathicity, and inhibition of interferon signaling. J Virol 87:10394–10400. https://doi.org/10.1128/JVI.00884-13
7. Garmashova N, Gorchakov R, Volkova E et al (2007) The Old World and New World alphaviruses use different virus-specific proteins for induction of transcriptional shutoff. J Virol 81:2472–2484. https://doi.org/10.1128/jvi.02073-06
8. Barraza Sanchez J, Montes M, Faraj S, Filomatori C (2025) Chikungunya virus protease expression in Escherichia coli Rosetta strain. Res Appl Med 3:2. https://doi.org/10.55634/3.3.2

Check for updates

Chapter 11

PAGE-Based Measurements of nsP2 Protease Activity

Mónica R. Montes and Santiago E. Faraj

Abstract

In this protocol, we describe a PAGE-based assay to evaluate the steady-state kinetics of the nsP2 protease. Using a recombinant GFP–TRX fusion substrate containing the nsP3/4 cleavage site, proteolytic activity is monitored through the appearance of cleavage products resolved by SDS-PAGE. Detection is achieved by Coomassie blue staining or GFP fluorescence emission, followed by densitometric analysis to quantify product formation and calculate initial reaction rates. The protocol provides a robust and accessible method for comparing enzyme activity under varying conditions, and is suitable for studies of substrate specificity, inhibitor screening, and the mechanistic characterization of protease function.

Keywords Chikungunya, nsP2 protease, Cysteine protease, PAGE-based assay, Steady-state kinetics

1 Introduction

The protease domain of the nonstructural protein nsP2 catalyzes the hydrolysis of the P1234 polyprotein through a series of sequential cleavage reactions. This protease contains a catalytic dyad composed of a histidine and a cysteine residue. The catalytic mechanism involves a nucleophilic attack by the cysteine thiol on the carbonyl carbon of the scissile peptide bond [1–3].

Measurements of nsP2 protease activity are used not only to assess the efficacy of candidate therapeutic compounds but also to elucidate the molecular determinants of substrate hydrolysis. Enzyme activity is typically monitored during the initial phase of the reaction, when substrate concentration remains almost constant and product accumulation does not yet affect the rate. Under these conditions, the appearance of product is proportional to time, enabling the characterization of enzyme-substrate properties and the comparison across different experimental conditions.

Two methods are commonly used to quantify nsP2 protease activity: the FRET-based assay, discussed in Chap. 12, and the PAGE-based assay, described here.

Claudia Filomatori and Eliana F. Castro (eds.), *Chikungunya Virus*, Methods in Molecular Biology, vol. 3067,
https://doi.org/10.1007/978-1-0716-5496-5_11,

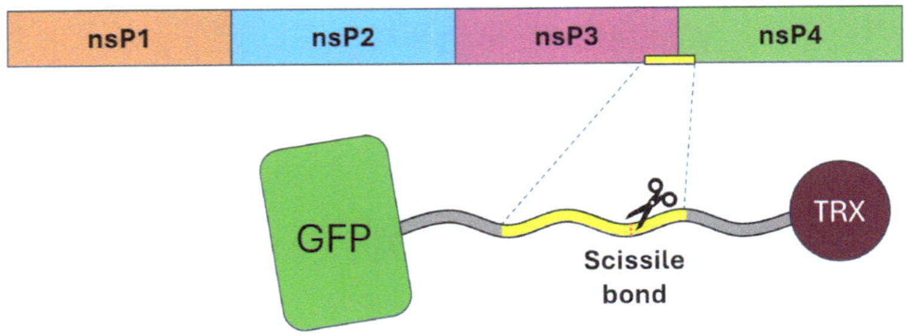

Fig. 1 nsP2 protease-catalyzed reaction. General scheme of the substrate, composed of GFP and TRX linked by a peptide containing the sequence of the nsP3/4 cleavage site

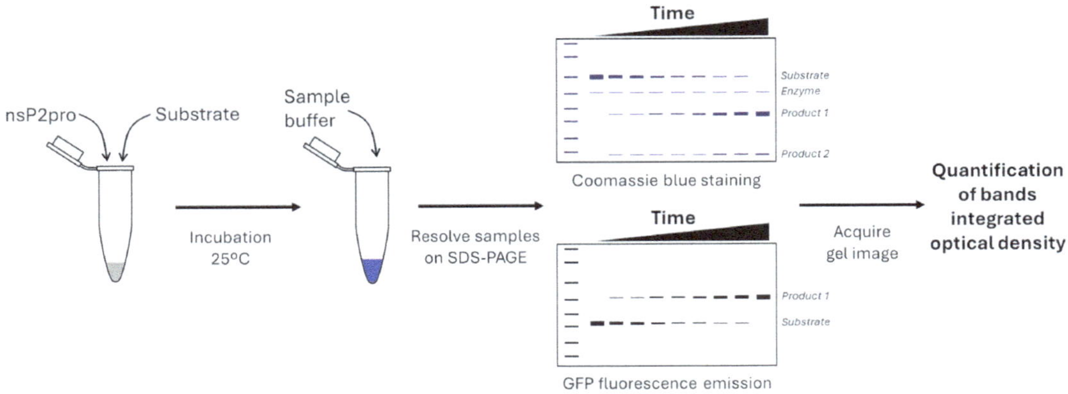

Fig. 2 Assessment of proteolytic reaction: scheme of the protocol for obtaining quantitative data. Enzymatic reactions are stopped by the addition of Sample buffer and resolved on SDS-PAGE. Bands may be visualized either by Coomassie blue staining or GFP fluorescence emission (*see* **Note 1**). Densitometric analysis provides the integrated optical density, which is used to calculate the amount of generated product

In the PAGE-based assay, hydrolysis of a peptide sequence inserted between two stable proteins is evidenced by the appearance of lower-molecular-weight species on the gel [4–7]. To assess the steady-state kinetics of the nsP2 protease domain (nsP2pro), we use the nsP3/4 scissile bond as a substrate for the reaction. This substrate consists of a recombinant protein composed of the superfolder green fluorescent protein (GFP) and *E. coli* thioredoxin (TRX) linked by a peptide with the sequence (ten amino acids prior and five amino acids after) of the cleavage site (Fig. 1) [7].

The reaction is initiated by mixing the substrate with the enzyme and stopped by the addition of sample buffer (Fig. 2). Proteolytic activity is monitored by the appearance of two lower-molecular-weight products on SDS-PAGE: *Product 1*, corresponding to the GFP moiety, and *Product 2*, corresponding to the TRX moiety. The protein bands are quantified by densitometry to calculate the amount of generated product and the initial reaction rate. We present two alternative methods for visualizing reaction products on the SDS-PAGE: Coomassie blue staining and fluorescence emission detection.

Using standard laboratory equipment, this method allows quantitative comparisons of enzyme activity for substrates with sequence variations, under varying experimental conditions and in the presence of inhibitory molecules.

2 Materials

All solutions should be prepared using ultrapure or deionized water.

2.1 Equipment

1. Hand-cast PAGE system and power supplies.
2. Water bath.
3. Stopwatch.
4. Vortex.
5. Gel documentation system (e.g., *ImageQuant* system or a digital camera).

2.2 Sodium Dodecyl Sulfate Polyacrylamide Gel Electrophoresis (SDS-PAGE)

1. 10% SDS solution. Store at room temperature.
2. Stacking gel buffer: 1.0 M Tris–HCl, pH 6.8 at 25°C.
3. Separating gel buffer: 1.0 M Tris–HCl, pH 8.8 at 25°C.
4. Acrylamide solution: 30% acrylamide/bisacrylamide (29:1).
5. APS: 10% ammonium persulfate solution, freshly prepared.
6. TEMED: tetramethylethylenediamine.
7. Isopropyl alcohol OR water-saturated 2-butanol (alternatively, plain water can be used).
8. Tris–glycine running buffer: 25 mM Tris base, 192 mM glycine, 0.1% SDS (do not adjust pH, which should be about 8.3).
9. Staining solution: 0.05% Coomassie brilliant blue R-250 solution in 40% (v/v) methanol (or ethanol) and 10% (v/v) acetic acid.
10. Destaining solution: 40% (v/v) methanol (or ethanol) and 10% (v/v) acetic acid.

2.3 Enzyme Reaction

1. Recombinant nsP2pro in 50 mM Tris–HCl, p pH 7.5 at 4 H 7.5 at 4°C, 50 mM NaCl, and 20% glycerol (see Chap. 10).
2. Recombinant substrate in 10 mM Tris–HCl, pH 7.5 at 4°C.
3. HEPES buffer: HEPES–KOH, pH 7.5 at 25°C (*see* **Note 2**).
4. 5× Sample buffer (Laemmli buffer): 300 mM Tris–HCl, pH 6.8 at 25°C, 10% SDS, 50% glycerol, and 0.05% bromophenol blue. If bands on SDS–PAGE are to be visualized by GFP fluorescence emission, replace SDS with 2.5 M NaCl in the sample buffer.

5. 1.0 M dithiothreitol in water (DTT).
6. 0.5 M ethylenediaminetetraacetic acid (EDTA) solution in water, pH 7.5 adjusted with KOH.

3 Methods

3.1 SDS-PAGE

Prepare the gel according to the Laemmli procedure [8]. This recipe is for a 1.0 mm gel.

1. Set up the glass plates for SDS-PAGE. Thoroughly clean the glass plates and assemble them in the casting frame with their bottoms aligned, inserting spacers along the sides if required. Seal the bottom of the assembly with tape or the system's sealing device, then place the plates vertically in the gel-casting stand. Check for leaks by adding a small amount of water, and adjust if needed.
2. Prepare solutions for the gels in separate containers. Separating gel (10%): mix 1.8 ml of water, 2.2 ml of Separating gel buffer, 2.0 ml of Acrylamide solution, 60 µl of SDS. Stacking gel (5%): mix 2.1 ml of water, 0.4 ml of Stacking gel buffer, 0.5 ml of Acrylamide solution, and 30 µl of the 10% SDS solution.
3. Add 60 µl of APS and 4 µl of TEMED to the separating gel solution. Immediately pour the mixture between the glass plates, leaving approximately 2 cm of space at the top for the stacking gel. Carefully overlay the gel solution with 0.5 ml of isopropyl alcohol, water-saturated 2-butanol or water to exclude oxygen and to create a flat, even interface for the stacking gel. Allow the separating gel to polymerize for 20–30 min.
4. Pour off the overlaying solvent by gently tilting the gel casting plate and rinse the gel surface with water. Wick away any remaining liquid with a piece of blotting paper.
5. Add 30 µl of APS and 3 µl of TEMED to the stacking gel solution. Immediately pour the mixture to fill the remaining space above the separating gel. Insert the comb carefully to form wells, ensuring it is level and properly positioned. Allow the stacking gel to polymerize for 20–30 min.

3.2 Enzyme Activity Assays

The following protocol allows the assessment of nsP2 protease activity under steady-state conditions. Final concentrations are summarized in Table 1 (*see* **Note 3**).

Reactions are performed in a volume of 18 µl (*see* **Note 4**). To obtain the enzymatic reaction time course, eight identical reactions are set up, each allowed to proceed for different time intervals (Table 2). In addition, a nonenzymatic hydrolysis (NEH) control is included, consisting of substrate incubated in the absence of enzyme to assess potential spontaneous hydrolysis (*see* **Note 5**).

1. Thaw the enzyme and substrate. To remove potential aggregates, centrifuge at ≥10,000 × *g* for 30 min at 4°C, and transfer supernatants to clean tubes. Measure protein concentration by recording absorbance at 280 nm. Calculate concentrations using extinction coefficients of 51,255 and 37,360 M^{-1} cm^{-1} for the enzyme and substrate, respectively.
2. Prepare the following solutions:
   - 2× Enzyme solution: 0.25 μM recombinant nsP2pro in 20 mM HEPES buffer, 0.5 mM EDTA.
   - 2× Substrate solution: 10 μM recombinant substrate in 20 mM HEPES buffer, 4 mM DTT.

Table 1

Final concentration of the components of the enzymatic reaction. The reaction is performed at 25°C

Reagents	Final concentration
Enzyme	0.125 μM
Substrate	5.0 μM
HEPES, pH 7.5	20 mM
DTT	2.0 mM
EDTA	0.25 mM

Table 2

Detailed composition of each tube in a typical set up of the experiment. The reaction is measured for different time intervals (tubes 1–8) to obtain the initial rate. Tube 9 is included to check for nonenzymatic hydrolysis

Tube		Incubation time (min)	2× Enzyme solution (μl)	2× Substrate solution (μl)	*Buffer solution* (μl)
1	Enzymatic reaction time course	15	9	9	—
2		10	9	9	—
3		5	9	9	—
4		4	9	9	—
5		3	9	9	—
6		2	9	9	—
7		1	9	9	—
8		0	9	9	—
9	NEH control	16	—	9	9

- Buffer solution: 20 mM HEPES buffer.
- 4× Sample buffer (reducing): mix four volumes of 5× Sample buffer with one volume of 1.0 M DTT (or one volume of β-mercaptoethanol).

3. Prepare nine tubes (considering a ten-lane SDS-PAGE will be used).
4. Add 9 μl of 2× Enzyme solution to tubes 1–8 (Table 2).
5. Add 9 μl of Buffer solution to tube 9 (Table 2).
6. Add 6 μl of 4× Sample buffer to tube 8 to inactivate the enzyme (time = 0).
7. Preheat tubes 1–9 and the 2× Substrate solution in a water bath at 25°C for 5 min prior to starting the reaction.
8. Start reactions by adding 9 μl of 2× Substrate solution to each tube, one at a time, at 30-s intervals, as shown in Table 3 (*see* **Note 6**). After each addition, mix the reaction gently but thoroughly, and immediately return the tube to the water bath.
9. Stop reactions after the corresponding incubation time by adding 6 μl of 4× Sample buffer (Table 3).
10. If quantifying Coomassie blue-stained bands, boil the samples for 10 min. Omit this step if GFP fluorescence emission is to be read (*see* **Note 7**).
11. Set up the SDS-PAGE and add Tris–glycine running buffer to the anode and cathode chambers of the gel unit. Load 15 μ l from each reaction tube into separate lanes of the gel. Run

Table 3
Example of a timing plan for reaction initiation and stop

Tube	Incubation time (min)	Addition of 2× Substrate solution (min)	Addition of 4× Sample buffer (min)
9	16	0:00	16:00
1	15	0:30	15:30
2	10	1:00	11:00
3	5	1:30	6:30
4	4	2:00	6:00
5	3	2:30	5:30
6	2	3:00	5:00
7	1	3:30	4:30
8	0	4:00	*Before initiating the time course*

the gel (about 1 h at 25 mA/gel) and stop electrophoresis just before the dye front reaches the bottom of the gel.

12. Stain the gel with Staining solution, then destain thoroughly using Destaining solution until no background staining is visible. If GFP fluorescence emission is to be read, omit this step and proceed directly to the following step.
13. Acquire a high-quality photograph of the SDS-PAGE using a gel documentation system in the normal mode (Coomassie blue staining) or fluorescent mode, as appropriate. If the gel is stained with Coomassie blue, a digital photograph with a good camera—ensuring uniform lighting and avoiding glare, shadows, or reflections that could interfere with band processing—is a viable option. A typical SDS-PAGE photograph is shown in Fig. 3a.
14. Quantify the Integrated Optical Density (IOD) of the bands corresponding to the *Substrate* and *Product 1* by densitometric analysis using image analysis software (such as *ImageJ*, *GelAnalyzer*, or similar). Ensure the image is properly calibrated, select each band carefully, and subtract background signal before recording band intensities (Fig. 3b) (*see* **Note 8**).
15. Band densitometry values, IODs, obtained from SDS-PAGE must be converted into product concentration values [P], in order to calculate the substrate hydrolysis rate (*see* **Note 9**). Use Eq.1 to express product concentration in μM units:

$$\text{Coomassie blue staining}: [\mathrm{P}] = \frac{\mathrm{IOD}_{P1}}{0.7\,\mathrm{IOD}_S + \mathrm{IOD}_{P1}}[\mathrm{S}]_0 \qquad (1)$$

$$\text{GFP fluorescence}: [\mathrm{P}] = \frac{\mathrm{IOD}_{P1}}{1.3\,\mathrm{IOD}_S + \mathrm{IOD}_{P1}}[\mathrm{S}]_0$$

where IOD_S and IOD_{P1} are the integrated optical density values obtained for the bands corresponding to the *Substrate* and *Product 1*, respectively, and $[\mathrm{S}]_0$ is the substrate in μM concentration. For Coomassie blue staining, the factor of 0.7 is calculated based on the ratio between the MW of *Product 1* (28.4 kDa) and the *Substrate* (41.1 kDa), and accounts for the fact that *Product 2* cannot be quantified under these conditions (*see* **Note 1**). For GFP fluorescence, the factor of 1.3 corrects for the 30% higher fluorescence of *Product 1* relative to the *Substrate* under our experimental conditions (see **Note 9**).

16. Plot product concentration versus time (Fig. 3c). The activity of the enzyme is determined from the slope of product-versus-time plots during the linear phase of the reaction.

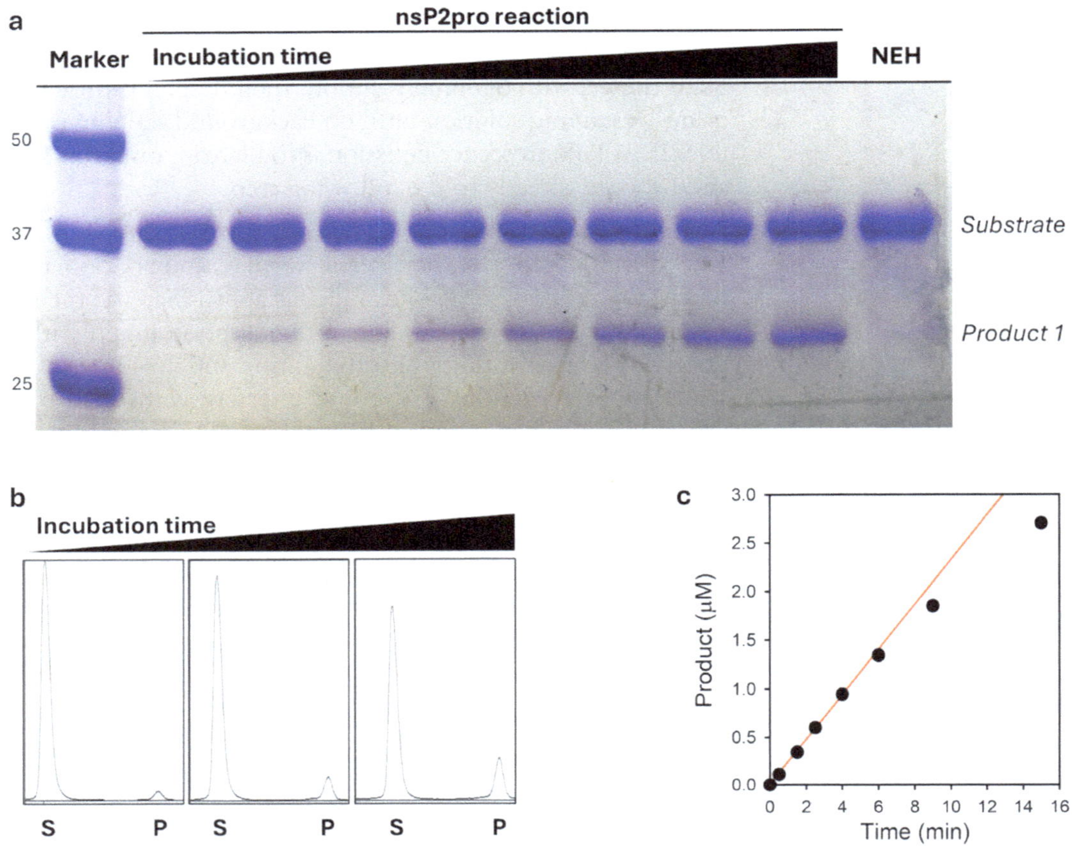

Fig. 3 Reaction time course. (**a**) Coomassie blue-stained SDS-PAGE of hydrolysis reactions. *Product 1* corresponds to the GFP moiety (*see* **Note 1**). Note that no detectable nonenzymatic hydrolysis (NEH) is observed. (**b**) Example densitograms obtained for some of the lanes of the gel shown in panel a. (**c**) Plot of [Product] vs. time. Activity value is obtained from the slope of the best-fitting line to the data during the initial linear phase of the reaction. In the experiment shown, the calculated activity is 44 nmolP. min^{-1}. $mgEz^{-1}$

4 Notes

1. The hydrolysis reaction yields two products: *Product 1*, corresponding to the GFP moiety, and *Product 2*, corresponding to the TRX moiety. However, *Product 1* is the only product that can be reliably detected by SDS-PAGE, as only a small amount of the substrate is converted under the initial reaction conditions and the low mass of TRX results in very faint bands when stained with Coomassie blue. In addition, TRX typically migrates with the running front in a 10% SDS-PAGE. Naturally, GFP fluorescence emission does not allow the detection of *Product 2*.

2. Alternatively, Tris buffer may be used instead of HEPES, although we have found lower activity values when using this buffer.
3. If the composition of the reaction mixture needs to be altered—for example, to test the effect of inhibitors or activators, or to vary pH, salt concentration— bear in mind that the reaction is highly sensitive to ionic strength, which may markedly decrease activity. Therefore, ionic strength must be kept constant in all assays to be compared, taking into account contributions from salts, buffers, and additives, and the effect of temperature. To ensure consistent ionic strength across experiments, we routinely apply the Debye–Hückel formalism, iteratively calculating the ionic strength of reaction mixtures and correcting apparent p*Ka* values, to estimate the fraction of ionized species and adjust ionic strength with NaCl as required. Under the conditions described in this chapter, the reaction mixture has an ionic strength of approximately 12 mM.
4. The reaction volume can be safely adjusted if desired, although smaller volumes may increase variability in data. Furthermore, the protocol can be modified by preparing a single large reaction mixture and withdrawing aliquots at different time points, which are then stopped with Sample buffer.
5. For control and reference purposes, a solution of recombinant nsP2pro might be analyzed on the SDS-PAGE. Note that, given the enzyme concentration used, it is not detectable by SDS-PAGE.
6. The enzyme concentration used, and the reactions' incubation time, should be adjusted according to the enzyme's activity, which can vary considerably between different preparations or with enzyme aging. Highly active preparations require shorter incubation times to ensure a proportional relationship between product formation and time, whereas less active preparations may need longer incubations to produce measurable amounts of product.
7. GFP fluorescence is lost when heated. Therefore, samples must not be heat-denatured (boiled) if results are to be analyzed by GFP fluorescence emission. Note that under this condition GFP retains significant structure even in the presence of SDS. Therefore, migration on SDS-PAGE is anomalous: *Product 1* (GFP) migrates ahead of the *Substrate* (GFP+TRX).
8. Fluorescence detection yields clearer and better-resolved peaks in densitograms, resulting in more reproducible and precise measurements. Whenever available, GFP fluorescence emission is recommended over Coomassie blue staining.
9. The calculation of product concentration requires IOD values to be proportional either to the protein mass (Coomassie

blue staining) or to the number of fluorophores (GFP fluorescence) in a given band. The relative fluorescence of GFP in the *Substrate* and *Product 1* should also be assessed; in our hands *Product 1* exhibited approximately 30% higher fluorescence than the *Substrate.*

Acknowledgments

We are grateful to Prof. Andres Merits (University of Tartu) for kindly providing the construct for the expression of the substrate. This work was supported by Agencia Nacional de Promoción Científica y Tecnológica PICT 598/2020 (2022–2025) and Universidad de Buenos Aires UBACyT 20020220400192BA (2023–2026).

References

1. Strauss EG, Rice CM, Strauss JH (1983) Sequence coding for the alphavirus nonstructural proteins is interrupted by an opal termination codon. Proc Natl Acad Sci U S A 80: 5271–5275
2. Abu Bakar F, Ng LFP (2018) Nonstructural proteins of alphavirus-potential targets for drug development. Viruses 10(2):71
3. Rupp JC, Sokoloski KJ, Gebhart NN et al (2015) Alphavirus RNA synthesis and non-structural protein functions. J Gen Virol 96:2483–2500
4. Stapleford KA, Rozen-Gagnon K, Das PK et al (2015) Viral polymerase-helicase complexes regulate replication fidelity to overcome intracellular nucleotide depletion. J Virol 89:11233–11244
5. Das PK, Puusepp L, Varghese FS et al (2016) Design and validation of novel chikungunya virus protease inhibitors. Antimicrob Agents Chemother 60:7382–7395
6. Rausalu K, Utt A, Quirin T et al (2016) Chikungunya virus infectivity, RNA replication and non-structural polyprotein processing depend on the nsP2 protease's active site cysteine residue. Sci Rep 6:37124
7. Utt A, Das PK, Varjak M et al (2015) Mutations conferring a noncytotoxic phenotype on chikungunya virus replicons compromise enzymatic properties of nonstructural protein 2. J Virol 89:3145–3162
8. Laemmli UK (1970) Cleavage of structural proteins during the assembly of the head of bacteriophage T4. Nature 227:680–685

Chapter 12

Assessing Chikungunya Virus Protease Activity and Pharmacological Inhibition Using the acc-CHIK$_{15}$-dnp Fluorogenic Peptide Substrate

Muhammad Arslan Rahat, Sparsh Makhaik and Jeanne A. Hardy

Abstract

Chikungunya virus, a mosquito-borne alphavirus, poses a significant global health threat due to its rapid spread and lack of approved antiviral therapies. The viral nonstructural protein 2 protease (nsP2p) plays a critical role in processing the viral polyprotein to produce key viral enzymes, making it an attractive target for drug development. In this study, we describe the expression and purification of recombinant CHIKV nsP2 protease and establish a robust fluorogenic assay to measure its enzymatic activity using the acc-CHIK$_{15}$-dnp peptide substrate. The assay enables real-time monitoring of proteolytic cleavage and provides a platform for evaluating small-molecule inhibitors. We present optimized conditions for protein expression in *E. coli*, affinity purification, and tag removal, achieving high-purity nsP2 protease suitable for biochemical characterization. Using this system, we also provide a robust protocol for determining IC_{50} values for candidate inhibitors. This work provides a reproducible methodology for high-throughput screening of CHIKV protease inhibitors, facilitating the discovery of potential antiviral compounds against chikungunya virus.

Keywords CHIKVP, Chikungunya virus protease, Activity assay, Affinity purification, Alphavirus, Plate reader, Drug discovery, Hit assessment, Redox PAINS identification

1 Introduction

Chikungunya virus (CHIKV) is a debilitating arthropod-borne virus (family *Togaviridae*, genus Alphavirus) that is considered to have significant pandemic potential. The name chikungunya derives from the Makonde word meaning "that which bent" referring to the bent posture of individuals suffering from severe arthralgia associated with infection [1]. This virus is spread through the bites of the *Aedes aegypti* and *Aedes albopictus* mosquitoes [2–5]. The first case of CHIKV was reported in 1952 in Tanganyika (now Tanzania). Due to climate change, mosquito migration, and human travel, the virus has spread to over 118 countries in Asia, Africa, Europe, and

Claudia Filomatori and Eliana F. Castro (eds.), *Chikungunya Virus*, Methods in Molecular Biology, vol. 3067,
https://doi.org/10.1007/978-1-0716-5496-5_12,

the Americas [6]. Acute CHIKV infections cause fever, joint pain, headache, and rash. Chronic chikungunya arthritis is reported in 40% of patients, often lasting more than 3 months [7]. During the late stages of pregnancy, this virus has high transmission rates of up to 62% from infected mother to child [8]. In newborns, CHIKV infection can lead to abnormalities like cleft palate, stillbirth, microcephaly, and hyperpigmentation [9]. Two vaccines have been approved by the FDA for the prevention of CHIKV Vimkunya [10] and Ixchiq [11]; however, recently adverse effects of Ixchiq have been reported, including 21 hospitalizations and three deaths [12]. Regardless of the efficacy and safety of available vaccines, breakthrough infections will be common, underscoring the importance of efficacious antivirals. Unfortunately, no antivirals have been approved for use in humans for treating Chikungunya virus infection.

Chikungunya virus (CHIKV) is an enveloped virus approximately 70 nm in diameter. Its genome consists of a single-stranded positive-sense RNA of ~12 kb, with a 5′ cap structure (7-methylguanosine) and a 3′ polyadenylated tail [13]. The genome contains two open reading frames (ORFs) encoding two polyproteins: one for nonstructural proteins (nsPs), including the CHIKV protease nsP2p (CHIKVP) [14] and one for structural proteins. The nonstructural proteins nsP1 to nsP4, which are liberated from the polyprotein by CHIKVP, play crucial enzymatic roles in viral replication and transcription. nsP1 is involved in RNA capping. nsP2 comprises at least two different functions. Its N-terminal region (aa 1–470) performs nucleoside triphosphatase and RNA triphosphatase activities, which are crucial for helicase function [15]. The C-terminal region (aa 471–798), which includes the protease and methyl-transferase like (MTL) domains (nsP2p, also called CHIKVP), is responsible for proteolytic processing of the nonstructural polyprotein [16]. The proteolytic activity derives from the cysteine 478 and histidine 548 catalytic dyad [16] (Fig. 1a). The nsP3 performs ADP ribosyl hydrolase activity and nsP4 is an RNA polymerase that is essential for viral RNA synthesis. Of the virally encoded enzymes, the nsP2 protease is a significant alphaviral drug target.

Due to their essential roles in polyprotein processing, proteases perform a crucial role in the life cycle of viruses, which makes them excellent drug targets. In addition, several key host restriction factors have been found to be targeted by viral proteases [17–19], which is an additional role that can be pharmacologically targeted [20]. Proteases from several viruses have been targeted successfully to develop antivirals, including Telaprevir for hepatitis C [21], Paxlovid (nirmatrelvir) for COVID-19 [22], and Saquinavir and others for HIV [20]. Similarly, CHIKV reliance on the activity of nsP2 protease (CHIKVP) makes it promising for antiviral development. To develop an effective inhibitor for CHIKVP, a reliable assay is essential to accurately measure its activity.

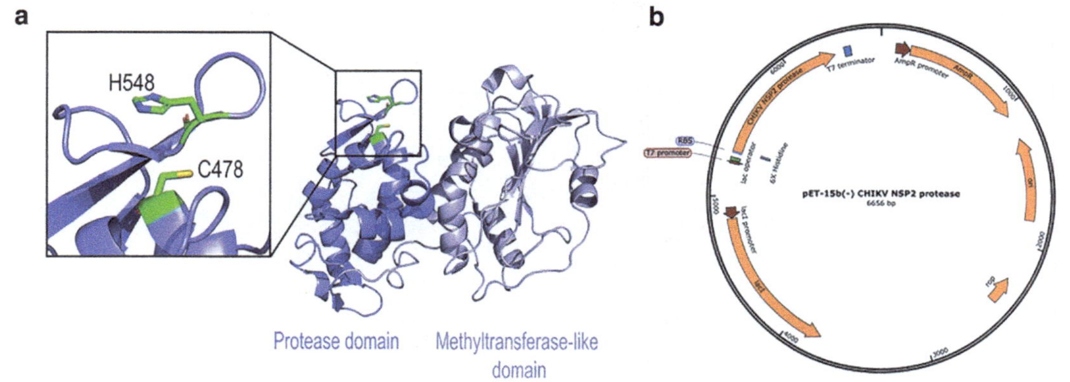

Fig. 1 (**a**) CHIKVP (PDB ID 4ZTB) is composed of the protease domain (dark blue) and a methyltransferase-like domain (light blue). The catalytic cysteine–histidine dyad (C478 and H548, green) is in the protease domain. (**b**) The CHIKVP expression construct expresses both the protease and the methyltransferase-like domains of CHIKVP with an N-terminal, Thrombin-cleavable His_6 tag from the pET-15b plasmid

Initially, protease assays often relied on gel-based substrate cleavage, a labor-intensive, end-point measurement [23], which lacks real-time resolution. Fluorogenic substrates have emerged as dynamic and sensitive alternatives [24]. Some proteases can cleave synthetic peptide substrates that lack C-terminal overhangs, enabling single fluorophore substrates to be developed. In contrast, the alphaviral proteases have thus far required longer substrate peptides containing both N- and C-terminal overhangs for efficient recognition and cleavage. This length dependence has necessitated use of a Förster resonance energy transfer (FRET) donor–quencher pair, which emits a measurable fluorescence signal upon peptide cleavage and enables continuous monitoring of the CHIKVP activity. Peptide substrates of various lengths derived from the sequences present in the CHIKV nsP1/nsP2, nsP2/nsP3, and nsP3/nsP4 polyprotein cleavage junctions have been reported that enable measurement of the activity of the CHIKVP [25–29]. The reported k_{cat}/K_M for these substrates differ (Table 1). The peptide substrate acc-$CHIK_{15}$-dnp exhibited significantly improved k_{cat} compared with previously reported substrates and demonstrated selectivity for CHIKVP over other alphaviruses, despite the high degree of sequence conservation among them. Most importantly, the vastly improved signal-to-noise relative to other fluorogenic substrates enabled us to run a small-molecule high-throughput screen on CHIKV that had been impossible with prior substrates [25]. Here, we describe the expression and purification of recombinant CHIKVP, a robust activity assay of CHIKVP, and method to assess the inhibitory properties of small molecules by measuring IC_{50} using acc-$CHIK_{15}$-dnp as a substrate specifically designed for CHIKVP [25]. This purified CHIKVP and inhibition assay can be robustly applied to perform high-throughput screening

Table 1
Kinetic parameters are given for substrates of CHIKVP

Substrate	Cleavage junction	K_M (μM)	k_{cat} (s^{-1})	k_{cat}/K_m ($M^{-1}s^{-1}$)	Ref
acc-βAla-DELRLDRAGG*YIFSS-Lys-dnp	nsP3/nsP4	1.36×10^{2}	3.4×10^{-1}	2.4×10^{3}	[25]
acc-βAla-DRAGG*YIFSS-Lys-dnp	nsP3/nsP4	1.80×10^{2}	2.8×10^{-2}	1.5×10^{2}	[25]
dabcyl-RAGG*YIFS-edans	nsP3/nsP4	1.46×10^{1}	3.4×10^{-3}	2.3×10^{2}	[26]
dabcyl-DRAGG*YIFSE-edans	nsP3/nsP4	1.02×10^{1}	4.1×10^{-3}	4.08×10^{2}	[27]
dabcyl-DELRLDRAGG*YIFSS-edans	nsP3/nsP4	2.5	4.2×10^{-2}	1.6×10^{4}	[28]
Nma-RAGA*GIIETk(dnp)-OH	nsP1/nsP2	3.6×10^{1}	3.3×10^{-3}	9.1×10^{-5}	[29]
Nma-RAGC*APSYRk(dnp)-OH	nsP2/nsP3	4.5	2.9×10^{-3}	6.54×10^{-4}	[29]
Nma-RAGG*YIFSSk(dnp)-OH	nsP3/nsP4	7.4	1.1×10^{-3}	1.58×10^{-4}	[29]

and lead optimization of inhibitors for CHIKVP in the context of pharmaceutical development.

2 Materials

2.1 Consumables

1. Electroporation cuvettes
2. 1 cm plastic cuvettes
3. Aluminum foil
4. 1.5 mL Eppendorf tubes
5. 50 mL conical tubes
6. Micropipettors and tips
7. Luria Bertani (LB) agar plates
8. Serological pipettes
9. 384-Well flat-bottom black microplate
10. 96-Well flat-bottom microplate
11. Low-binding 0.22-μm syringe filter
12. 5 mL nickel affinity column
13. 16% acrylamide 1.5 mm-thick gels for SDS-PAGE
14. 10 kDa MWCO dialysis tubing 34 mm dry flat width
15. 10 kDa MWCO centrifugal filtration units
16. 0.22 μm vacuum filters
17. Disposable Kim wipes

2.2 Equipment

1. pH meter
2. Stationary incubator for growth of bacterial colonies on plates
3. Shaking floor incubator for growth of bacterial cultures
4. 399 Electroporation System
5. Spectrophotometer
6. Vortex
7. Microfluidizer
8. Superspeed Floor Centrifuge and accompanying rotors
9. Multimode Plate Reader
10. Chromatography system
11. Gel imaging system

2.3 Chemicals

1. LB medium
2. BL21(DE3) electrocompetent *E. coli* cells
3. Ampicillin

4. Isopropyl-β-D-thiogalactopyranoside (IPTG)
5. TRIS base [Tris(hydroxymethyl)aminomethane]
6. NaCl
7. NaOH
8. HCl
9. Dithiothreitol (DTT)
10. Coomassie Brilliant Blue R concentrate
11. Bromophenol blue
12. Methanol
13. Acetic acid
14. CHIKVP acc-CHIK$_{15}$-dnp substrate
15. Unstained Protein Standard Broad Range
16. Bicinchoninic acid (BCA) assay reagent

2.4 Reagents

All solutions are made with Nanopure water.

1. 100 mg/mL ampicillin. Dissolve 200 mg of ampicillin in 2 mL of water. Mix and store at −20 °C.
2. 1 M IPTG. To prepare 10 mL of 1 M IPTG, dissolve 2.383 g of IPTG into 10 mL of water. Filter using a 0.22-μm syringe filter and store at −20 °C. Although it is not typical outside our lab, we make this solution on the day of the growth and induction.
3. LB growth media. Dissolve 20 g of LB powder in 1 L of water in a baffled Erlenmeyer flask. Cover the flask loosely with aluminum foil and steam sterilize by autoclaving for 45 min at 121 °C and 16 PSI pressure.
4. 1 M DTT. Dissolve 154 mg of DTT in 1 mL (total volume) of water and store at −20 °C.
5. 4 M NaCl. Dissolve 234 g of NaCl into 1 L (total volume) of water and store at room temperature.
6. 1 M Tris-HCl. Dissolve 121.14 g of Tris into 1 L (total volume) water and store at room temperature.
7. 5 M NaOH. Dissolve 200 g of NaOH in 1 L of water (total volume) and store it at room temperature.
8. Lysis buffer: 50 mM Tris-HCl pH 7.5, 500 mM NaCl, 5% glycerol. To prepare the 2 L of lysis buffer, combine 100 mL of 1 M Tris, 250 mL of 4 M NaCl, and 100 mL of glycerol in a 2-L breaker and add 1000 mL of nanopure H_2O. Adjust the pH of the solution to 7.5 using 5 M NaOH or 1 M HCl while stirring. After adjusting the pH, adjust the final volume to 2 L. Filter the solution using a 0.22-μm vacuum filtration

apparatus, then degas the solution for 1 h by attaching house vacuum to a vacuum degassing cap tightly secured on a 2-L reagent bottle with stirring. Degassing helps prevent bubble accumulation during chromatographic Steps 3.2 and 3.5. Store the solution at 4 °C.

9. Elution buffer: 50 mM Tris-HCl pH 7.5, 500 mM NaCl, 5% glycerol, 1 M Imidazole. To prepare the 1 L of elution buffer, combine with stirring 50 mL of 1 M Tris, 125 mL of 4 M NaCl, 68.07 g of imidazole and 50 mL of glycerol in a 1-L breaker and add 500 mL of nanopure H_2O. Adjust the pH of the solution using the pH meter to 7.5 using 1 M HCl while stirring. After adjusting the pH, adjust the final volume to 1 L. Filter the solution using 0.22 μm vacuum filtration apparatus and then degas the solution for 1 h by attaching house vacuum to a vacuum degassing cap tightly secured on a 1-L reagent bottle with stirring. Degassing helps prevent bubble accumulation during chromatography Steps 3.2 and 3.5. Store it at 4 °C.
10. Dialysis buffer: 50 mM Tris-HCl pH 7.5, 500 mM NaCl, 5% glycerol, 0.5 mM DTT. Prepare the dialysis buffer using the same material as in lysis buffer and add 154 mg of DTT at the time of dialysis (*see* **Note 1**).
11. 1% w/v bromophenol blue. Dilute 10 mg bromophenol blue in 1 mL methanol.
12. 3× gel-loading buffer. To make 30 mL of solution mix 0.9 mL of 1% w/v bromophenol blue, 9 mL of glycerol, 1.8 g of SDS, 11.25 mL of 0.5 M Tris-HCL pH 6.8, and 8.9 mL of water. Mix and store at −20 °C.
13. Fixing solution. Mix 400 mL of methanol, 100 mL of acetic acid, and 500 mL of water with stirring. Store at room temperature.
14. Staining solution. Mix one part Coomassie Brilliant Blue R concentrate with five parts water. Store in a flammable cabinet at room temperature.
15. Destaining solution. Dissolve 10 mL of acetic acid and 90 mL of water while stirring. Store it in a 100-mL glass bottle.
16. Activity assay buffer: 1 mM CHAPS, 25 mM HEPES. To make 50 mL of activity assay buffer, dissolve 30.7 mg CHAPS, 298 mg HEPES, and 7.7 mg of DTT in 30 mL of water. After adjusting the pH to 7.5 with 5 M NaOH solution while stirring, adjust the final volume to 50 mL. Filter the solution using a 0.22-μm vacuum filtration apparatus and store it at 4 °C.

3 Methods

3.1 Expression of CHIKVP

This manuscript describes expression, purification, and assessment of CHIKVP activity in a way that is applicable for drug development. An overview of the approaches used in this chapter are shown (Fig. 2a).

1. Retrieve a 100-μL aliquot of BL21(DE3) electrocompetent *E. coli* cells and an aliquot of the CHIKVP expression plasmid containing the coding sequence for the CHIKVP protease (Fig. 1b) from the −80 °C freezer and place them on ice to thaw (*see* **Note 2**).
2. Spray 70% ethanol on the workstation. Light a flame for sterilization and utilize proper sterile technique for molecular biology approaches.
3. Transfer 1 μL of CHIKV plasmid (45 ng/μL) into the 100 μL of thawed BL21(DE3) electrocompetent *E. coli* (10^8 CFU/μg DNA).
4. Using a micropipette, transfer the cell:plasmid mixture to an electroporation cuvette and apply an electrical shock of 1490 V.
5. Allow the cells to recover by adding 300 μL of LB media. After mixing, transfer all cells to a sterile 1.5 mL Eppendorf tube. Tape the tube in a horizontal position to the platform of an orbital shaker in a 37 °C incubator. Incubate with shaking at 230 rpm for 30 min.
6. After 30 min, transfer 20 μL of the cells onto LB plates containing 100 μg/mL ampicillin. Spread the cells in a circular pattern using a sterilized stainless-steel spreading rod. Incubate the plate at 37 °C overnight to allow growth of colonies that have been transformed with the CHIKVP expression plasmid.
7. After 16 h, assess the success of the transformation by monitoring the growth of colonies on the plate. If colonies are observed, they should be propagated immediately. It is also possible to store the plate for future use for up to 1 week by wrapping the outer edge of plate with parafilm (to prevent contamination and desiccation) and transfer the wrapped plate to a 4 °C refrigerator.
8. Using flame sterilization, add 50 μL of 100 mg/mL ampicillin to 50 mL LB broth in a 250-mL Erlenmeyer flask, which will be used as a seed culture.
9. Pick one colony from the LB agar-amp plate containing the transformed cells and use it to inoculate the seed culture flask. Cover the flask with sterilized aluminum and incubate with shaking at 230 rpm and 37 °C overnight.

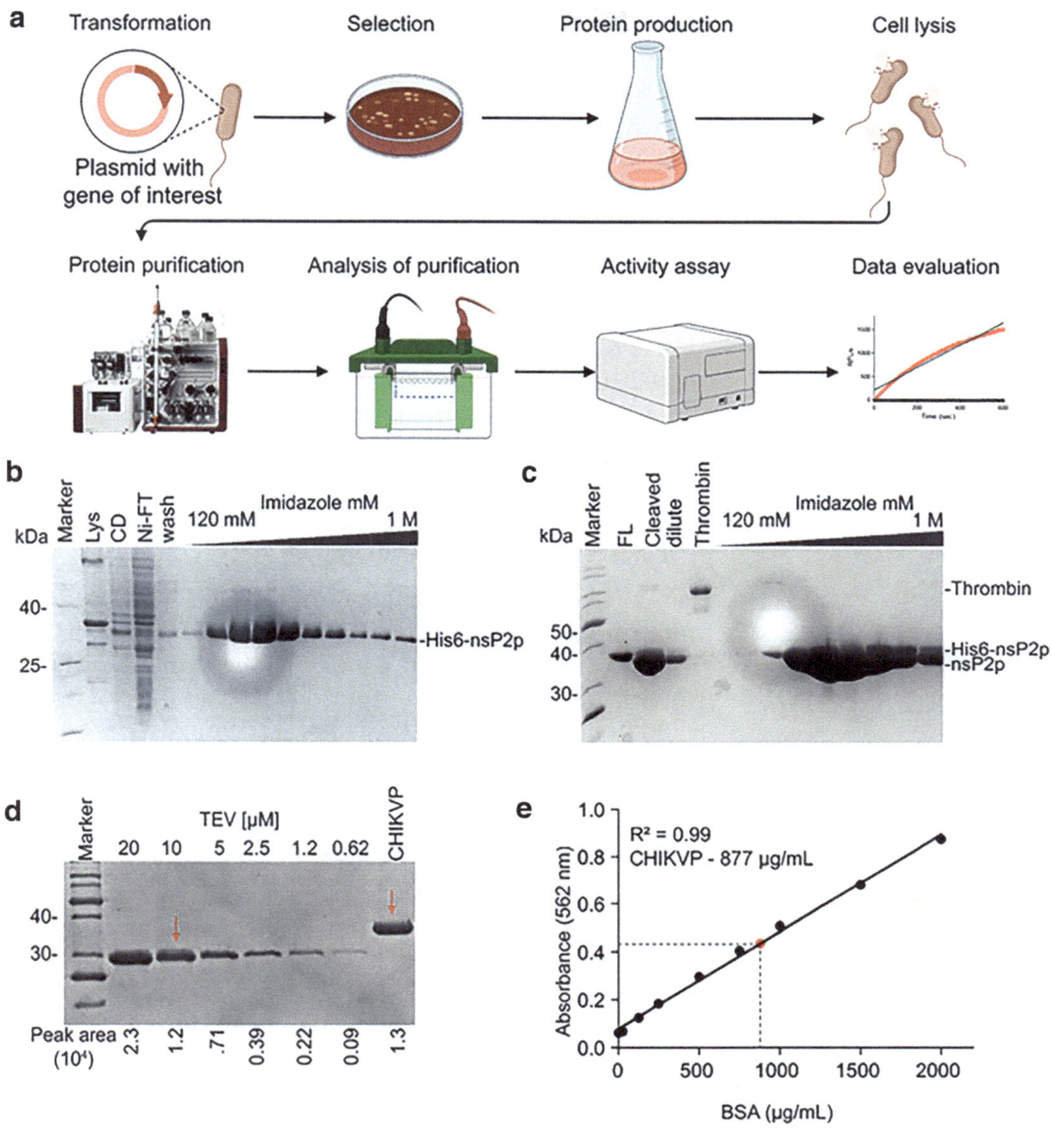

Fig. 2 (**a**) Production, purification, and assessment of CHIKVP activity using acc-CHIK$_{15}$-dnp. (**b**) SDS-PAGE during purification by Ni-affinity. Fractions include cell lysate following overexpression (Lys), pellet of cell debris (CD), flow-through upon loading to the Ni-charged column (Ni-FT), wash with 80 mM imidazole, and fractions from the applied imidazole gradient to elute CHIKVP. (**c**) SDS-PAGE of purification by Ni-affinity following thrombin cleavage to remove the His$_6$tag. Fractions include the full-length (FL) CHIKVP prior to thrombin digestion, CHIKVP following thrombin digestion (cleaved), cleaved protein following dilution (dilute), thrombin, and fractions from the imidazole gradient. (**d**) Gel-based estimation of 20-fold-diluted CHIKVP concentration using a twofold serial dilution of 20 µM purified Tobacco etch virus (TEV) protease as a standard. The arrow indicates the lane estimated to have the same concentration as TEV. The quantification of band intensity is listed below each lane. (**e**) Standard curve of BSA dilutions (black dots) for estimating concentration of CHIKVP (red dot)

10. The next morning, add 1 mL 100 mg/mL ampicillin into each of eight Fernbach flasks, each containing 1 L of sterilized LB media.
11. Sterilely remove and set aside 1 mL of media to serve as a blank for monitoring culture growth.
12. Using a sterilizing flame, inoculate each 1 L of sterilized media with 5 mL of seed culture grown overnight using 5 mL sterile serological pipette. Incubate the culture at 37 °C with vigorous aeration 230 rpm.
13. After 2 h of incubation, check the optical density (OD_{600}) every 20 min. Once the OD_{600} reaches 0.6 (approximately 4 h after inoculation), induce expression in the culture by sterile addition of 1 mL of 1 M IPTG solution to a final concentration of 1 mM.
14. Continue to incubate the cells at 25 °C for 4 h post induction with vigorous shaking at 230 rpm.
15. Harvest the cells by centrifugation in a superspeed floor centrifuge in 1000 mL bottles at 5000 rpm (4685 rcf) for 10 min. We expect to see a healthy-looking, uniformly radially distributed cell pellet of 3–4 cm in diameter (*see* **Note 3**). The pellet should be collected by manual scraping with a weighing spoon and placed into a tared 50 mL conical tube. The mass of the wet pellet should be recorded for use in calculating the total yield of protein. The harvested pellet should be stored at −80 °C until purification. The freezing and subsequent thawing process is important for lysis of the cell membrane, so cell pellets should be frozen even if purification will be conducted on the same day as harvest.

3.2 Preparation of CHIKVP by His_6-Mediated Affinity Purification

1. Add 50 mL of room-temperature lysis buffer to the frozen cell pellet. Incubate the cells in a 37 °C water bath for 15 min or until thawed. To make a homogeneous suspension, vortex the cells to fully suspend them. Visually confirm that the cells are uniformly suspended and free of clumps before lysing them using four passages through a microfluidizer at 15,000 psi.
2. Clarify the cell lysate by centrifuging at 27,000 rcf for 50 min at 4 °C in a superspeed floor centrifuge. Collect the supernatant and reserve the pellet to assess the aggregated or insoluble protein through SDS-PAGE gel.
3. Retain 20 μL samples of the cell lysate and the cell debris (pellet) for SDS-PAGE analysis in labeled 1.5 mL Eppendorf tubes. Scratch the cell debris pellet with a pipette tip to obtain a sample of an approximately 1 mm sphere.
4. Using the chromatography system, load the supernatant at 2 mL/min flow rate onto a 5-mL nickel affinity column that

has been pre-equilibrated with lysis buffer. Monitor absorbance at 280 nm (A_{280}).

5. Wash the unbound protein from the column with an 8% elution buffer:92% lysis buffer mixture until the A_{280} returns to baseline.
6. Elute the protein using a linear gradient from 10% elution buffer:90% lysis buffer to 100% elution buffer at 1 mL/min. Expect to observe CHIKVP begin to elute at 15% elution buffer with the maximal elution at 45% elution buffer.
7. Assess the purity of the protein obtained using SDS-PAGE (Fig. 2b, c).

3.3 Assessing Purity of CHIKVP by SDS-PAGE

1. Due to the presence of cysteines in CHIKVP, it is important to analyze CHIKVP using reducing SDS-PAGE. To prepare the appropriate gel-loading buffer, add 125 μL of 1 M DTT to 875 μL of 3× gel-loading buffer in an Eppendorf tube.
2. Dilute a 20-μL sample of the lysate with 60 μL of lysis buffer.
3. Dilute the 1 mm sphere from the cell debris pellet from Step 3.2.3 into 60 μL 2 M urea to determine the content of CHIKVP in the insoluble pellet. Observing CHIKVP in the pellet may be indicative of incomplete lysis or of CHIKVP that has been unfolded during the lysis process. We do not typically observe high levels of CHIKVP in the cell debris pellet (Fig. 2 b, c).
4. Mix 10 μL of 3X-dye premixed with DTT and 20 μL of diluted lysate, cell pellet, or collected fractions in labelled Eppendorf tubes.
5. Heat the dye-containing sample at 95 °C for 10 min to denature the proteins.
6. Prepare and assemble a 16% acrylamide 15-well SDS gel in running buffer.
7. Load 8 μL of each of the samples to be analyzed in individual wells of an SDS gel using a micropipette with a gel loading tip. Load 5 μL of protein marker into one well of the SDS page gel.
8. Electrophorese the gel at 70 V, 400 mA for 10 min. Then increase the voltage to 170 V and run for an additional 55 min, until the dye front reaches the bottom of the gel.
9. Remove the gel from the plates and place it into a rectangular plastic container with a fitted lid. Add enough fixing solution to cover the gel and microwave for 30 s. Decant the fixing solution into the waste bottle.
10. Pour in a sufficient volume of staining solution to cover the gel and microwave for 30 s. Immediately remove the staining solution, which may be saved for reuse.

11. Destain the gels twice with destaining solution. For each round of destaining, microwave for 30 s, then immediately decant the destain solution into an appropriately labeled waste bottle.
12. Again, add destaining solution and microwave for 30 s. Add two disposable wipes, which have each been twisted and tied in a knot, and incubate on a tilting platform at 50 tilts/min until the gel is fully destained.
13. Image the destained gel using a gel imaging system to assess purity.

3.4 Removal of the His_6 Tag and Separation from CHIKVP

1. Pool fractions of purified CHIKVP in a beaker. Remove a sample for SDS-PAGE analysis.
2. Add 108 µM thrombin protease to the CHIKVP sample (CHIKVP-to-thrombin ratio 50:1). Thrombin will proteolytically remove the His_6tag that was used for prior steps of the purification.
3. Prepare a 6-cm-long piece of 10 kDa MWCO dialysis tubing by folding and refolding and then securing one end of the dialysis tube with an appropriate length clip.
4. Assure that the dialysis tubing is free of leaks by filling it with a small amount of water.
5. Transfer the pooled thrombin-treated CHIKVP fractions to the prepared dialysis tubing. Seal the tubing with a second dialysis clip.
6. Submerge the filled dialysis tube into the dialysis buffer, add DTT to final concentration of 0.5 mM and leave overnight stirring at 4 °C.

3.5 Second Affinity Purification by Imidazole Gradient to Remove Cleaved His_6 Tag from CHIKVP

Although the His_6tag has been removed, CHIKVP still binds to His-trap nickel affinity column, presumably through ionic interactions, making it necessary to run an imidazole gradient to separate CHIKVP from the cleaved His_6 tag.

1. Retrieve the CHIKVP sample from the dialysis beaker (*see* **Note 4**) and filter it with 0.22 µM syringe filter (*see* **Note 5**).
2. Remove a 20 µL sample for SDS-PAGE analysis as in Step 3.3.
3. Load the dialysate at a 1 mL flow rate onto a 5 mL nickel affinity column that has been pre-equilibrated with lysis buffer. Wash the column with five column volumes of lysis buffer.
4. Elute the cleaved CHIKVP using a linear gradient of imidazole from 5% to 100% while collecting 2 mL fractions. Expect to observe CHIKVP begin to elute at 42% elution buffer with the maximal elution at 62% elution buffer.

5. Assess the purity of the protein obtained using the SDS-PAGE as in Step 3.3.
   Pool the fractions containing CHIKVP, which appears at 38 kDa on the SDS-PAGE gel. Use the Beer–Lambert law: absorbance = extinction coefficient ε × concentration in molar × pathlength in cm to calculate the concentration from the 280 nm absorbance of a 1-μL drop at using a NanoDrop spectrophotometer. The extinction coefficient (ε) for CHIKVP is 50,880 $M^{-1}cm^{-1}$. The yield of the CHIKVP preparation should be determined to report mg CHIKVP/gram of cell pellet, which is assessed in Step 3.1.16 (*see* **Note 6**).
6. If the protein is at a lower concentration than desired, CHIKVP can be concentrated using a 10 kDa MWCO centrifugal filter. It is important to wet the filter with nanopure water and perform one cycle of centrifugation prior to addition of purified CHIKVP. We have observed that initial CHIKVP concentrations greater than 40 μM allow CHIKVP to be robustly concentrated without significant losses during the concentration process.
7. Once a suitable concentration has been attained, prepare 100 μL aliquots of CHIKVP and store at −80 °C (*see* **Note 7**).
8. For assessment of k_{cat} (Step 3.7), it is essential that the total enzyme (E_T) concentration be known with utmost accuracy since $k_{cat} = V_{max}/E_T$. Once a suitable concentration of CHIKVP has been attained, it is important to very accurately evaluate concentration. To ensure accurate assessment of the total enzyme (E_T) concentration, we recommend using the average of the results from three independent assays: (1) Beer–Lambert law from Absorbance at 280 nm as described in Step 3.5.5, (2) BCA, and (3) gel-based assessment of concentration (Fig. 2d). The BCA assay should be performed according to the manufacturer's instructions. To quantify the band intensity from a destained SDS gel, use a quantify tool on the gel imaging system or the ImageJ software. In this example, the 280 nm absorbance by NanoDrop suggested a CHIKVP concentration of 212 μM. The gel-based assay estimated the CHIKVP concentration to be 216 μM. The BCA assay reported 224 μ M. The calculated average CHIKVP concentration from these three estimates is therefore 217 μM.

3.6 Single-Point Activity Assay

A single-point initial activity assay should be performed on the day of purification to establish the baseline activity of the preparation. The single-point activity assay can then be run on a routine basis to confirm that the activity of the enzyme has not diminished during storage or routine freeze–thaw cycles.

1. Prepare to measure the fluorescence of the acc-CHIK$_{15}$-dnp substrate (Fig. 3a, b) in a fluorescence plate reader by setting these parameters:
   - Temperature—25 °C
   - Excitation wavelength—340 nm
   - Auto cutoff—475 nm
   - Emission wavelength—490 nm
   - PMT optics—medium
   - six flashes
   - Read from top
   - Shake once at the start for 7 s.
   - Read fluorescence every 5 s for 600 s.
2. Retrieve a frozen 100 μL aliquot of the CHIKVP and incubate for ~5 min on ice to thaw.
3. Gently mix by pipetting after thawing.
4. For the enzymatic activity assay, prepare a 5.26 μM solution of CHIKVP in Activity Assay Buffer and mix gently.
5. To prepare the substrate, dilute the 10 mM acc-CHIK15-dnp substrate in DMSO to 400 μM in activity assay buffer (*see* **Note 8**). Specifically, to prepare 200 μL of 400 μM acc-CHIK$_{15}$-dnp solution, to a 1.5-mL Eppendorf tube, first add 8 μL of 10 mM acc-CHIK15-dnp dissolved in DMSO. Next, add 192 μL of activity assay buffer and mix gently to prevent precipitation that sometimes occurs due to vigorous agitation. Because the acc-CHIK$_{15}$-dnp substrate is photosensitive, wrap the vial with aluminum foil to avoid light exposure and store on ice prior to use. It is best to make this solution fresh each day that the assay is run due to acc-CHIK$_{15}$-dnp solubility; however, this solution can be stored at −20 °C for short periods provided that freeze–thaw cycles are minimized.

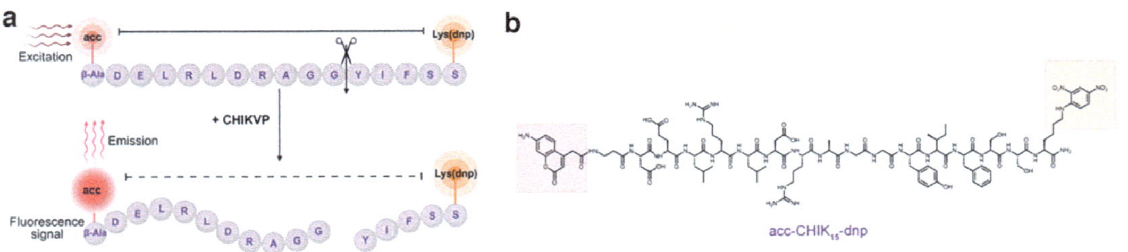

Fig. 3 (**a**) Schematic diagram showing acc-CHIK$_{15}$-dnp in which dnp quenches acc until cleavage by CHIKVP. (**b**) Chemical structure of acc-CHIK$_{15}$-dnp

6. The final 20 μL assay mixture will contain 5 μM CHIKVP and 20 μM acc-CHIK$_{15}$-dnp substrate.
7. Add 1 μL of the 400 μM fluorogenic peptide acc-CHIK$_{15}$-dnp substrate to each well of a 384-well black flat-bottom plate that is to be assayed. At least two wells should be reserved as controls to measure background signal.
8. Place the substrate-loaded plate on a bench top as close as possible to the plate reader to minimize transfer time once substrate has been added. Pipet 19 μL of the 5.26 μM CHIKVP solution into substrate-loaded wells of the experimental group and 19 μL of activity assay buffer into the substrate-loaded wells of the control group.
9. Immediately insert the plate into the plate reader and initiate the data collection protocol. The data collection protocol initially shakes the plate for 7 s to mix the substrate–enzyme mixture and then starts reading. The initial reading is set to zero. It is critical to start the reading as quickly as possible because addition of the substrate to CHIKVP will immediately initiate the peptide cleavage process (*see* **Note 9**). The initial linear activity, which represents initial velocity, V_0, is expected to be approximately 3.3 RFU/s (with the plate reader settings in Step 3.6.1) in the experimental group, with no activity in the control group.
10. For each of the individual samples, the initial velocity must be fit to the initial linear region of the progress curve. Typically, we observe a linear response for at least the first 100–150 s, whereas fitting of the data obtained over longer periods of time typically produces poorer fits (Fig. 4a–d) and R^2 values (*see* **Note 10**). The initial velocity at a single-substrate concentration provides an initial indication of enzyme activity.

3.7 Substrate Titration to Assess K_M and k_{cat}

To ensure that the enzyme preparation is of the expected quality, we suggest measuring K_M and k_{cat} by performing a substrate titration. The procedure follows the single-point activity assay (Sect. 3.6) with the following additions.

1. Prepare a twofold serial dilution of acc-CHIK$_{15}$-dnp substrate in DMSO starting from 700 μM. This is important so that all wells contain the same total volume of DMSO.
2. In each well of a 384-well flat-bottomed black plate add 1 μL of substrate in DMSO. Next, rapidly 19 μL of 5.26 μM CHIKVP solution. Measure the activity for 600 s as in Step 3.6.6 and 3.6.7. Fit V_0 for each substrate concentration as in Step 3.6.8.
3. Convert the V_0 in RFU/s to μM/s using a standard curve as in Fig. 4e. The standard curve can be made from free dnp. In this case, we prepared the standard curve by fully digesting

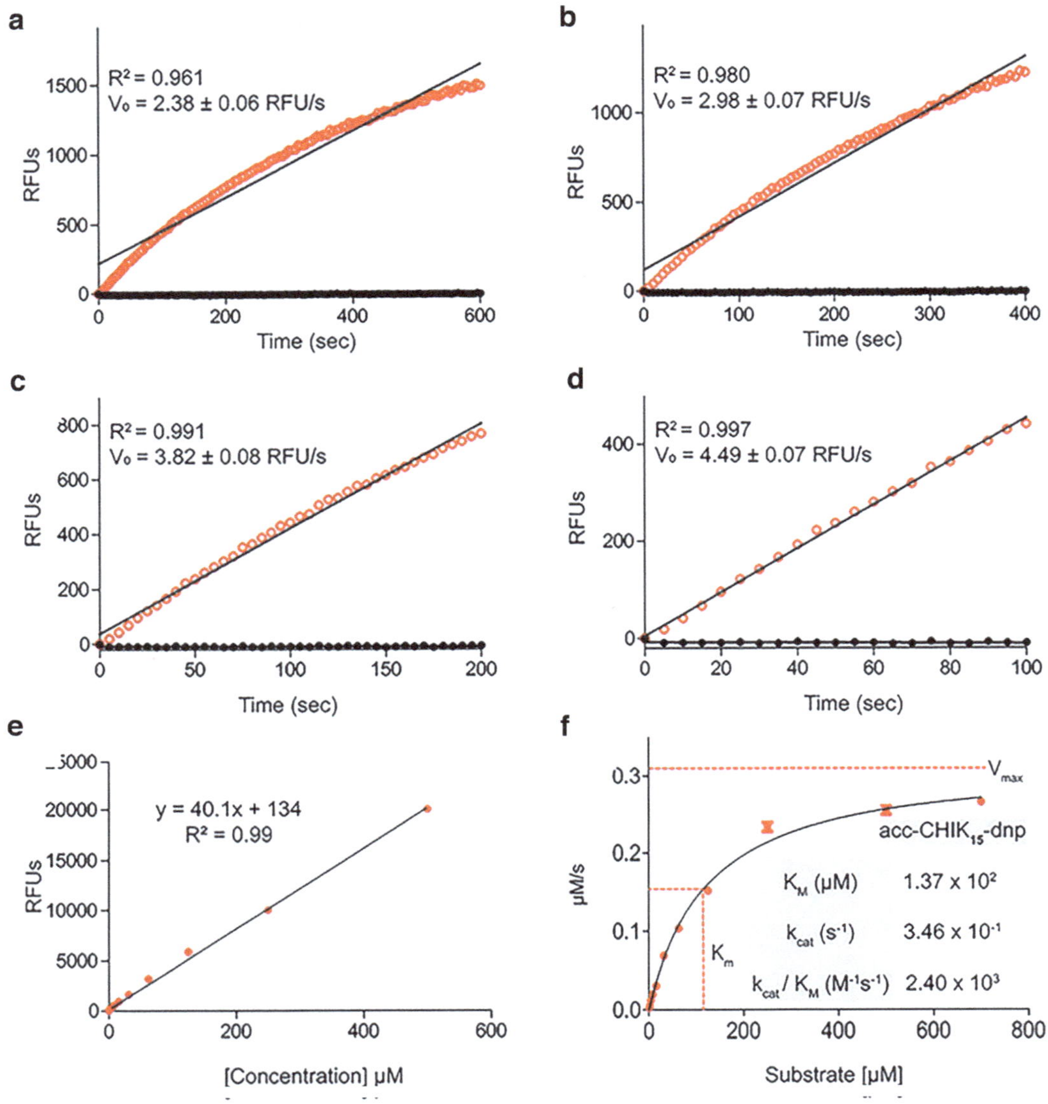

Fig .4 (**a–d**) Progress curves for cleavage of acc-CHIK$_{15}$-dnp (relative fluorescence units; RFU) are fit to the data for (**a**) 600 s, (**b**) 400 s, (c) 200 s, or (**d**) 100 s. The best fit for V_0 as judged by R^2 is for 100 s. (**e**) A dnp standard curve, which is required to convert RFU to concentration of dnp, is shown. (**f**) The acc-CHIK$_{15}$-dnp substrate is titrated from 0 to 800 μM in the presence of 5 μM CHIKVP for Michaelis–Menten assessment of CHIKVP. Observed V_{max} and K_M values are indicated

1 mM of acc-CHIK$_{15}$-dnp using trypsin (substrate:trypsin ratio 100:1), serially diluting the fully cleaved substrate in DMSO and measuring the fluorescence as in Step 3.6.1.

4. Plot the titration curve of [substrate] vs V_0 in μM/s using GraphPad Prism Software to obtain K_M and V_{max} as in Fig. 4f.

3.8 Screening a Compound Library to Identify Inhibitors

A key application of this assay is the identification of inhibitors of CHIKVP, as has been reported previously [25]. Here we provide an example of a small-scale screen we performed on in-house-synthesized compounds derived from prior screening. This approach (Fig. 5a) can be utilized for screening of any library of choice.

1. To identify small-molecule inhibitors of CHIKVP, add 1.5 μ L of inhibitor of choice at mM concentration suspended in DMSO to the bottom of a conical 96-well plate. Add 1.5 μL of DMSO to each well that will serve as a vehicle control.
2. Prepare 5.6 μM CHIKVP in activity assay buffer and store on ice. Minimize the time on ice as the protein loses activity

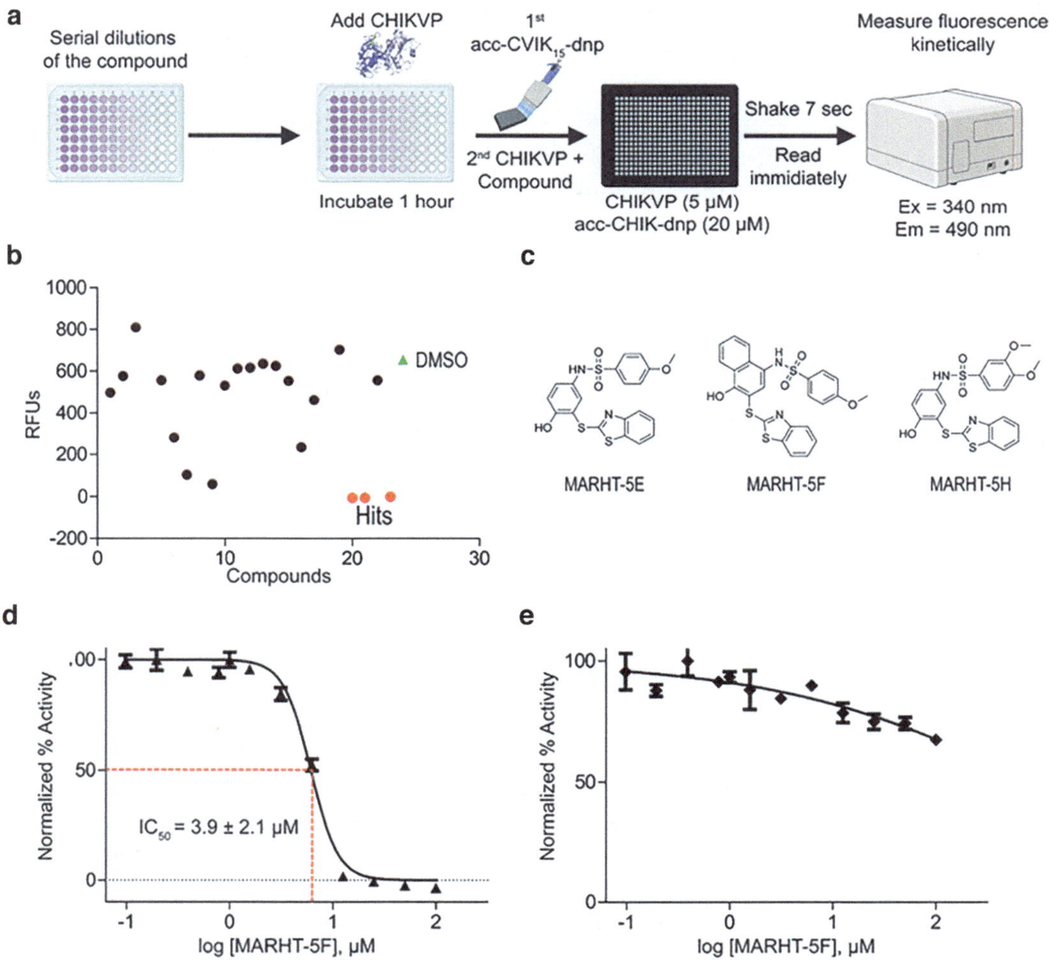

Fig. 5 (**a**) Schematic of activity assay. (**b**) Assessment of inhibition of a 20-compound library derived from a prior high-throughput screen of CHIKVP inhibitors [25]. The DMSO control (green) and hits (red) are identified. (**c**) Structures of the three most inhibitory compounds. (**d**) Apparent IC_{50} of MARHT-5F inhibition of 5 μM CHIKVP. (e) The observed inhibition of MARHT-5F is reversed by the presence of 1 mM DTT, suggesting that this compound acts by nondrug-like mechanisms and may have pan-assay interfering properties

once it thaws (*see* **Note 11**). Vigorously add 23.5 μL of 5.6 μM CHIKVP and incubate at room temperature for 1 h to allow compound binding. It is important to add the protein vigorously as it improves mixing. It is critical to place the compound suspended in DMSO in the well plate first and then add the aqueous solution onto the DMSO solution as this method greatly improves the solubility of most compounds, due to the contact rate of the compound to the aqueous solution. Adding DMSO-suspended compounds to aqueous solution tends to increase the observation of insolubility of compounds during screening.

3. To prepare for the kinetic measurement, add 1 μL of 400 μM acc-CHIK$_{15}$-dnp substrate to desired wells in a black 384-well plate, as described in Steps 3.6.6–3.6.9.
4. Transfer 19 μL of the compound-CHIKVP mixture to the well containing the acc-CHIK$_{15}$-dnp substrate. Immediately shake for 7 s, then measure the fluorescence kinetically every 5 s for 10 min using the parameters described in **Step 3.6**.
5. Identify compounds with the most significantly depressed reaction rates (lowest RFU/s) as compared to the DMSO control as hits. Using this method, three hits were identified from a library of 20 compounds (Fig. 5b, c).

3.9 Determination of IC_{50} of Chikungunya Virus Protease Inhibitors

To measure the IC_{50} of identified CHIKVP inhibitors, treat 5 μM of CHIKVP with a twofold serially diluted inhibitor from 100 μM to 97 nM and measure activity with 20 μM acc-CHIK$_{15}$-dnp substrate (Fig. 5a).

1. Prepare 5.6 μM CHIKVP in activity assay buffer and store on ice (*see* **Note 11**).
2. Prepare a 3.5-mM stock solution of the test compound in DMSO.
3. Add 10 μL of DMSO to wells A1–A12 of a 96-well conical bottom plate.
4. Add and mix by repeated pipetting 10 μL of the 3.5 mM compound solution to well A1, followed by twofold serial dilutions by transferring 10 μL sequentially from well A1 to A2, A2 to A3, and so on, through well A11, using a manual pipette. Well A12, containing only DMSO, will serve as the vehicle control.
5. From each dilution, transfer 3 μL to a separate well.
6. Add 47 μL of 5.6 μM CHIKVP to each compound-containing well and mix gently with a manual pipettor.
7. Incubate the mixture at room temperature (20–25 °C) for 1 h.

8. For each compound to be tested, add 1 µL of 400 µM of acc-CHIK$_{15}$-dnp substrate to 24 wells of a black 384-well assay plate to allow for assessment of the activity of technical replicates.
9. Set the fluorescence plate reader parameters as described in Step 3.6.
10. Using a multichannel pipette, transfer 19 µL of the protein–inhibitor mixture in duplicate to the wells containing the substrate, and measure fluorescence immediately using the fluorescence plate reader as in Steps 3.6.6–3.6.9. Wells with the highest concentration of the inhibitor are expected to have the lowest V_0 and the DMSO control will have the highest V_0. The obtained IC_{50} of three biological replicates for compound MARHT-5F was 3.9 ± 2.1 µM (Fig. 5d).

3.10 Processing Kinetic Data for IC_{50} Assessment

While data from activity assays can be fit manually and using many different software tools, we find it convenient to use the pre-set tool suite that is available in the GraphPad Prism software and for this reason include guidance on this approach for data analysis.

1. In the GraphPad Prism software under *New Table & Graph*, select the type XY. Select the option for two replicates using the *mean and error* and the *SEM* options.
2. Insert compound concentrations and the corresponding V_0 values into the GraphPad Prism spreadsheet.
3. Under the *Analysis* functions, normalize the enzyme activity by assigning the lowest value (blank or positive inhibitor control such as RA0002034 [30]) to 0% and highest value (DMSO control) to 100%.
4. Transform the normalized values of the data to log [compound].
5. To find the IC_{50} of the compound, perform the log [inhibitor] vs normalized response with a variable slope. This analysis will provide an estimated IC_{50} of the compound, which is a useful measure for comparing the inhibition properties of the compounds for a particular protein.

3.11 Example of PAINS Assessment

In inhibitor development, particularly that aimed at pharmacological endpoints, as is the case for CHIKVP, it is essential that identified inhibitors function through tractable, drug-like mechanisms. Pan-assay interference compounds (PAINS) are defined as those that show inhibition through nonspecific modes of action, including chemical aggregation, singlet oxygen production, redox activity and cysteine oxidation [31–33]. We provide an example of an identified hit, MARHT-5F that exhibits non-drug-like properties. A literature search suggested that compound 6b, which is structurally similar to

MARHT-5F, showed errant redox behavior against histone acetyltransferase Rtt109 [32]. To determine if the MARHT-5F inhibition of CHIKVP was due to redox activity and/or cysteine oxidation, the compound was tested for inhibition in a buffer containing the reducing agent DTT [32, 33]. We anticipate that this approach, and related approaches for assessing other mechanisms, such as aggregation and micelle formation, may be useful for other compounds discovered through high-throughput screening.

1. Mix 990 µL of the activity assay buffer with the 10 µL of 100 mM DTT solution.
2. Perform the IC_{50} assay as in Step 3.9 and compare the IC_{50} values of the compound in non-reducing activity assay buffer and buffer containing DTT. If the compound does not possess these two redox-related PAINS behaviors, the IC_{50} should be similar.

For example, the obtained IC_{50} of the MARHT-5F was significantly higher at 895 µM in buffer containing DTT as compared to 3.93 µM buffer without DTT (Fig. 5e), which indicates that MARHT-5F has nonspecific, redox-related modes of inhibition. Compounds with PAINS properties should be discontinued from further development.

4 Notes

1. Always add fresh DTT from the 1 M stock (fresh or stored at −20 °C) to solutions just before use.
2. The protein sequence for N-terminally His_6-tagged CHIKVP used in these studies is: MAHHHHHHLVPRGSGMAGICSHQMTFDTFQNKANVCWAKSLVPILETAGIKLNDRQWSQIIQAFKEDKAYSPEVALNEICTRMYGVDLDSGLFSKPLVSVYYADNHWDNRPGGKMFGFNPEAASILERKYPFTKGKWNINKQICVTTRRIEDFNPTTNIIPVNRRLPHSLVAEHRPVKGERMEWLVNKINGHHVLLVSGYNLALPTKRVTWVAPLGVRGADYTYNLELGLPATLGRYDLVVINIHTPFRIHHYQQCVDHAMKLQMLGGDSLRLLKPGGSLLIRAYGYADRTSERVICVLGRKFRSSRALKPPCVTSNTEMFFLFSNFDNGRRNFTTHVMNNQLNAAFVG. The thrombin-recognition site is LVPRGSG, with cleavage following LVPR. Thus, a residual GSG tripeptide is left at the CHIKVP N-terminus following thrombin cleavage.
3. Healthy cells form a radially distributed pellet of ~3–4 cm diameter. A pellet with an overly shiny and slimy appearance, with black flecks in the center of the pellet and with cell debris

extending toward the top of the centrifuge bottle is suggestive that the culture has been contaminated with bacteriophage. In that case, all culture glassware should be treated for phage contamination by soaking in a 10 M KOH bath. All incubators should be sterilized by UV irradiation overnight at 254 nm.

4. Wet dialysis tubing is slippery, so place the dialysis tubing in a beaker to protect it from contamination due to dropping or other mishandling.
5. Filter the CHIKVP protein using 22 μm syringe filter after thrombin cleavage. This helps to remove aggregated protein, which can clog the column.
6. Typically, we obtain 14–16 mL of 30–50 μM CHIKVP from 8 L of culture, which produces 45–50 g of wet pellet.
7. 100 μL aliquots are appropriate for most assays. This small aliquot size prevents waste and the need for repeated freeze–thaw cycles.
8. Due to its increased solubility in DMSO, acc-$CHIK_{15}$-dnp should be stored as a 10 mM stock in DMSO and be freshly diluted into activity assay buffer on the day that the activity assay should be run.
9. Addition of the protein to substrate will immediately initiate the peptide cleavage process, therefore, it is critical to start measuring the fluorescence immediately upon substrate addition.
10. We attribute the loss in linearity to a combination of substrate consumption and product inhibition.
11. CHIKVP is sensitive to time after thawing. We routinely observe a loss of approximately 20%–25% activity per hour of incubation at room temperature. Incubation on ice and the presence of reductants like DTT ameliorate to a very small degree this loss in activity, but activity diminishes as a function of time even with DTT and on ice.

Acknowledgments

We thank Tom Hobman, University of Alberta for providing us with the plasmids for Chikungunya virus protease expression and Olivier Julien for development of this process. MAR was supported by a UMass BTP fellowship associated with NIH T32 GM135096. SM was funded by a Translational Graduate Student Assistantship from the Center for Bioactive Delivery in the Institute for Applied Life Sciences, UMass Amherst. This work was supported by Canadian

Institute of Health Research PS 162417. ChatGPT was used for repharasing during some portions of the writing process and to check for grammatical errors.

References

1. Robinson MC (1955) An epidemic of virus disease in Southern Province, Tanganyika Territory, in 1952-53. I. Clinical features. Trans R Soc Trop Med Hyg 49:28–32. https://doi.org/10.1016/0035-9203(55)90080-8
2. Ross RW (1956) The Newala epidemic: III. The virus: isolation, pathogenic properties and relationship to the epidemic. Epidemiol Infect 54:177–191. https://doi.org/10.1017/S0022172400044442
3. Chikungunya fact sheet. https://www.who.int/news-room/fact-sheets/detail/chikungunya?utm_source=chatgpt.com. Accessed 21 Mar 2025
4. Higgs S, Vanlandingham D (2015) Chikungunya virus and its mosquito vectors. Vector-Borne Zoonotic Dis 15:231–240. https://doi.org/10.1089/vbz.2014.1745
5. Tsetsarkin KA, McGee CE, Volk SM et al (2009) Epistatic roles of E2 glycoprotein mutations in adaption of chikungunya virus to *Aedes albopictus* and *Ae. aegypti* mosquitoes. PLoS One 4:e6835. https://doi.org/10.1371/JOURNAL.PONE.0006835
6. Zeller H, Van Bortel W, Sudre B (2016) Chikungunya: its history in Africa and Asia and its spread to new regions in 2013–2014. J Infect Dis 214:S436–S440. https://doi.org/10.1093/INFDIS/JIW391
7. Amaral JK, Bilsborrow JB, Schoen RT (2020) Chronic chikungunya arthritis and rheumatoid arthritis: what they have in common. Am J Med 133:e91–e97. https://doi.org/10.1016/J.AMJMED.2019.10.005
8. Ferreira FCPADM, Filippis AMBde, Moreira MEL et al (2024) Perinatal and neonatal chikungunya virus transmission: a case series. J Pediatric Infect Dis Soc 13:576–584. https://doi.org/10.1093/JPIDS/PIAE102
9. Sagay AS, Hsieh SC, Dai YC et al (2024) Chikungunya virus antepartum transmission and abnormal infant outcomes in a cohort of pregnant women in Nigeria. Int J Infect Dis 139:92–100. https://doi.org/10.1016/J.IJID.2023.11.036
10. VIMKUNYA™ (Recombinant) | A Chikungunya Vaccine | For Travelers. https://vimkunya.com/?gad_source=1&gad_campaignid=22505135260&gbraid=0AAAAA_c1E7jnRfbzbODcnwUwJtCTn-V3k_&gclid=CjwKCAjwiNXFBhBKEiwAPSaPCZINvS8EPzqKRHb-92Cogmukq-K_OoKNQFFb7rXSxdgwmxy9Qtc_OhoCC78QAvD_BwE. Accessed 31 Aug 2025
11. For Healthcare Providers | IXCHIQ® (chikungunya vaccine, live) HCP. https://ixchiqhcp.com/. Accessed 8 Apr 2025
12. FDA Update on the Safety of Ixchiq (Chikungunya Vaccine, Live). FDA Suspends Biologics License: FDA Safety Communication | FDA. https://www.fda.gov/safety/medical-product-safety-information/fda-update-safety-ixchiq-chikungunya-vaccine-live-fda-suspends-biologics-license-fda-safety. Accessed 12 Jan 2026
13. Metz SW, Gardner J, Geertsema C et al (2013) Effective chikungunya virus-like particle vaccine produced in insect cells. PLoS Negl Trop Dis 7:e2124. https://doi.org/10.1371/JOURNAL.PNTD.0002124
14. Strauss JH, Strauss EG (1994) The alphaviruses: gene expression, replication, and evolution. Microbiol Rev 58(3):491–562
15. Karpe YA, Aher PP, Lole KS (2011) NTPase and 5′-RNA triphosphatase activities of chikungunya virus nsP2 protein. PLoS One 6:e22336. https://doi.org/10.1371/JOURNAL.PONE.0022336
16. Pastorino BAM, Peyrefitte CN, Almeras L et al (2008) Expression and biochemical characterization of nsP2 cysteine protease of Chikungunya virus. Virus Res 131:293–298. https://doi.org/10.1016/J.VIRUSRES.2007.09.009
17. Hill ME, Kumar A, Wells JA et al (2018) The unique cofactor region of Zika virus NS2B-NS3 protease facilitates cleavage of key host proteins. ACS Chem Biol 13:2398–2405. https://doi.org/10.1021/ACSCHEMBIO.8B00508
18. Raj S, Irfan M, Ranjith-Kumar CT (2025) Chikungunya virus non-structural protein 2 (nsP2) inhibits RIG-I and TLR-mediated immune response. Virology 616:110777. https://doi.org/10.1101/2025.04.30.651505
19. Bouraï M, Lucas-Hourani M, Gad HH et al (2012) Mapping of chikungunya virus interactions with host proteins identified nsP2 as a highly connected viral component. J Virol 86:3121–3134. https://doi.org/10.1128/JVI.06390-11
20. Majerová T, Konvalinka J (2022) Viral proteases as therapeutic targets. Mol Aspects Med 88:101159. https://doi.org/10.1016/J.MAM.2022.101159

21. Tompa DR, Immanuel A, Srikanth S, Kadhirvel S (2021) Trends and strategies to combat viral infections: a review on FDA approved antiviral drugs. Int J Biol Macromol 172:524–541. https://doi.org/10.1016/j.ijbiomac.2021.01.076
22. Bermejo H, Luis J, Patiño C et al (2024) The design, synthesis and mechanism of action of paxlovid, a protease inhibitor drug combination for the treatment of COVID-19. Pharmaceutics 16:217. https://doi.org/10.3390/PHARMACEUTICS16020217
23. Das PK, Puusepp L, Varghese FS et al (2016) Design and validation of novel chikungunya virus protease inhibitors. Antimicrob Agents Chemother 60:7382–7395. https://doi.org/10.1128/AAC.01421-16
24. Smith AJ, Rahat MA, Scaramozza LJ et al (2025) Inhibition of Zika virus protease by modulating NS2B-NS3 interactions. Biochemistry 65(1):64-76. https://doi.org/10.1021/acs.biochem.5c00644
25. Makhaik S, Rut W, Choudhary S et al (2025) A robust fluorogenic substrate for chikungunya virus protease (nsP2) activity. Protein Sci 34. https://doi.org/10.1002/pro.70069
26. Singh H, Mudgal R, Narwal M et al (2018) Chikungunya virus inhibition by peptidomimetic inhibitors targeting virus-specific cysteine protease. Biochimie 149:51–61. https://doi.org/10.1016/J.BIOCHI.2018.04.004
27. Narwal M, Singh H, Pratap S et al (2018) Crystal structure of chikungunya virus nsP2 cysteine protease reveals a putative flexible loop blocking its active site. Int J Biol Macromol 116:451–462. https://doi.org/10.1016/J.IJBIOMAC.2018.05.007
28. Rausalu K, Utt A, Quirin T et al (2016) Chikungunya virus infectivity, RNA replication and non-structural polyprotein processing depend on the nsP2 protease's active site cysteine residue. Sci Rep 6:37124. https://doi.org/10.1038/srep37124
29. Saisawang C, Sillapee P, Sinsirimongkol K et al (2015) Full length and protease domain activity of chikungunya virus nsP2 differ from other alphavirus nsP2 proteases in recognition of small peptide substrates. Biosci Rep 35(3):e00196. https://doi.org/10.1042/BSR20150086
30. Merten EM, Sears JD, Leisner TM, et al (2024) Identification of a cell-active chikungunya virus nsP2 protease inhibitor using a covalent fragment-based screening approach. Proc Nat Acad Sci 121. https://doi.org/10.1073/pnas.2409166121
31. McGovern SL, Caselli E, Grigorieff N, Shoichet BK (2002) A common mechanism underlying promiscuous inhibitors from virtual and high-throughput screening. J Med Chem 45(8):1712–1722. https://doi.org/10.1021/jm010533y
32. Dahlin JL, Nissink JWM, Strasser JM et al (2015) PAINS in the assay: chemical mechanisms of assay interference and promiscuous enzymatic inhibition observed during a sulfhydryl-scavenging HTS. J Med Chem 58:2091–2113. https://doi.org/10.1021/JM5019093
33. Baell JB, Nissink JWM (2017) Seven year itch: pan-assay interference compounds (pains) in 2017—utility and limitations. ACS Chem Biol 13:36–44. https://doi.org/10.1021/ACSCHEMBIO.7B00903

Chapter 13

Epitope Tagging and Coimmunoprecipitation to Identify Viral Protein Interactors

Zhenlan Yao and Melody Li

Abstract

Affinity purification-mass spectrometry (AP-MS) is a powerful proteomic approach for dissecting the interaction network between virus and host. Traditional AP-MS employs overexpression of viral proteins as baits to enrich host interactors. However, overexpressed viral proteins may mislocalize to inappropriate cellular compartments and trigger endoplasmic reticulum stress by overwhelming the protein-folding machinery, which leads to false identification of host factors. To overcome these limitations, we introduce an AP-MS strategy based on direct infection with an epitope-tagged chikungunya virus (CHIKV/myc-E2), which we used to successfully uncover two new antiviral factors in CHIKV cellular reservoirs—macrophages. In this protocol, we will describe this technique step by step: (1) design and construction of myc-tagged virus by advanced multi-fragment assembly, (2) in vitro transcription and preparation of infectious myc-tagged virus stocks, and (3) immunoprecipitation of myc-tagged viral protein and its interactome for mass spectrometry analysis. This strategy enables accurate identification of viral interactors in a physiologically relevant context, providing a framework for future proteomic studies using tagged viruses.

Keywords Epitope tagging, Multi-fragment assembly, Tagged virus, AP-MS, Chikungunya virus, Macrophage

1 Introduction

Affinity purification-mass spectrometry (AP-MS) is an elegant proteomic approach for enriching bait proteins and identifying interactors with high confidence [1, 2]. AP-MS has been widely used in studies of virus–host interactions making great contributions to the discoveries of SARS-CoV-2 host factors during the pandemic [3, 4]. However, most virus–host interaction studies overexpress viral bait proteins in cell lines to recruit host interactors. Solo expression of certain viral proteins out of the infection context may lead to viral proteins localizing to different cellular compartments, and overexpression potentially triggers endoplasmic reticulum (ER) stress by overloading the protein-folding machinery [5, 6]. These will

Claudia Filomatori and Eliana F. Castro (eds.), *Chikungunya Virus*, Methods in Molecular Biology, vol. 3067,
https://doi.org/10.1007/978-1-0716-5496-5_13,

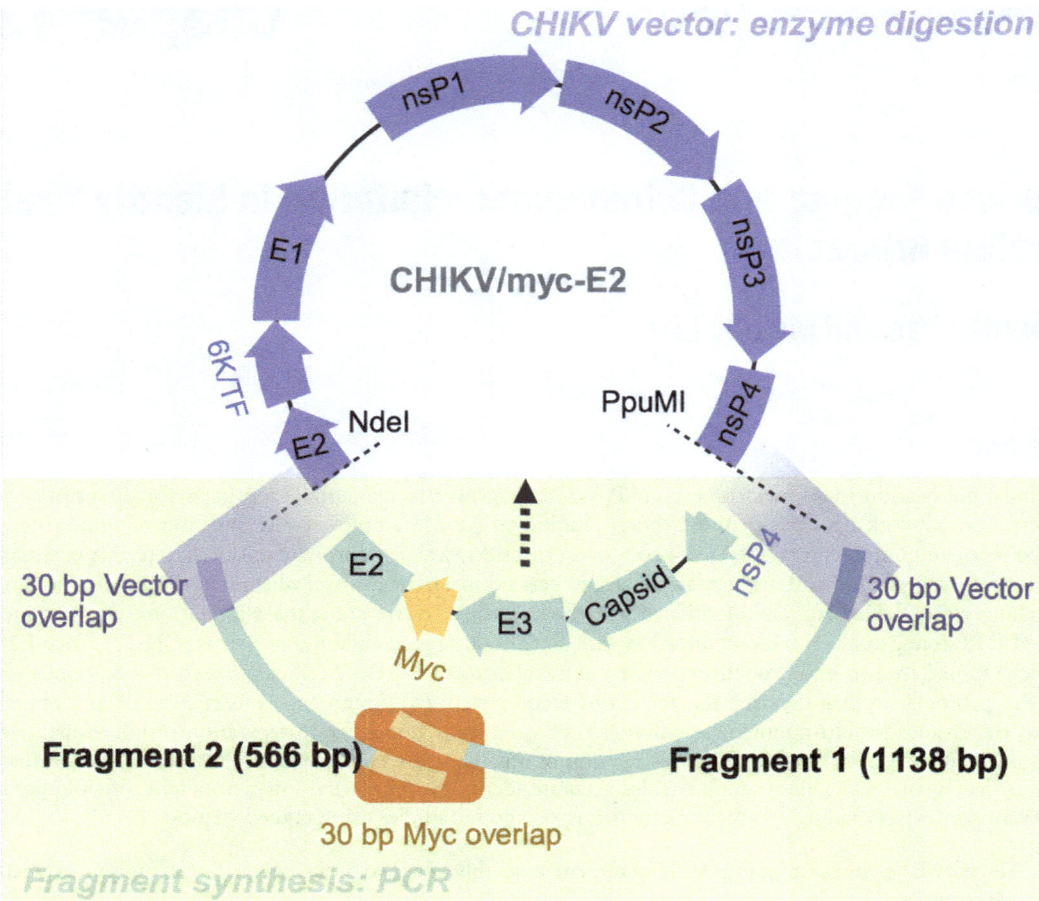

Fig. 1 Schematic of three-fragment assembly for generating CHIKV/myc-E2. The CHIKV vector was prepared by double digestion of the CHIKV 181/clone 25 cDNA plasmid with PpuMI and NdeI. Fragments 1 and 2 were PCR-amplified with 30 bp overhangs: one side overlapping with the CHIKV vector and the other containing the Myc sequence. The CHIKV vector, Fragment 1, and Fragment 2 were assembled to reconstitute the full CHIKV genome and introduce a Myc tag between the E3 and E2 coding regions

lead to false identification of host factors, which normally do not interact with the viral bait protein in the context of infection. In this protocol, we will introduce our strategy to employ epitope-tagged CHIKV to enrich virus–host protein complexes from infected cell lysates for MS analysis.

Chikungunya virus (CHIKV) is a highly pathogenic mosquito-borne alphavirus which has been causing epidemics worldwide in past decades and reemerge as a new outbreak striking more than 7000 cases in southern China in 2025 [7–10]. CHIKV infection is characterized by rash, fever, and severe joint pains, which may develop into incapacitating arthralgia for several months. Macrophages are recognized as one of the CHIKV cellular reservoirs for chronic infection and inflammation in virus-induced arthralgia [11–13]. Studies showed evidence of active CHIKV replication by detection of viral proteins and replication intermediates (negative-sense viral RNA) in macrophages in synovial joints of nonhuman primates up to 3 months post CHIKV infection [12]. We previously determined viral glycoprotein interactors through AP-MS analysis on macrophages infected with CHIKV expressing myc-tagged E2 glycoprotein [14]. Instead of using a viral protein overexpression system, pulldown of myc-tagged CHIKV E2 protein that engages with host and viral interactors throughout the course of infection leads to efficient identification of host factors with authentic interactions with the viral glycoprotein and functional impact on the infection.

Here, we dissect the process of exploiting myc-tagged CHIKV to identify macrophage host factors through AP-MS, which includes: (1) design and construct E2-tagged CHIKV (CHIKV/myc-E2) through multi-fragment ligation; (2) prepare CHIKV/myc-E2 stocks for infecting macrophages; and (3) enrich for myc-tagged E2 along with host and viral interactors through immunoprecipitation for downstream MS analysis. The goal of this protocol is to provide knowledge and reference for (1) strategically labeling alphavirus on specific viral proteins; (2) applying tagged virus in cutting-edge proteomic approaches, such as AP-MS and the emerging proximity-labeling MS technologies.

2 Materials

2.1 Molecular Cloning of CHIKV/myc-E2

1. Primers for Fragment 1:
   F-PpuMI-E3-myc: 5′-CCAGATCCAACTTCGAGAAGCT-3′.
   R-PpuMI-E3-myc: 5′-CAGATCTTCTTCGCTAATAAGTTTTTGTTC-(Myc tag)-ACTACTGCGTCGCTGGCGG-3′.
2. Primers for Fragment 2:

   F-myc-E2: 5′-GAACAAAAACTTATTAGCGAAGAAGATCTG-(Myc tag)-AGTAAGGACAACTTCAATGTCTATAAAG-3′.
   R-E2-myc: 5′-AATGTGCGATCTGGGGTG-3′.

1. NdeI
2. PpuMI
3. 10× rCutSmart
4. Purple 6× gel loading dye
5. SYBR safe DNA gel stain

6. 1% agarose gel: dissolve 1 g of agarose powder in 100 mL of 1× TAE buffer by heating until completely melted. Pull 30–40 mL of melted agarose and add SYBR Safe in a 1:10,000 ratio for a small gel format (8 × 10 cm).
7. 50×Tris/acetic acid/EDTA (TAE) buffer
8. Zymoclean Gel DNA Recovery Kit (Zymo Research)
9. Q5® Hot Start High-Fidelity 2× Master Mix
10. NEBuilder® HiFi DNA Assembly Master Mix
11. NEB® 10-beta competent *E. coli* (high efficiency)
12. Ampicillin/Carbenicillin
13. Lab heat block or water bath
14. Thermocycler for PCR amplification
15. Gel electrophoresis apparatus

2.2 Production of Parental CHIKV 181/Clone 25 and CHIKV/myc-E2

1. NoteI-HF
2. SP6 RNA polymerase
3. 10× SP6 reaction buffer
4. 25 mM ribonucleotide solution mix
5. m7G(5')ppp(5')G RNA Cap Structure Analog
6. 10 mM dithiothreitol (DTT): dissolve 15.4 mg DTT in 10 mL water and save the aliquots at −20°C
7. RNaseOUT™ recombinant ribonuclease inhibitor
8. RNA gel loading dye (2×)
9. Quick-RNA Miniprep Kit (Zymo Research)
10. Minimum essential medium (MEM)
11. Fetal bovine serum (FBS)
12. Phosphate-buffered saline (PBS)
13. 0.25% trypsin–EDTA
14. BHK-21 fibroblasts
15. ECM 830 Square Wave Electroporation System
16. BTX electroporation cuvettes, 2 mm
17. 1.5 mL Eppendorf (eppi) tubes, DNase/RNase-free
18. 2 mL screw-cap microcentrifuge tubes, DNase/RNase-free
19. 50 mL conical tubes, DNase/RNase-free
20. 150 mm Petri dishes
21. Refrigerated centrifuge with swing bucket for mammalian cell centrifugation

2.3 Immunoprecipitation for Mass Spectrometry

1. RPMI 1640 Medium (ATCC modification).
2. MEM non-essential amino acids solution (100×).
3. β-Mercaptoethanol: dilute b-mercaptoethanol in water to make 1 M stock.
4. Phorbol 12-myristate 13-acetate (PMA): 1 mg/mL PMA stock in DMSO.
5. THP-1 cells.
6. Human serum from human male AB plasma.
7. 1 M Tris-HCl, pH8: dissolve 15.76 g of Tris-HCl in 100 mL water to make 1 M solution stock. Adjust the solution pH to 8.
8. 0.5 M EDTA, pH 8.
9. 3 M NaCl: dissolve 17.55 g of NaCl in 100 mL water.
10. Glycerol.
11. Nonidet® P 40 Substitute (NP-40), Proteomics Grade.
12. 100 mM phenylmethanesulfonyl fluoride solution (PMSF) in ethanol.
13. cOmplete™ Protease Inhibitor Cocktail.
14. EZview™ Red Anti-c-Myc Affinity Gel.
15. 2× or 4× Laemmli sample buffer.
16. Urea.
17. Acetone.
18. NP-40 lysis buffer: 150 mM NaCl, 0.1% (v/v) NP-40, 5% (v/v) glycerol, 100 mM Tris–HCl, 5 mM EDTA. Store at 4°C for later use.
19. Cell lysis working buffer: 1 mM PMSF, 1 tablet/10 mL complete protease inhibitor cocktail, 1 mM DTT, in prepared NP-40 lysis buffer. Calculate the amounts of NP-40 lysis buffer required for immunoprecipitation.
20. Elution buffer: 8 M urea in 100 mM Tris-HCl. Dissolve 4.8 g urea crystals in ~6 mL water, add 1 mL of 1 M Tris–HCl (pH 8.0), then adjust the final volume to 10 mL with water and mix until fully dissolved.
21. Macrophage differentiation medium: 10% (v/v) inactivated human AB serum (*see* **Notes 1and 2**), 1× NEAA, 50 ng/mL PMA, in RPMI 1640 ATCC modification medium.
22. Macrophage resting medium: 10% (v/v) inactivated human AB serum, 1× NEAA, in RPMI 1640 ATCC modification medium.
23. Virus binding buffer (VB): 1 % (v/v) inactivated human AB serum in PBS.
24. Cell scrappers.

25. Tube rotator/revolver.
26. Benchtop centrifuge with fixed-angle rotor.
27. Eppendorf ThermoMixer.

3 Methods

3.1 Design and Construct of E2-Tagged CHIKV cDNA Plasmid Through Multi-fragment Ligation (Fig. 1)

The CHIKV genome is composed of two open reading frames (ORFs) encoding the nonstructural and structural polyproteins. The nonstructural polyprotein is processed into nonstructural proteins 1–4 (nsP1–4) and is important for viral replication and genome synthesis. The structural polyprotein is processed into capsid, E3, E2, 6 K/TF, and E1 where E3, E2, and E1 are glycoproteins and are important for viral entry, assembly, and egress. The myc tag can be introduced after the furin cleavage site between E3 and E2 [15]. The myc tag at the N-terminal of E2 will be preserved in uncleaved long precursors (E3-myc-E2-6K-E1), partially cleaved p62 (E3-myc-E2) and fully matured myc-E2 (*see* **Note 3**). The NEBuilder® HiFi DNA Assembly Kit can be used for introducing the myc tag into the CHIKV vaccine strain 181/clone 25 infectious clone cDNA plasmid through three-fragment ligation. The NEBuilder® HiFi DNA Assembly involves three key enzymes (5' to 3' exonuclease, DNA polymerase, DNA ligase) and enables seamless ligation of up to six fragments in a single reaction.

1. Generate the CHIKV vector for three-fragment ligation through digestion of restriction sites PpuMI at the end of nsP4 and NdeI in the middle of E2 in the CHIKV genome. Set up the double digestion in 100 μL reaction on ice:

Components	100 μL Reaction
CHIKV 181/clone 25 plasmid	5 μg
10× CutSmart Buffer	10 μL
PpuMI	4 μL
NdeI	4 μL
H_2O	Up to 100 μL

2. Mix the components well and incubate the reaction at 37°C for 4 h to facilitate complete digestion.
3. Optional: Stop the digestion by heat inactivation at 80°C for 20 min (*see* **Note 4**).
4. Mix the 100 μL reaction with 20 μL of 6× Purple Gel Loading Dye. Load the reaction and DNA ladder to 1% agarose gel

stained with SYBR Safe. Separate the digested fragment from the CHIKV vector through electrophoresis in 1× TAE buffer.

5. Cut off the gel piece containing the CHIKV vector (12331 bp) under UV light. Purify the CHIKV vector through Zymoclean Gel Recovery Kit and elute the CHIKV vector with 15 μL DNase-free water to ensure a high concentration (at least 50 ng/μL) for ligation (*see* **Note 5**). Save the purified vector at −20°C for later use.

6. PCR amplification to generate Fragment 1 (Fig. 2).
   Add the components in a sterile microcentrifuge tube on ice:

Component	Volume	Final concentration
Q5® Hot Start High-Fidelity 2× Master Mix, NEB	12.5 μL	1×
10 μM F-PpuMI-E3-myc	1.25 μL	0.5 μM
10 μM R-PpuMI-E3-myc	1.25 μL	0.5 μM
CHIKV 181/clone 25 plasmid	20 ng	0.8 ng/μL
H_2O	Up to 25 μL	

7. Perform the PCR amplification (1138 pb) according to thermocycling conditions:

Step	Temperature	Time
Initial denaturation	98°C	30 s

CCAGATCCAACTTCGAGAAGCTCAGAG|GACCCGTCATAACTTTGTACGGCGGTCC
Overlapping overhang PpUMI
TAAATAGGTACGCACTACAGCTACCTATTTTGCAGAAGCCGACAGCAGGTACCTA
AATACCAATCAGCCATAATGGAGTTTATCCCAACCCAAACTTTCTACAATAGGAG
GTACCAGCCTCGACCTTGGACTCCGCGCCCTACTATCCAAGTTATCAGACCCAGA
CCGCGTCCGCAAAGGAAAGCCGGGCAACTTGCCCAGCTGATCTCAGCAGTTAAT
AAACTGACAATGCGCGCGGTACCTCAACAGAAGCCGCGCAAGAATCGGAAGAAT
AAGAAGCAAAAGCAAAAGCAGCAGGCGCCACGAAACAACATGAATCAAAAGAA
GCAGCCCCCTAAAAAGAAACCGGCTCAAAAGAAAAAGAAGCCGGGCCGTAGAG
AGAGAATGTGCATGAAAATCGAAAATGATTGCATCTTCGAAGTCAAGCATGAAG
GTAAGGTAACAGGTTACGCGTGCTTGGTAGGGGACAAAGTAATGAAGCCAGCAC
ACGTAAAGGGGACCATCGATAATGCGGACCTGGCCAAATTGGCCTTCAAGCGGT
CATCTAAGTACGACCTTGAATGCGCGCAGATACCCGTGCACATGAAGTCCGACGC
TTCGAAGTTCACCCATGAGAAACCGGAGGGGTACTACAACTGGCACCACGGAGC
AGTACAGTACTCAGGAGGCCGGTTCACCATCCCTACAGGTGCGGGCAAACCAGG
GGACAGCGGTAGACCGATCTTCGACAACAAGGGGCGCGTGGTGGCCATAGTTTT
AGGAGGAGCTAATGAAGGAGCCCGTACAGCCCTCTCGGTGGTGACCTGGAACAA
AGACATCGTCACGAAAATCACCCCTGAGGGGGCCGAAGAGTGGAGTCTTGCCAT
TCCAGTTATGTGCCTGCTGGCAAATACCACGTTCCCCTGCTCCCAGCCCCCTTGCA
CACCCTGCTGCTACGAAAAAGAGCCGGAGAAAACCCTGCGCATGCTAGAAGACA
ACGTCATGAGCCCCGGGTACTATCAGCTGCTACAAGCATCCTTAACATGTTCTCC
CCGCCGCCAGCGACGCAGTAGTGAACAAAAACTTATTAGCGAAGAAGATCTG
Myc tag overhang

Fig. 2 Fragment 1–genomic region from the end of nsP4 to the end of E3 with 5′ overhang overlapping CHIKV vector (highlighted in cyan) around PpuMI site and 3′ overhang containing myc tag (highlighted in yellow). The overlapping region with the vector and the myc tag at the ends of the primers were color-coded (*see* **Notes 6 and 7**)

Step	Temperature	Time
32 repeated cycles	98°C	10 s
	68°C	30 s
	72°C	35 s
Final extension	72°C	2 min
Hold	10°C	

8. PCR amplification to generate Fragment 2 (Fig. 3). Set up the PCR reaction:

Component	Volume	Final concentration
Q5® Hot Start high-fidelity 2× master mix, NEB	12.5 μL	1×
10 μM F-myc-E2	1.25 μL	0.5 μM
10 μM R-E2-myc	1.25 μL	0.5 μM
CHIKV 181/clone 25 plasmid	20 ng	0.8 ng/μL
H_2O	Up to 25 μL	

9. Perform the PCR amplification (566 bp) according to thermocycling conditions:

Step	Temperature	Time
Initial denaturation	98°C	30 s
32 repeated cycles	98°C	10 s
	61°C	30 s
	72°C	18 s
Final extension	72°C	2 min
Hold	°C	

GAACAAAAACTTATTAGCGAAGAAGATCTGAGTAAGGACAACTTCAATGTCTATAAAGCCAT
Myc tag overhang

AAGACCGTACCTAGCTCACTGTCCCGACTGTGGAGAAGGGCACTCGTGCCATAGTCCCGTAGCG
CTAGAACGCATCAGAAACGAAGCGACAGACGGGACGCTGAAAATCCAGGTTTCCTTGCAAATC
GGAATAAAGACGGATGATAGCCATGATTGGACCAAGCTGCGTTACATGGACAATCATATGCCAG
CAGACGCAGAGAGGGCCAGGCTATTTGTAAGAACGTCAGCACCGTGCACGATTACTGGAACAA
TGGGACACTTCATCCTGGCCCGATGTCCGAAAGGAGAAACTCTGACGGTGGGATTCACTGACGG
TAGGAAGATCAGTCACTCATGTACGCACCCATTTCACCACGACCCTCCTGTGATAGGCCGGGAA
AAATTTCATTCCCGACCGCAGCACGGTAGAGAACTACCTTGCAGCACGTACGCGCAGAGCACCG
CTGCAACTGCCGAGGAGATAGAGGTACA|TATGCCCCCAGACACCCCAGATCGCACATT
Nde I Overlapping overhang

Fig. 3 Fragment 2—genomic region of partial E2 with 5′ overhang containing myc tag and 3′ overhang overlapping with the CHIKV vector around the NdeI site. The overlapping region and the myc tag in the primers were color-coded and annotated as previously described in Fragment 1

10. Perform the three-fragment ligation using NEBuilder HiFi Assembly Kit: To obtain optimized ligation result, the three-fragment ligation shall be carried out according to the three-fragment assembly rules recommended by NEB: 1:2 Vector:Insert DNA molar ratio, 0.03–0.2 pmol total amount of fragments, and 10 μL NEB HiFi Assembly Master Mix in a 20-μL total volume. Briefly, set up the NEBuilder HiFi DNA assembly with the recommended amounts of CHIKV vector, Fragment 1, and Fragment 2 (*see* **Note 8**).

	Length	Mass	Moles
CHIKV Vector (digested)	12331 bp	50 ng	6.583 fmol
Fragment 1	1138 bp	9.229 ng	13.166 fmol
Fragment 2	566 bp	4.590 ng	13.166 fmol
Total amounts	32.915 fmol (0.032915 pmol)		
Add H_2O up to 10 μL			

Add 10 μL 2× NEB HiFi assembly master mix to the 10 μL vector–fragment mix. Incubate the ligation mixture at 50°C for 20 min. Save the ligation product in −20°C for competent bacterial cell transformation.

Since the assembled CHIKV 181/clone 25 E2-myc is larger than 10 kb, high-efficiency *E. coli* strain for cloning large plasmids, such as NEB 10-beta, is recommended for transformation.

3.2 Transformation of Ligation Product into Competent E. coli *with High Efficiency*

1. Thaw NEB 10-beta competent cells on ice and add 10 μL NEBuilder HiFi ligation product to 50 μL competent cells. Flick the tube five times to mix the ligation product and competent cells well.
2. Incubate the mixture on ice for 10 min.
3. Heat shock the mixture at 42°C for 30 s and place it on ice for 5 min to cool down.
4. Add 900 μL of NEB 10-beta/stable outgrowth medium into the mixture and pipette it well.
5. Shake the cells in the outgrowth medium at 250 rpm at 37°C for 1 h.
6. Spin down the competent cells at 2500 rcf at room temperature for 3 min.
7. Resuspend the competent cells in 100 μL outgrowth medium and spread the cells onto ampicillin/carbenicillin agar plate (*see* **Note 9**).
8. Incubate the ampicillin agar plate at 37°C for 18 h.

9. Pick three colonies from the ampicillin/carbenicillin agar plate and inoculate them into 4 mL of liquid LB containing ampicillin/carbenicillin. Incubate the bacterial culture at 37°C for 12 h.
10. Extract the plasmids and send them for sequencing for confirmation.

3.3 Production of CHIKV 181/Clone 25 E2-myc Virus Stocks

1. Linearization of the viral cDNA plasmid with NotI restriction site, which is introduced after the polyA tail in CHIKV/myc-E2. To linearize the viral cDNA plasmid as a transcription template, set up the digestion as follows:

Components	100 μL reaction
CHIKV 181/clone 25 E2-myc plasmid	5 μg
10× CutSmart buffer	10 μL
NotI-HF	4 μL
H_2O	Up to 100 μL

Incubate the reaction at 37°C for 4 h to facilitate complete digestion. Stop the digestion by heat inactivation at 65°C for 20 min (optional).

2. Purify the linearized CHIKV 181/clone 25 E2-myc with Zymoclean Gel Recovery Kit and elute the viral cDNA in 20 μL RNase-free H_2O. Save the eluted cDNA in −20°C for later use or proceed directly with transcription.
3. In vitro transcription of the viral RNA: The CHIKV 181/clone 25 E2-myc plasmid has an SP6 promoter upstream of its 5'UTR and a polyA tail downstream of its 3'UTR. Transcription system using SP6 polymerase can be applied to synthesize the viral RNA with a polyA tail from the linearized CHIKV 181/clone 25 E2-myc plasmid. Set up 50 μL reaction:

Components	50 μL reaction	Final concentration
10× SP6 reaction buffer (NEB)	5 μL	1×
10 mM DTT (NEB)	5 μL	1 mM
25 mM ribonucleotide solution mix (NEB)	1 μL	0.5 mM
10 mM Cap [m^{7}G(5')ppp(5')G (NEB)] (*see* **Note 10**)	2.5 μL	1 mM
Linearized viral cDNA template	1 μg	
RNAseOUT (Invitrogen)	1 μL	
SP6 RNA polymerase (NEB)	2.5 μL	
Add H_2O up to 50 μL		

Incubate the reaction mix for 1.5 h at 41°C.

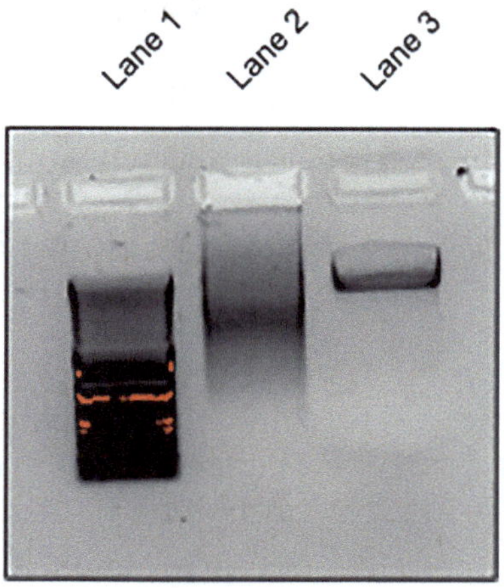

Fig. 4 Quality control of in vitro transcribed viral RNA (Lane 2) in comparison to linearized viral cDNA plasmid control (Lane 3). The band of transcribed viral RNA is a little bit smeared in 1% native agarose gel and moves faster than the linearized DNA template control of a similar size. Lane 1: 10 kb DNA ladder.

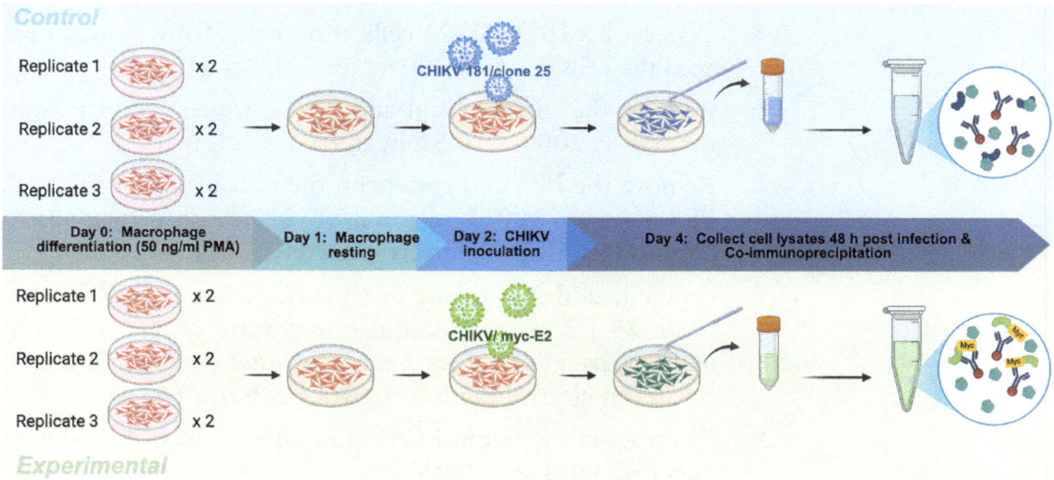

Fig. 5 Workflow of macrophage differentiation, CHIKV infection, and co-immunoprecipitation. THP-1 cells were seeded and differentiated into macrophages with 50 ng/mL PMA in twelve 150-mm dishes on Day 0. On Day 1, cells were switched to PMA-free complete medium and rested for 24 h before infection. On day 2, six dishes were infected with CHIKV 181/clone 25 (control), and six dishes were infected with CHIKV/myc-E2 (experimental). On day 4, macrophages were harvested and lysed for co-immunoprecipitation using anti-c-myc agarose affinity gels. For each group, lysates from two dishes were pooled together one biological replicate for co-immunoprecipitation.

To check the transcription result, run 2μL of reaction product on a 1% agarose gel (*see* **Note 11**).

3.4 Quality Control of In Vitro Transcribed CHIKV/myc-E2 RNA

1. Mix 2 μL of transcription product with 2 μL of the RNA gel loading dye (NEB).

2. Run the transcription product together with linearized CHIKV 181/clone 25 E2-myc as control on a 1% agarose gel stained with SYBR Safe at 135 V for 5–10 min.
3. Check the transcribed RNA quality: The RNA band should be a strong band that runs faster than the linearized template control (*see* **Note 12**) (Fig. 4). If the transcribed viral RNA has good yield, save the remaining viral RNA at −80°C for electroporation.

3.5 Electroporate Viral RNAs into BHK-21 cells [16]

1. Purify the transcribed viral RNA through Zymo Quick-RNA Miniprep Kit and elute the RNA in 20 μL RNase-free H_2O (*see* **Note 13**).
2. Cell preparation: Culture the BHK-21 cells in MEM with 7.5% FBS in a 150 mm dish until the cell confluency reaches 80%.
3. Gently rinse the cells with PBS and trypsinize them at 37°C for 3–4 min.
4. Add complete culture medium to inactivate trypsin and gently pipette the cells up and down to remove clumps.
5. Transfer 2×10^7 BHK-21 cells into a new 50 mL conical tube. Spin the cells down at 200 rcf for 5 min at 4°C.
6. Remove the medium and add 30 mL ice-cold PBS to wash the cells at 200 rcf for 5 min at 4°C (*see* **Note 14**).
7. Remove the PBS and resuspend the cells in 1 mL of ice-cold PBS. Aspirate 400 μL of the BHK-21 cells into two electroporation cuvettes (2 mm) respectively (*see* **Note 15**).
8. Carefully add 2 μg of the in vitro transcribed CHIKV 181/clone 25 E2-myc RNA into one cuvette of the BHK cells without making bubbles. Leave the other cuvette of BHK-21 cells as an electroporation control (*see* **Note 16**).
9. Electroporation with ECM 830: five pulses, 1 s between pulses, at 99 μs and 960 V.
10. Let the cells rest for 5 min at room temperature after the electroporation. In the meantime, prepare two 150 mm dishes containing 25 mL MEM with 7.5% FBS in each.
11. Transfer the electroporated BHK-21 cells from one cuvette to one dish and gently shake the dishes to spread the cells evenly in the medium. Label the dishes of cells electroporated with 1) CHIKV 181/clone 25 E2-myc RNA or 2) no RNA that will serve as an electroporation control.
12. Place the dishes in the incubator and culture the cells for 28 h (*see* **Note 17**).

3.6 Harvest the CHIKV 181/Clone 25 E2-myc Virus and Titrate the Infectious Viral Particles

Check the cell viability and confluency 28 h after electroporation. The cells of electroporation control should have approximately 100% confluency, while there should be significant cell death in the dish of cells electroporated with CHIKV 181/clone 25 E2-myc RNA due to active viral replication and cytopathic effects.

1. Harvest the BHK-21 cells supernatant and transfer it to a 50-mL conical tube. Spin the supernatant at 200 rcf at 4°C for 5 min to remove the cell debris and dead cells.
2. Transfer the clarified supernatant to a new 50-mL conical tube. Aliquot the supernatant into 2 mL screw-cap microtubes and store the virus stock aliquots at −80°C (*see* **Note 18**).
3. Measure the infectious myc-tagged CHIKV through plaque assay as previously mentioned. This protocol usually produces myc-tagged CHIKV stocks with a titer over 1×10^8 pfu/mL.

3.7 Purify myc-Tagged CHIKV E2 Along with Host and Viral Interactors Through Immunoprecipitation (Fig. 5)

3.7.1 Culture and Maintenance of THP-1 Monocytes

The THP-1 monocytes are cultured in RPMI 1640 ATCC modification medium which contains 10% FBS, 1× NEAA and 0.05 mM β -mercaptoethanol. The THP-1 density shall be maintained around 3.5×10^5 to 1×10^6 cells/mL.

3.7.2 Differentiating THP-1 Monocytes to Macrophages by PMA Stimulation

1. For the myc-tagged protein complex co-immunoprecipitation step in AP-MS, two sets of macrophages are required at the scale of 150 mm dishes: THP-1 macrophages infected with CHIKV/myc-E2 and THP-1 macrophages infected with parental untagged CHIKV 181/clone 25 (serves as negative control for AP-MS).
2. Transfer total amounts of 4.8×10^8 THP-1 cells into twelve 50 mL conical tubes (4×10^7 cells in each tube).
3. Spin down the cells at 200 rcf at 4°C for 5 min to remove the medium.
4. Resuspend the cells in 50 mL of macrophage differentiation medium and seed 25 mL of the cell suspension evenly into a 150 mm dish (2×10^7 cells/dish).
5. Culture the THP-1 cells in the differentiation medium for 24 h.
6. After 24-h incubation with the differentiation medium, the THP-1 cells should have differentiated into macrophage-like cells and firmly attached to the dishes. Replace the culture medium with PMA-free macrophage resting medium.

Table 1
CHIKV inoculation

Virus stock	Titer (pfu/mL)	MOI	Virus (μL)/ dish	VB buffer (μL)/ dish
Formulation[a]			**(Cell numbers per dish × MOI)/ pfu × 1000**	**5000–Virus (μL)/dish**
CHIKV/ E2-myc	pfu1	10	$((2\times10^7\times10)/\text{pfu1})\times1000$	Add VB buffer up to 5 mL
Parental CHIKV	pfu2	10	$((2\times10^7\times10)/\text{pfu2})\times1000$	Add VB buffer up to 5 mL

[a] To calculate the required volume of virus stock for each culture dish, we first determine the total number of infectious viral particles needed. This is calculated using the multiplicity of infection (MOI), which represents the number of viral particles added per cell. Because terminally differentiated macrophages do not proliferate, the number of cells at infection should be equal to the initial seeding density. Therefore, the total infectious units needed per dish can be calculated as: Total viral particles = Number of cells per dish × MOI Here, each dish contains 2×10^7cells and the MOI is 10, then: 2×10^7 cells $\times 10 = 2 \times 10^8$ infectious particles Next, to convert this to the volume of virus stock needed, we use the viral titer expressed as plaque-forming units per milliliter (pfu/mL):
$\text{Required volume} \quad (\text{mL}) = \frac{\text{Total infectious particles}}{\text{Virus stock titer} \quad (\text{pfu / mL})}$b Finally, for experimental convenience, convert milliliters to microliters by multiplying the volume by 1000 for practical pipetting.

3.7.3 Infecting THP-1-Derived Macrophages with myc-Tagged CHIKV

1. Thaw the CHIKV/myc-E2 virus stocks at room temperature.
2. Prepare the virus and VB buffer mixture based on the MOI used for infection according to Table 1.
3. Remove the medium from the fully differentiated macrophages and add 5 mL of the virus inoculum (virus–VB buffer mixture) into each dish.
4. Place the dishes of macrophages with CHIKV/E2-myc or parental CHIKV in the incubator for 1 h. During the incubation, shake the plate every 15 min to ensure the cells are evenly exposed to the virus inoculum (*see* **Note 19**).
5. Add 20 mL of RPMI 1640 ATCC modification medium with 10% inactivated human AB serum in each dish and let the infection continue for another 48 h.

3.7.4 Co-immunoprecipitation of myc-Tagged E2 and E2 Precursors with Interactors for Mass Spectrometry Analysis

1. Add 5 mL ice-cold PBS into each dish to rinse the macrophages.
2. Discard the PBS and add 2 mL cell lysis working buffer into each dish. Place the dishes on ice and let the cell lysis go on for 15 min.
3. Use a cell scraper to thoroughly detach the macrophages from the dish (*see* **Note 20**). Collect the cells from two dishes under the same treatment into a pre-cooled 15 mL conical tube on ice (combine cells from two dishes into one conical tube as a biological replicate).

4. Place the conical tubes on a tube rotator and rotate the cell lysates up and down at 4°C for 20 min.
5. To clarify the cell lysis, spin down the cell debris with a fixed-angle rotor at 14,000 rcf for 15 min at 4°C. Transfer supernatant to a new pre-cooled 15 mL conical tube.
6. While waiting for cell lysis to complete, aspirate 210 μL EZview Red anti-c-myc agarose affinity gel [210 μL affinity gel/per biological replicate (two dishes)] into a new 1.5-mL microcentrifuge tube. Add 1.5 mL NP-40 lysis buffer to wash the beads and get rid of the equilibration buffer. Spin down the beads at 1800 rpm at 4°C for 2 min. Discard the supernatant and repeat the washing step for another two times.
7. Resuspend the agarose affinity gel in 0.5 mL NP-40 lysis buffer and move the resin into a new pre-cooled 15 mL conical tube.
8. Continue from Step 5 for preparation of input controls: Aspirate 45 μL of clarified supernatant from each tube as whole cell lysate (WCLs) for later comparison. Add 15 μL 4× Laemmli sample buffer containing 5% β-mercaptoethanol to each WCL. Denature the WCLs at 99°C for 10 min. Save the denatured WCLs at −80°C for later use.
9. Continue from Step 5 for immunoprecipitation: Add the remaining clarified supernatants into the 15 mL conical tubes containing agarose affinity gels and bring the total volume up to 10 mL/ tube by adding NP-40 lysis buffer. Load the tubes to the rotator and rotate the cell lysate–affinity gel mixture at 4°C for 4 h.
10. Optional: After immunoprecipitation, save 45 μL flowthrough from each tube for checking the depletion efficiency of myc-tagged E2 and E2 precursors. Add 15 μL 4× Laemmli sample buffer containing 5% β-mercaptoethanol to each flowthrough. Denature the flowthrough at 99°C for 10 min. Save the denatured flowthrough at −80°C for western blot analysis later.
11. Add 3 mL NP-40 lysis buffer to resuspend the affinity gel in each tube and aliquot the resin into three 1.5 mL microcentrifuge tubes (1 mL/ tube). Invert the tubes ten times and spin down the affinity gel at 1800 rpm at 4°C for 2 min. Remove the supernatant and add new NP-40 lysis buffer into each tube (1 mL/tube).
12. Repeat the washing step (Step 18) three times.
13. Protein elution from the affinity gel: Thoroughly remove the supernatant after the last washing step, add 50 μL of the elution buffer (8 M urea, 100 mM Tris–HCl pH 8) into each tube (~60–70 μL beads/tube). Shake the tubes on Eppendorf ThermoMixer at room temperature for 15 min at 1200 rpm.

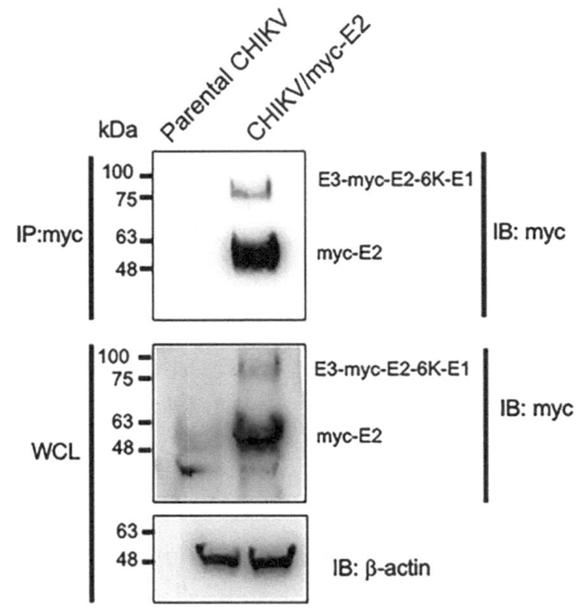

Fig. 6 Validation of myc-tagged E2 and E2 precursors from macrophages infected with CHIKV/myc-E2. Proteins immunoprecipitated by anti-myc affinity gel and corresponding whole cell lysates (WCLs) as input control from macrophages either infected with parental CHIKV or CHIKV/ myc-E2 were probed by myc antibody. Both myc-E2 and uncleaved precursor E3-myc-E2-6K-E2 were detected by myc antibody

Collect the supernatant into a new microcentrifuge tube and place on ice.

14. Add another 50 μL of elution buffer/tube and shake at room temperature. Combine the second-round elution with the first one.
15. Take 10 μL of the pooled protein elution and mix the elution with 10 μL 2× Laemmli sample buffer containing 5% β-mercaptoethanol for western blot confirmation (*see* **Note 21**) (Fig. 6).
16. Carefully transfer the eluted proteins to a new tube and precipitate them by addition of four volumes of −20°C acetone (360 μL per tube). Incubate at 4°C overnight.
17. Next day: Precipitate the proteins in acetone at 16,100 rcf for 30 min at 4°C. Wash the protein pellets with −20°C acetone for an additional time. Remove the acetone and save the protein pellet at −80°C for MS analysis.

4 Notes

1. Using human serum to differentiate THP-1 into macrophages will confer stronger and longer attachment of the macrophages. Human serum from blood type AB donors is recommended, as it will not react with cells bearing A or B antigens.
2. The inactivation procedure for human AB serum is the same as for FBS: heat at 56 °C for 30 min, then allow it to cool to room temperature before use.
3. Before full maturation of CHIKV E2, the long precursor E3–E2–6K/TF-E1 will undergo proteolytic processing in ER and be cleaved at E2 | 6 K/TF and 6 K/TF | E1 by ER signal peptidases. The partially matured precursors in ER will form into p62/E1 heterodimer in which p62 is uncleaved E3–E2 polyprotein. Matured E2 will be released from p62 precursor after furin cleavage to remove E3 in Golgi compartment, and E2 will interact with E1, forming heterodimer again [17]. Adding myc tag at the N-terminal of E2 in the viral genome will label matured E2 and also uncleaved precursors, including E3–E2-6K/TF-E1 and p62. Thus, immunoprecipitating E2 and precursors will co-immunoprecipitate viral proteins strongly interacting with them as well, such as E3 and E1.
4. Heat inactivation to stop the digestion is optional, as the agarose dissolving buffer (ADB) from Zymoclean Gel Recovery Kit and other gel purification kit containing high salt concentrations will inhibit enzymatic activities.
5. Since the fragments and vector first need to be assembled in a small volume (10 μL) of water, enhancing the vector concentration as much as possible will allow for assembly with a higher number of fragments (usually with lower concentrations). Ideally, high concentration of the digested vector from the gel recovery can be achieved by eluting with 6–20 μL water.
6. The original amino acid sequences around the E3–E2 furin cleavage site is: (E3)...PRRQRR | SIKDNF...(E2). The furin cleavage site is annotated as "|". To introduce myc tag at the N terminal of E2 and keep the furin cleavage site, the myc sequence "EQKLISEEDL" replaces the second amino acid isoleucine in E2 and is flanked with two serines to increase flexibility. As such, the sequence of E3-myc-E2 is: (E3)... PRRQRR | SSEQKLISEEDLSKDNF...(E2).
7. According to NEB instruction, 15–20 bp overlap regions between each fragment/vector for 2–3 fragment ligation is enough to achieve optimal ligation efficiency. Nevertheless, we extend the overlap region to 30 bp to contain the myc tag.

We found the prolonged overlap regions to not affect ligation efficiency.

8. The NEBioCalculator (https://nebiocalculator.neb.com/#!/ligation) can be used for calculating mass and moles of fragments and vectors for the ligation.
9. Same with ampicillin, carbenicillin also belongs to the penicillin family and confers resistance to the *AmpR* gene. Since carbenicillin is more stable than ampicillin in growth media, it is recommended to give stronger selection.
10. The ratio of Cap analog [m7G(5′)ppp(5′)G] to GTP determines both capping efficiency and in vitro transcription yield [18]. Substituting a large proportion of GTP with cap analog increases the capping rate but reduces RNA yield. A commonly used ratio is 4:1 (cap analog:GTP), which provides a balance between adequate RNA yield and sufficient capping. In this protocol, we used a 2:1 ratio for in vitro transcription of viral RNA. For applications requiring higher capping efficiency, higher ratios can be tested to optimize conditions.
11. DNA ladder is not compatible for estimating the in vitro transcribed viral RNA size. The transcribed single-stranded viral RNA usually forms complicated secondary structures and has smaller molecular weight which makes them move faster than the DNA of the same size. If evaluating the actual size of transcribed viral RNA is warranted, it is recommended to use an RNA ladder.
12. Running the transcribed RNA integrity on a native 1% agarose gel is sufficient to have a quick and raw estimation of the RNA integrity, which usually presents as a thick intense band smaller than the linearized DNA template. If accurate evaluation is warranted, running the transcribed RNA on denaturing agarose gel with RNA ladder is recommended.
13. Using a smaller volume of water to condensate RNA for higher concentration helps to introduce less water into the 400 μL electroporation mixture at a later step.
14. Pre-cool the PBS on ice. Ice-cold PBS will inhibit the RNase activity.
15. Pre-cool the cuvettes at 4°C to inhibit RNase activity.
16. It is recommended to use brand-new filtered RNase-free tips to avoid RNA degradation through the entire electroporation process.
17. Incubating the electroporated cells for a longer period increases viral yield but also raises the risk of mutations in progeny genomes. To balance yield with genetic stability, harvesting at 28 h post-electroporation is recommended for CHIKV stock generation.

18. It is important to store virus stocks in screw-cap tubes, which can prevent accidental aerosol release or leakage. In addition, screw caps withstand freeze–thaw stress better than the snap caps, which may pop open under pressure.
19. Frequent shaking of the dishes during incubation helps ensure even distribution of viruses across the cells.
20. Macrophages are resistant to enzymatic dissociation. We tested both trypsin and Accutase treatments of THP-1–derived macrophages at room temperature and 37°C, but neither achieved complete detachment. Therefore, physical detachment with a cell scraper is recommended as the most effective method.
21. Urea is a chaotropic agent that disrupts non-covalent interactions and unfolds protein structures. Proteins eluted in urea buffer are already sufficiently denatured, so additional heat denaturation is unnecessary.

Acknowledgments

This work was supported by NIH grant R01AI158704, 2023 UCLA DGSOM W. M. Keck Foundation Junior Faculty Award, and a JCCC–BSCRC Ablon Scholars Award to M.M.L.

References

1. Morris JH, Knudsen GM, Verschueren E et al (2014) Affinity purification–mass spectrometry and network analysis to understand protein-protein interactions. Nat Protoc 9:2539–2554
2. Wu S, Zhang S, Liu C-M et al (2024) Recent advances in mass spectrometry-based protein interactome studies. Mol Cell Proteomics 24:100887
3. Bouhaddou M, Reuschl A-K, Polacco BJ et al (2023) SARS-CoV-2 variants evolve convergent strategies to remodel the host response. Cell 186:4597–4614. e26
4. Gordon DE, Jang GM, Bouhaddou M et al (2020) A SARS-CoV-2 protein interaction map reveals targets for drug-repurposing. Nature 583:459–468
5. Wang X, Wang W, Wang T et al (2023) SARS-CoV-2 ORF8 protein induces endoplasmic reticulum stress-like responses and facilitates virus replication by triggering calnexin: an unbiased study. J Virol 97(3):e0001123
6. Mazel-Sanchez B, Iwaszkiewicz J, Bonifacio JPP et al (2021) Influenza A viruses balance ER stress with host protein synthesis shutoff. Proc Nat Acad Sci 118:e2024681118
7. Zhang Y, Zhang Y, Fang Z et al (2025) China's chikungunya virus outbreak is a wake up call 645:41–41
8. Ribeiro dos Santos G, Jawed F, Mukandavire C et al (2025) Global burden of chikungunya virus infections and the potential benefit of vaccination campaigns. Nat Med 31:2342–2349
9. Lima Cavalcanti TYV, de, Pereira MR, Paula SOde et al (2022) A review on chikungunya virus epidemiology, pathogenesis and current vaccine development. Viruses 14:969
10. Chen R, Puri V, Fedorova N et al (2016) Comprehensive genome scale phylogenetic study provides new insights on the global expansion of chikungunya virus. J Virol 90(23):10600–10611
11. Srivastava P, Chaudhary S, Malhotra S et al (2023) Transcriptome analysis of human macrophages upon chikungunya virus (CHIKV) infection reveals regulation of distinct signaling and metabolic pathways during the early and late stages of infection. Heliyon 9:e17158
12. Labadie K, Larcher T, Joubert C et al (2010) Chikungunya disease in nonhuman primates involves long-term viral persistence in macrophages. J Clin Invest 120:894–906

13. Zarrella KM, Sheridan RM, Ware BC et al (2025) Chikungunya virus persists in joint-associated macrophages and promotes chronic disease. Nat Microbiol 11:1302–1317
14. Yao Z, Ramachandran S, Huang S et al (2024) Interaction of chikungunya virus glycoproteins with macrophage factors controls virion production. EMBO J 43:4625–4655
15. Jose J, Tang J, Taylor AB et al (2015) Fluorescent protein-tagged Sindbis Virus E2 glycoprotein allows single particle analysis of virus budding from live cells. Viruses 7:6182–6199
16. Bick MJ, Carroll J-WN, Gao G et al (2003) Expression of the zinc-finger antiviral protein inhibits alphavirus replication. J Virol 77:11555–11562
17. Brown RS, Wan JJ, Kielian M (2018) The alphavirus exit pathway: what we know and what we wish we knew. Viruses 10:89
18. Thermo Fisher Scientific. Methods of IVT mRNA capping. Available at: https://www.thermofisher.com/blog/behindthebench/methods-of-ivt-mrna-capping/ (accessed 29 July 2026).

Check for updates

Chapter 14

Plaque Reduction Neutralization Test Against Chikungunya Virus Using Liver Homogenate

Thais Alkifeles Costa and Betânia Paiva Drumond

Abstract

We adapted the plaque reduction neutralization test (PRNT) to detect chikungunya virus (CHIKV) neutralizing antibodies in liver homogenates instead of serum. This approach enables serological investigation using tissues collected from animal carcasses as nonhuman primate ones, during arbovirus surveillance, expanding opportunities for CHIKV monitoring and strengthening epidemiological surveillance efforts.

Keywords PRNT, Adapted viral neutralization test, Chikungunya virus, Viral neutralization, Neutralizing antibodies, Arbovirus surveillance, Serological surveillance

1 Introduction

The plaque reduction neutralization test (PRNT) is recognized as the gold standard for detecting and quantifying neutralizing antibodies against several viral pathogens [1]. Traditionally, the assay is conducted using sera; however, obtaining this type of material is not always feasible in certain sampling contexts. For example, the yellow fever surveillance program in Brazil is mostly based on passive surveillance of nonhuman primate (NHP) carcasses, when liver and other organs are usually sampled for viral investigation [2]. These NHP carcasses collected during yellow fever surveillance could be used for research into other arboviruses, not only using molecular analysis but also serological tests. Given that, we adapted PRNT to be used with liver homogenates instead of sera, in order to expand the scope of epidemiological surveillance of zoonotic arboviruses, such as chikungunya virus (CHIKV).

This protocol describes the adaptation of PRNT for the detection of neutralizing antibodies against CHIKV from liver homogenates. The use of solid tissues in serological tests has been previously described in the literature, with reports of antibody detection

Claudia Filomatori and Eliana F. Castro (eds.), *Chikungunya Virus*, Methods in Molecular Biology, vol. 3067,
https://doi.org/10.1007/978-1-0716-5496-5_14,

in samples from defrosting fluids and muscle tissue trituration [3, 4]. Based on the conventional PRNT methodology, which uses serum samples, the protocol described here was adapted to use liver homogenates obtained from maceration and subsequent centrifugation of liver tissue. This protocol is based on previous studies of Costa et al. [5], which demonstrated the feasibility of using liver homogenates to investigate antibodies against arboviruses as CHIKV, YFV, and Mayaro virus (MAYV) in experimentally (mice) or naturally (nonhuman primates) animals. Given the epidemiological importance of CHIKV and the potential of serological tests to improve epidemiological surveillance, this study describes the anti-CHIKV PRNT protocol adapted for the use of liver homogenate as an alternative sample to serum.

2 Materials

All materials used for the manipulation of cells, culture medium, liver samples, liver homogenates, serum, or viruses must be sterile.

Procedures must be performed in a class II biological safety cabinet, unless otherwise specified, with the use of appropriate personal protective equipment (PPE), including gloves, a mask, a lab coat, and a disposable head cap.

2.1 Preparation of VERO Cell Plates

1. VERO cells (CCL-81) cultured in tissue culture flasks.
2. MEM 5% FBS: Eagle's minimum essential medium (MEM) supplemented with 5% fetal bovine serum (FBS) and antibiotics: 200 U/mL penicillin, 40 μg/mL streptomycin, and 2 μg/mL amphotericin B.
3. 0.25% Trypsin solution.
4. 1× Phosphate-buffered saline (PBS).
5. 5 and 10 mL graduated serological pipettes.
6. Micropipettes with volume ranges of 0.5–10 μL and 100–1000 μL.
7. Micropipette tips compatible with 0.5–10 μL and 100–1000 μ L pipettes.
8. 12-well flat-bottom polystyrene tissue culture plates.
9. Cell counter (Neubauer chamber or automatic counter).
10. Cell incubator at 37°C with 5% CO_2 atmosphere.
11. Sterile plastic biohazard waste container.
12. Water bath set to 37°C.
13. 70% (v/v) ethanol.

2.2 Preparation of Liver Homogenate

1. Fresh or frozen liver tissues (at least 50 mg) to be used for liver homogenate preparation. These include liver samples to be investigated and liver samples from noninfected mice to be used as control.
2. 2.0 mL screw-cap microtubes (*see* **Note 1**).
3. 1.5 mL microtubes.
4. 2.0 mm sterile borosilicate beads (*see* **Note 2**).
5. Scalpel with No. 22 blade.
6. Sterile polystyrene Petri dishes.
7. Analytical balance.
8. Benchtop tissue homogenizer.
9. MEM 0% FBS: Eagle's minimum essential medium supplemented with antibiotics: 200 U/mL penicillin, 40 μg/mL streptomycin, and 2 μg/mL amphotericin B.
10. Microcentrifuge compatible with 1.5 and 2.0 mL microtubes.
11. Micropipette with a volume range of 100–1000 μL.
12. Micropipette tips compatible with 100–1000 μL pipettes.
13. 0.22 μm, 30 mm polyethersulfone (PES) syringe filter.
14. 3.0 mL syringe.
15. 70% (v/v) ethanol.
16. Sterile tweezers.
17. Sterile plastic biohazard waste container.

2.3 Adapted PRNT

1. Liver homogenate samples diluted 1:10 in MEM 0% FBS, filtered through a 0.22 μm membrane and heat-inactivated (56°C for 30 min).
2. CHIKV stock previously titrated (*see* **Note 3**).
3. Control sera: positive and negative sera controls.
4. HEPES buffer (4-[2-hydroxyethyl]-1-piperazineethanesulfonic acid), 1 M solution (*see* **Note 4**).
5. MEM 2% FBS: Eagle's minimum essential medium supplemented with 2% FBS, 200 U/mL penicillin, 40 μg/mL streptomycin, and 2 μg/mL amphotericin B.
6. MEM 2% FBS/HEPES solution: Eagle's minimum essential medium supplemented with 2% FBS, 200 U/mL penicillin, 40 μg/mL streptomycin, 2 μg/mL amphotericin B, and 2% HEPES-KOH.
7. 2× Medium 199 (*see* **Note 5**).
8. 3.0% (w/v) Carboxymethylcellulose (CMC) (*see* **Note 6**).
9. CMC/199 solution (*see* **Note 7**).

10. 3.7% (v/v) formaldehyde (*see* **Note 8**).
11. 1% (w/v) crystal violet.
12. 1.5 mL microtubes.
13. 50 mL conical tubes.
14. Micropipettes with volume ranges of 20–200 µL and 100–1000 µL.
15. Micropipette tips compatible with 20–200 µL and 100–1000 µL pipettes.
16. Multichannel pipettes with volume ranges of 30–300 µL and 100–1000 µL.
17. Multichannel reagent reservoir.
18. 1.2 mL 96-well deep-well plates.
19. Lids for 96-well deep-well plates.
20. Cell incubator at 37°C with 5% CO_2 atmosphere.
21. Dry-bath thermoblocks set to 56°C.
22. Negatoscope.
23. Water bath set to 37°C.
24. Sterile plastic biohazard waste container.
25. 70% (v/v) ethanol.

3 Methods

To ensure optimal cell performance during the assay, use VERO cells that have undergone less than 30 passages. VERO cells should be maintained using routine culture protocols to prepare 175 cm^2 culture flasks that will be used to prepare the cell plates for the experiment.

Flasks should contain a nearly confluent monolayer and must have been passaged within the previous 48 h. Cell plates should be prepared one day prior to the adapted PRNT assay.

All materials must be sterile. All procedures should be conducted in a class II biological safety cabinet with appropriate personal protective equipment (PPE), including gloves, a mask, a lab coat, and a disposable head cap.

3.1 Preparation of VERO Cell Plates

Prior to starting the experiment, disinfect all necessary equipment with 70% (v/v) ethanol. Warm the MEM 5% FBS, trypsin solution, and PBS in a 37°C water bath before use.

1. Remove the culture medium from 175 cm^2 flasks containing VERO cell monolayers using a pipette or by inverting the flask into a sterile waste container.

2. Wash the cell monolayer by gently adding 3 mL of PBS prewarmed at 37°C. Gently rock the flask to wash the cell monolayer and then remove the PBS using a pipette or by inverting the flask into a sterile waste container.
3. Repeat the previous step three times and remove any residual PBS with a serological pipette.
4. Add 3 mL of 0.25% trypsin to the cell monolayer and gently homogenate it to cover the cell monolayer. Incubate the flask in a 5% CO_2 incubator at 37°C for 3 min (*see* **Note 9**).
5. Add 5 mL of MEM 5% FBS to neutralize the trypsin and gently resuspend the cells by pipetting up and down. Transfer the cell suspension to a 50 mL conical tube.
6. Prepare a 1:10 dilution of the cell suspension in PBS in a microtube for cell counting.
7. Count the cells using a Neubauer chamber (or an automatic cell counter). Based on the cell count, calculate the number of cells required for seeding 12-well plates (1.8×10^5 cells per well). The goal is to obtain 85%–95% monolayer confluence overnight.
8. Seed 1.8×10^5 cells per well and adjust the final volume to 1 mL with MEM 5% FBS. Gently homogenize cells and medium by carefully rocking the plate.
9. Incubate the 12-well plates overnight at 37°C in a 5% CO_2 incubator.

3.2 Preparation, Dilution, and Filtration of Liver Homogenates

Prior to starting the experiment, disinfect all necessary equipment with 70% (v/v) ethanol.

1. Label the 2.0 mL screw-cap microtubes according to the samples to be processed. In each screw-cap microtube, add four 2.0 mm sterile borosilicate beads using sterile tweezers and then add 200 μL of MEM 0% FBS with a micropipette.
2. If the tissue is frozen, thaw it and place it in a sterile Petri dish. Using a pipette tip, immobilize the tissue (like a pin) while cutting a piece of the tissue with a scalpel. Weigh 50 mg of liver tissue using an analytical balance and transfer the sample to the corresponding microtube containing borosilicate beads and MEM 0% FBS. If fresh tissue is used, place it directly in a sterile Petri dish and proceed with the same cutting and weighing steps.
3. Ensure that all microtubes are properly capped before proceeding. Place the microtubes in the benchtop tissue homogenizer, ensuring that they are balanced.

4. Macerate the liver tissue for 1 min, pause for 30 s, and then homogenize it for 1 min. The pause prevents overheating of the sample during maceration.
5. For supernatant clarification, centrifuge the microtubes at 16,000 × *g* for 10 min.
6. Carefully remove the tubes from the centrifuge and transfer the supernatant to a new, appropriately labeled 1.5 mL microtube.
7. Repeat the clarification (Steps 5 and 6) four times or until the supernatant is clear and no pellet is visible at the bottom of the microtube after centrifugation (*see* **Note 10**).
8. After clarification of the liver homogenate, dilute it 1:10 in MEM 0% FBS to a final volume of 1 mL.
9. Saturate a 0.22 μm filter attached to a 3 mL syringe with MEM 0% FBS. Pass 500 μL of MEM 0% FBS through the filter and discard it (*see* **Note 11**).
10. Using the saturated filter (Step 9), filter the diluted homogenate into a labeled 1.5 mL microtube. Take notes on the volume of liver homogenate recovered after filtration. At least 800 μL is required to perform the neutralization assay (*see* **Note 12**).
11. Place the microtube containing the diluted and filtered homogenate in a dry-block heater to inactivate the complement system at 56°C for 30 min. After this step, the sample can be stored at −20°C until use in the neutralization test.

3.3 Adapted PRNT

Before starting the experiment, disinfect all necessary equipment with 70% (v/v) ethanol. Ensure that at least 800 μL of each homogenate sample is prepared, diluted, filtered, and heat-inactivated. Positive serum, negative serum, and serum/liver controls must also be diluted and heat-inactivated in advance. The viral stock to be used must be titrated beforehand.

The 96-deep-well plates used in this protocol do not have dedicated lids; therefore, lids from standard 12-well flat-bottom polystyrene plates can be used. Ensure that the selected lid fits the deep-well plate properly.

3.3.1 Preparation of Positive and Negative Controls

1. Prepare the negative control, diluting the negative serum 1:10 in MEM 0% FBS to a final volume of 1 mL. Heat-inactivate the complement at 56°C for 30 min. Store the inactivated serum at −20°C until use.
2. Prepare the positive control, diluting the positive anti-CHIKV serum 1:10 in MEM with 0% FBS to a final volume of 1 mL. Heat-inactivate the complement at 56°C for 30 min. Store at −20°C until use.

3. Prepare the serum/liver homogenate control, mixing equal volumes of positive anti-CHIKV serum (100 μL) and liver homogenate from uninfected mice (100 μL). Dilute the mixture 1:10 in MEM 0% FBS to a final volume of 1 mL. Heat-inactivate the complement at 56°C for 30 min. Store at −20°C until use. This control is used to test possible background matrix effects from the homogenate.

3.3.2 Dilution of Samples and Controls

The following steps describe the preparation of the sample dilution plate. A 96-deep-well plate will be used for performing serial dilutions of the 1:10 diluted liver homogenates described above. Positive and negative controls will also be diluted in the same plate.

1. Add 20 mL of MEM 2% FBS/HEPES solution to a multichannel reagent reservoir.
2. Using a multichannel pipette, dispense 350 μL of MEM 2% FBS/HEPES from the reservoir into wells 1 to 9 of rows B through F (Fig. 1a). Cover the plate and set aside.
3. Thaw the liver homogenate samples (≥800 μL) and controls were previously diluted 1:10, filtered, and heat-inactivated. Centrifuge the microtubes for 10 min at 16,000 × *g* and transfer the supernatant to a new 1.5-mL microtube (*see* **Note 13**).

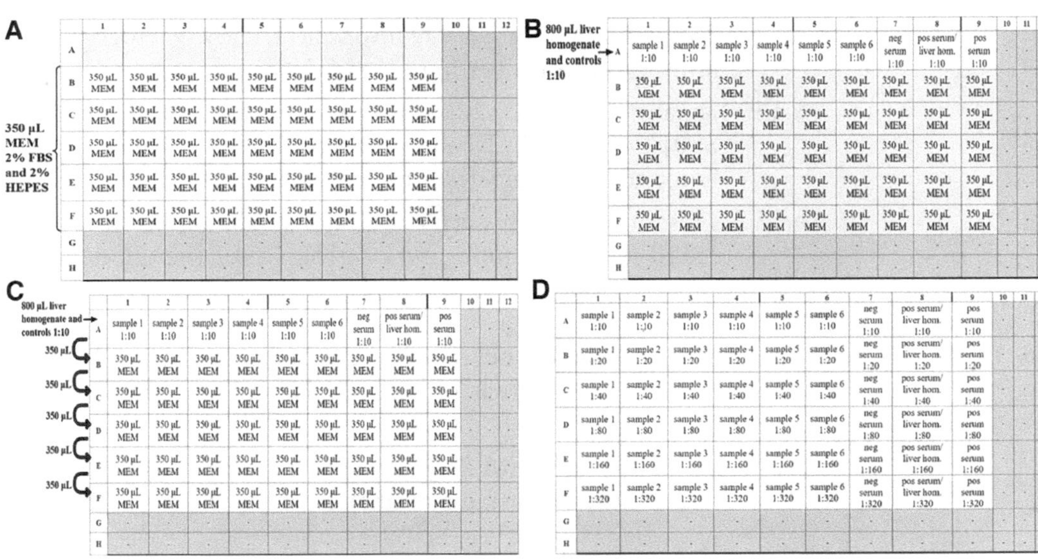

Fig. 1 (**a**) Layout of the sample dilution plate after addition of 350 μL of MEM 2% FBS/HEPES for sample and control dilutions. (**b**) Layout of the sample dilution plate after addition of liver homogenate samples and positive and negative controls. (**c**) Layout of the sample dilution plate during serial dilution of samples and controls. (**d**) Final layout of the sample dilution plate after serial dilution. MEM: Eagle's minimum essential medium; FBS: Fetal bovine serum

4. Define the position of each sample on the 96-deep-well plate map (Fig. 1b).
5. Using a micropipette, add 800 μL of each liver homogenate into wells A1–A6. Samples 1–6 will be added to wells A1–A6, respectively (Fig. 1b).
6. Add 800 μL of the negative serum control into well A7 (Fig. 1b).
7. Add 800 μL of the serum/liver homogenate control into well A8 (Fig. 1b).
8. Add 800 μL of the positive serum control into well A9 (Fig. 1b).
9. Homogenize the contents of wells A1–A9 by pipetting up and down three times.
10. Serially dilute (1:20) the samples and controls. Using a multichannel pipette, transfer 350 μL from wells A1–A9 into wells B1–B9, respectively. Change tips and homogenize by pipetting up and down three times (Fig. 1c) (*see* **Note 14**).
11. Serially dilute (1:40) the samples and controls. Transfer 350 μ L from wells B1–B9 into wells C1–C9. Change tips and homogenize by pipetting up and down three times (Fig. 1c).
12. Serially dilute (1:80) the samples and controls. Transfer 350 μ L from wells C1–C9 into wells D1–D9. Change tips and homogenize by pipetting up and down three times (Fig. 1c).
13. Serially dilute (1:160) the samples and controls. Transfer 350 μL from wells D1–D9 into wells E1–E9. Change tips and homogenize by pipetting up and down three times (Fig. 1c).
14. Serially dilute (1:320) the samples and controls. Transfer 350 μL from wells E1–E9 into wells F1–F9. Change tips and homogenize by pipetting up and down three times (Fig. 1c).
15. This procedure results in dilutions ranging from 1:10 to 1:320 for each sample and control. Cover the dilution plate and keep it at 4°C until use (Fig. 1d).

3.3.3 Preparation of the Assay Plate

1. Using a multichannel pipette, add 300 μL of MEM 2% FBS/HEPES into the wells of columns 10 and 11 (virus controls) (Fig. 2a).
2. Add 600 μL of MEM 2% FBS/HEPES to the wells of column 12 (cell controls) using a multichannel pipette (Fig. 2b).

3.3.4 Preparation of the Virus Solution

1. Serially dilute the viral stock 1:10 and prepare a virus working solution containing 100 PFU per 90 μL of MEM 2% FBS/HEPES. Prepare 30 mL of the viral working solution (11.1 PFU/μL).
2. Keep the prepared viral working solution on ice until use.

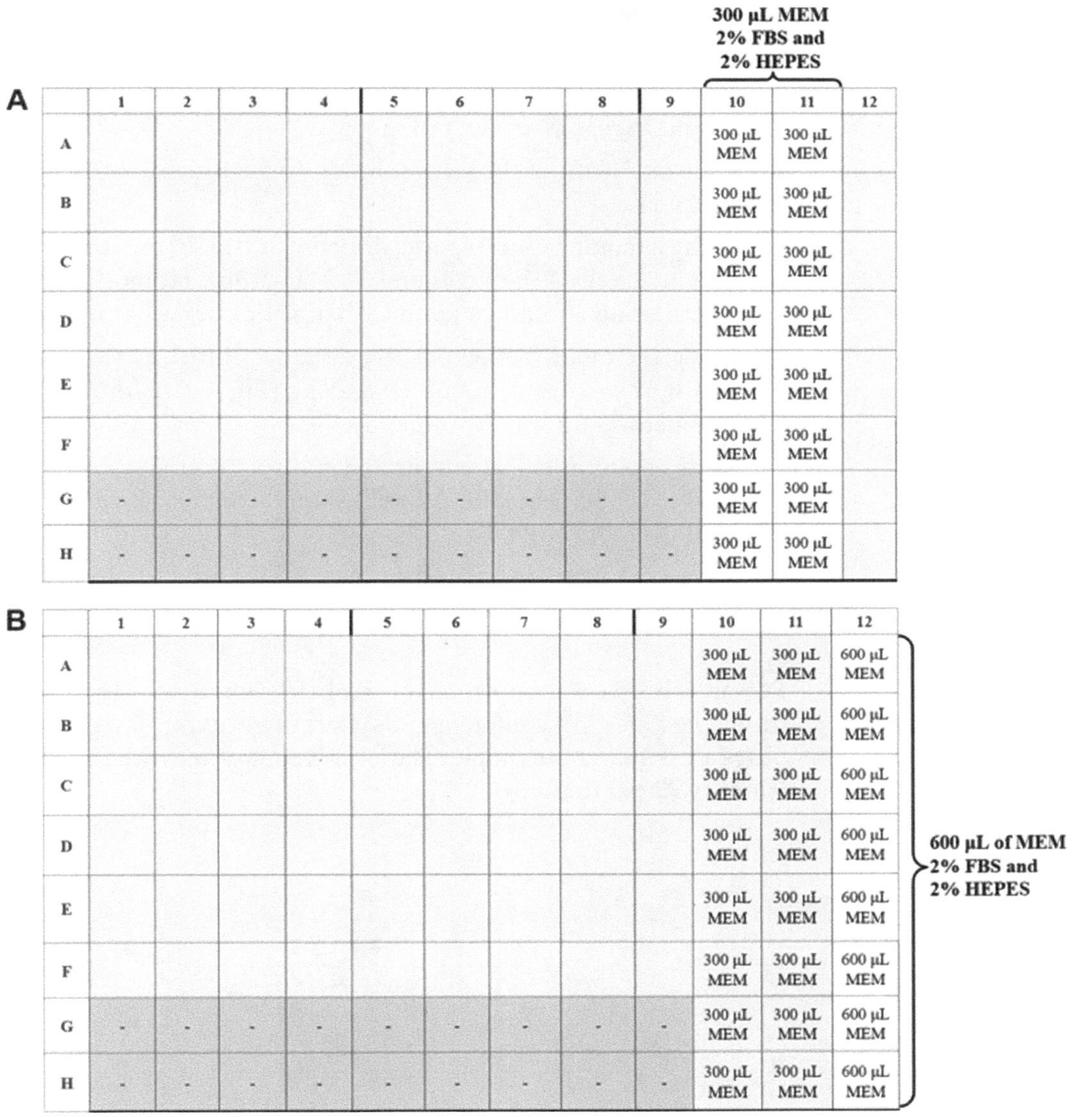

Fig. 2 (**a**) Layout of the assay plate after adding MEM 2% FBS/HEPES to virus control wells. (**b**) Layout of the assay plate after adding MEM 2% FBS/HEPES to cell control wells. MEM: Eagle's minimum essential medium; FBS: Fetal bovine serum

3.3.5 PRNT Assay

At this stage, the sample dilution plate, assay plate, and 30 mL of virus solution (11.1 PFU/µL) must all be ready.

The liver homogenates and controls in the dilution plate will be transferred to the assay plate and incubated with the virus.

1. Using a multichannel pipette, transfer 300 µL from each well in column 1 of the sample dilution plate into the corresponding well in column 1 of the assay plate. Change tips.

2. Repeat this process (Step 1) for columns 2–9, using fresh tips for each column.
3. After transferring all samples and controls, the assay plate should appear as shown in Fig. 3a.
4. Transfer 30 mL of the virus solution to a multichannel reagent reservoir.
5. Using a multichannel pipette, dispense 300 μL of the virus solution into wells A–F of columns 1–9 (Fig. 3b). Homogenize by pipetting up and down gently. Change tips between columns.
6. Using the multichannel pipette, dispense 300 μL of virus solution into wells of columns 10 and 11 (Fig. 3c). Homogenize by pipetting up and down gently.
7. At this point, the final sample and control dilutions range from 1:20 to 1:640 (Fig. 3d). All wells should contain a final volume of 600 μL (*see* **Note 15**).
8. Cover the plate and incubate in a 37°C, 5% CO_2 incubator for 1 h.

3.3.6 Infection of Monolayer Cells in a 12-Well Plate

Prepare VERO cell monolayers the day before infection. Monolayers should be 85%–95% confluent at the start of infection. Two 12-well plates are required per sample. The assay is compatible with triplicates (three wells per dilution).

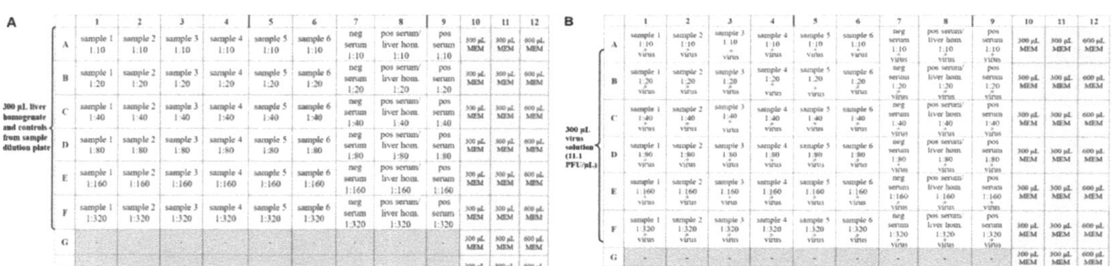

Fig. 3 (**a**) Final layout of the assay plate after transfer of samples from the sample dilution plate. (**b**) Final layout after mixing virus with samples and controls in the assay plate. (**c**) Final layout after adding virus solution to virus control wells in the assay plate. (**d**) Final dilutions of samples and controls used in the neutralization assay

1. Remove and discard the supernatant from the 12-well plate into a biohazard container.
2. Using a micropipette, add 180 µL of the assay-plate inoculum into each well of the 12-well plate (Fig. 4).
3. Gently rock the cell plate to ensure complete coverage of the monolayer.
4. Incubate the plates for 1 h at 37°C and 5% CO_2, gently rocking every 10 min using cross-movements (back and forth and from one side to the other). This prevents drying of the monolayer and ensures homogeneous infection.
5. After adsorption, gently overlay each well with approximately 1 mL of CMC/199 solution prewarmed at 37°C. Incubate at 37°C with 5% CO_2 for 48 h.
6. After 48 h of infection, in a fume hood, add approximately 2 mL of 3.7% formaldehyde solution to each well. Fix the cells for at least 2 h at room temperature (it can be incubated overnight as well).
7. Discard the formaldehyde-containing supernatant into a proper biohazard container. Wash the plates under running water until no CMC or formaldehyde remains (*see* **Note 16**).
8. Cover the monolayer with 1% (w/v) crystal violet solution and incubate for 30 min at room temperature.
9. Discard the staining solution, rinse the wells with running water, and air-dry the plates. The plates are now ready for PFU counting.

3.3.7 PFU Counting and Analysis of Results

1. Count PFUs in each well using a negatoscope, marking each plaque with a permanent marker.
2. For the experiment to be considered valid, the following conditions must be met: cell control wells must contain no PFUs; virus control wells must contain ~100 PFUs; negative control

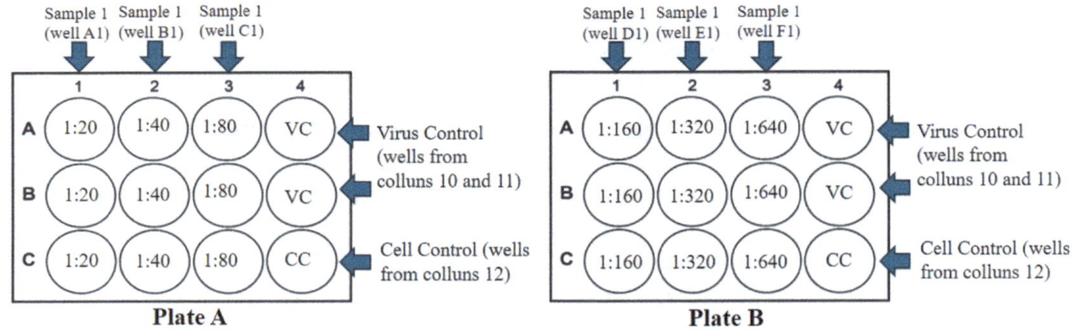

Fig. 4 Inoculating the contents of the assay plate into the VERO cell monolayer of the 12-well plate. VC: virus control; CC: cell control

dilutions must contain ~100 PFUs (indicating no neutralization); positive controls must show a decrease in PFU counting proportional to antibody titer in the control serum. At least one of the serum dilutions must neutralize 50% of viral inoculum. To calculate the neutralization percentage for each well, compare its PFU count with that of the virus control. Samples are considered positive when they show ≥60% neutralization at the 1:20 dilution and ≥40% neutralization at the 1:40 dilution [5]. For more information on analyzing results, see Costa et al. [5].

4 Notes

1. The use of screw-cap microtubes is required because it prevents liquid leakage during tissue maceration in benchtop homogenizers.
2. Borosilicate beads must be treated with nitric acid before use.
3. The virus stock used in the assay must be titrated, and ideally, the virus stock should be adjusted to achieve 10^4 PFU/mL to reduce variation in PFU counts due to viral dilution. Viral stock should be derived from a viral clone resulting in homogeneous lysis plaques. The viral stock must be stored at −70°C.
4. HEPES buffer should be prepared at a final concentration of 1 M, pH 7.5. The solution must be sterilized by filtration (0.22-μm filters), and aliquots are stored at −20°C.
5. Medium 199 should be prepared at a 2× concentration, as it will later be mixed with the 3.0% CMC solution to yield the final 1× concentration.
6. To prepare the 3% (w/v) CMC solution, dissolve 6.0 g of CMC powder in 200 mL of water and autoclave the mixture. Preheating the water before adding the CMC powder improves dissolution during autoclaving. After the autoclaving cycle is complete, allow the solution to cool and store it at 4°C until use.
7. To prepare the CMC/199 solution, combine 200 mL of 2× 199 medium with 200 mL of autoclaved 3% (w/v) CMC solution in a 1:1 (v/v) ratio. The final concentrations are 1.5% (w/v) CMC and 1× 199 medium. Supplement the mixture with 2% FBS, 200 U/mL penicillin, 40 μg/mL streptomycin, and 2 μg/mL amphotericin B, and store at 4°C.

8. To prepare 3.7% (v/v) formaldehyde, dilute commercial 37% (v/v) formaldehyde 1:10 in ultrapure water. Mix thoroughly and store at room temperature.
9. After 3 min of incubation with trypsin, gentle tapping of the flask can assist in detaching the cells. This step should be performed quickly, as prolonged exposure to trypsin is detrimental to cells.
10. The homogenate clarification step is critical for obtaining accurate and reliable PRNT results. Analyses of experimentally infected animal samples have shown basal neutralization in PRNT using homogenates [5], likely due to the unusual nature of the sample and the presence of tissue debris that interferes with PFU formation [6]. Removing as much debris as possible during clarification is therefore essential. Clarification is considered complete only when no pellet is visible and the supernatant is translucent. This typically occurs after the fifth centrifugation cycle, but some samples may require additional rounds.
11. Pre-wetting the filter with MEM 0% FBS is important to decrease sample loss when filtering the liver homogenate.
12. Filtering 1 mL typically yields approximately 800 μL of recovered sample. It is advisable to filter a larger volume if needed to run different experiments.
13. After thawing and complement inactivation, solid residues may appear; therefore, an additional clarification round is required.
14. It is not recommended to use a dilution lower than 1:20 due to the possibility of nonspecific neutralization.
15. After pipetting all samples and the virus solution into the assay plate, verify that each well contains 600 μL (sample + virus or control + virus).
16. Formaldehyde must be washed off carefully to prevent disruption of the monolayer by the water flow.

Acknowledgments

We do thank the Centers for Research in Emerging Infectious Diseases (CREID), CREID Pilot Research Program, Dr Nikos Vasilakis and Dr Kathryn Hanley from the Coordinating Research on Emerging Arboviral Threats Encompassing the Neotropics (CREATE-NEO). We also thank the Secretaria Municipal de Saúde de Belo Horizonte, Laboratório de Zoonoses, Centro de Controle de Zoonoses da Prefeitura de Belo Horizonte, Secretaria de Estado de Saúde de Minas Gerais. Finally, we thank our colleagues from Laboratório de Vírus/UFMG and the support of Pró-Reitorias de Graduação, de Pós-graduação, de Pesquisa/Universidade Federal de Minas Gerais/Brazil.

Funding The research reported in this publication was supported by the National Institute of Allergy and Infectious Diseases (NIAID) of the National Institutes of Health (NIH) under Award Number U01AI151378, through CREID Pilot Research Program 2021 (Pilot award number X.XXX.0217530 within CREATE-NEO grant U01 AI151807) of the Centers for Research in Emerging Infectious Diseases (grant 1U01AI15378). BPD and TAC research are supported by Conselho Nacional de Desenvolvimento Científico e Tecnológico do Ministério da Ciência e Tecnologia e Inovação (CNPq). The funders had no role in the design of the study; collection, analyses, or interpretation of data; writing of the manuscript; or in the decision to publish the results.

References

1. Roehrig JT, Hombach J, Barrett AD (2008) Guidelines for plaque-reduction neutralization testing of human antibodies to dengue viruses. Viral Immunol 21(2):123–132. https://doi.org/10.1089/vim.2008.0007
2. Ministério da Saúde, Secretaria de Vigilância em Saúde (2014) Guia de vigilância de epizootias em primatas não humanos e entomologia aplicada à vigilância da febre amarela. Disponível em. https://www.gov.br/saude/pt-br/assuntos/saude-de-a-a-z/f/febre-amarela/publicacoes/guia_vigilancia_epizootias_primatas_entomologia.pdf. Accessed 2 Nov 2025
3. Gamble HR, Patrascu IV (1996) Whole blood, serum, and tissue fluids in an enzyme immunoassay for swine trichinellosis. J Food Prot 59(11):1213–1217
4. Beck R, Beck A, Boskovic I, Mihaljevic Z, Marinculic A (2005) Evaluation of ELISA for detection of *Trichinella* antibodies in muscle juice samples of naturally infected pigs. Vet Parasitol 132(1–2):91–95
5. Costa TA, Arruda MS, Garcia-Oliveira GF, Reis EVdS, Guimarães ACDS, Moreira GD et al (2024) Detection of neutralizing antibodies against arboviruses from liver homogenates. PLoS Negl Trop Dis 18(12):e0012740. https://doi.org/10.1371/journal.pntd.0012740
6. Holland JJ, McLaren LC (1959) The mammalian cell–virus relationship. J Exp Med 109(5):487–504. https://doi.org/10.1084/jem.113.2.419

Chapter 15

Screening of Antiviral Agents Against CHIKV Using Reporter Virus

Eliana F. Castro and Diego Alvarez

Abstract

Chikungunya virus (CHIKV) causes a disease characterized by chronic musculoskeletal inflammation for which specific antivirals are not yet available. Currently, a supportive therapy to alleviate fever and pain is used, but it does not limit viral replication or the persistence of chronic arthritis symptoms. Thus, the identification and development of new active molecules against CHIKV is urgently needed. Here, we present a cell-based methodology that enables the implementation of a rapid and cost-effective strategy for high- and medium-throughput screening (HTS) of compounds, including repurposed drugs or novel molecules. This methodology allows for the identification of novel antiviral hits with a good activity and selectivity profile against CHIKV.

Keywords Antiviral activity, Antiviral screening, Reporter virus, Chikungunya, Antiviral hits

1 Introduction

Strategies to reduce CHIKV transmission and infection occur at the individual and community levels. Control of the mosquito vector is a strategy to limit the spread of CHIKV and other mosquito-borne infections. Currently, two chikungunya vaccines are licensed in different countries: a live-attenuated vaccine (IXCHIQ) and a virus-like particle vaccine (VIMKUNYA) [1]. However, there is no specific antiviral treatment for this potentially pandemic agent, and the current therapy is aimed at alleviating fever and pain. This supportive therapy does not limit viral replication or the persistence of chronic arthritis symptoms; thus, the identification of new molecules active against CHIKV replication is urgently needed.

Phenotypic screening is a valuable approach in the identification of antivirals and allows the selection of active compounds with varied mechanisms of antiviral action, including actives acting on viral factors and host factors. It involves *in vitro* cell-based testing of structurally varied compounds or a specific series of molecules belonging to the

Claudia Filomatori and Eliana F. Castro (eds.), *Chikungunya Virus*, Methods in Molecular Biology, vol. 3067,
https://doi.org/10.1007/978-1-0716-5496-5_15,

same chemical family to select an active hit, which can be developed into a lead compound with enhanced activity and/or selectivity. Here we describe a method for identifying potential antiviral compounds against CHIKV using an infectious CHIKV reporter virus carrying a ZsGreen expression cassette under the virus subgenomic promoter [2–4]. This reporter virus reproduces the complete viral cycle, allowing the detection of inhibitors that interfere with processes mediated by the structural and nonstructural proteins of the virus, including virus entry to the host cell, uncoating, genome replication, morphogenesis, and release of progeny viral particles. Green fluorescence can be used as a proxy for virus replication in infected cells, and antiviral activity is assessed based on the detection and automated numbering of fluorescent viral foci (Fig. 1). Importantly, our experimental setup allows monitoring compound cytotoxic effects on the same screening plate through crystal violet staining after viral fluorescent focus quantification. Notably, CHIKV-ZsGreen infection in Vero cells is not cytopathic at the time of analysis, and crystal violet staining allows estimating cytotoxicity in the cell culture. The dye binds to proteins and DNA, and therefore it stains adhered cells. Then, the recording of absorbance of the stained plate serves as a measure of cell viability.

Overall, we describe a phenotypic assay to test antiviral activity in a medium- to high-throughput format that allows rapid and low-cost detection of active compounds, as well as obtaining preliminary cytotoxicity data in the same test. Finally, this methodology allows for the identification of novel antiviral hits with a good selectivity

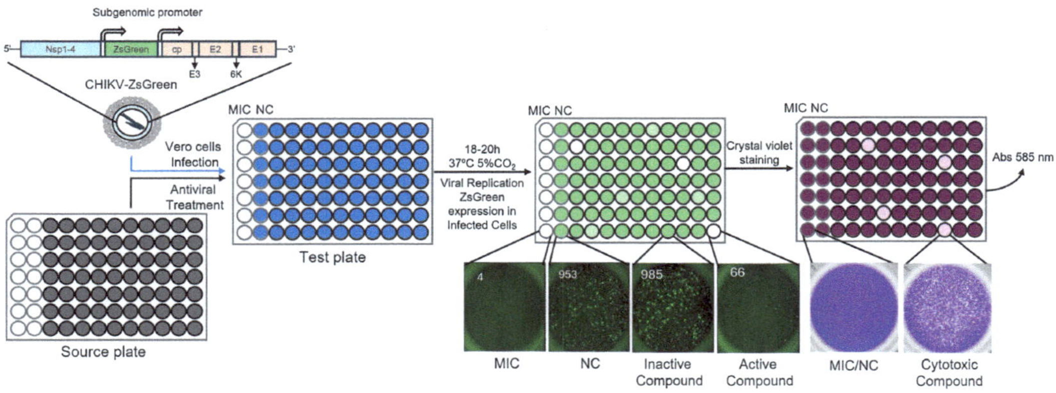

Fig. 1 Schematic representation of antiviral screening using the CHIKV-ZsGreen reporter virus. Vero cells in 96-well plates are infected (blue circles) and simultaneously treated with antiviral compounds, which are pre-spotted in a source plate (white circles: solvent, gray circles: compounds). Untreated infected cells (negative control, NC) and mock-infected cells (MIC) are included. When Vero cells are infected with CHIKV-ZsGreen, ZsGreen is expressed in the cell cytoplasm under the control of the virus subgenomic promoter. After 18–20 h of incubation at 37°C and 5% CO_2, cells are fixed and fluorescent foci are quantified. A representative image of automated numbering of fluorescent infectious foci using the ELISpot equipment and the ImmunoSpot 7.0.30.2 software (ImmunoSpot C.T.L, Germany) is shown. Finally, cell monolayers are stained with crystal violet, and absorbance at 585 nm is determined to estimate compounds' cytotoxicity

profile against CHIKV and could be adapted to any recombinant CHIKV that expresses a reporter fluorescent protein during its replication cycle [4].

2 Materials

Prepare all solutions using ultrapure water (prepared by purifying deionized water to attain a sensitivity of 18 MΩ-cm at 25°C) and analytical grade reagents. Sterilize and store all reagents at 4°C (unless indicated otherwise). Diligently follow all waste disposal regulations when disposing of waste materials.

2.1 Cell Culture

For cell culture, all materials and reagents must be sterilized.

1. Vero cells (ATCC CCL-81).
2. Culture medium: Dulbecco's Modified Eagle Medium, 10% fetal bovine serum (FBS), 100 IU/mL penicillin, and 100 mg/mL streptomycin.
3. Infection medium (IM): Dulbecco's Modified Eagle Medium, 2% FBS, 100 IU/mL penicillin, and 100 mg/mL streptomycin.
4. Infection medium–dimethyl sulfoxide (DMSO): IM with 2% DMSO.
5. Dulbecco's phosphate-buffered saline (DPBS, commercial): 2.7 mM (200 mg/L) KCl, 1.5 mM (200 mg/L) KH_2PO_4, 138.0 mM (8000 mg/L) NaCl, 8.1 mM (1150 mg/L) Na_2HPO_4 anhydrous.
6. 0.1% trypsin-EDTA.

2.2 Virus Stock Production

Recombinant CHIKV is derived from infectious cDNA clones, for example, CHIKV-ZsGreen [2]. The obtention of infectious viral particles from cDNA clones is described in Chap. 7. Virus stocks should be prepared in large quantities to avoid (any) batch effects.

2.3 Fixation and Staining

1. Phosphate-buffered saline (PBS): 2.7 mM KCl, 1.5 mM KH_2PO_4, 145 mM NaCl, 8.1 mM Na_2HPO_4, deionized H_2O, pH 7.4.
2. 4% paraformaldehyde in PBS (PFA 4%) (*see* **Note 1**).
3. 0.1% (w/v) crystal violet in 20% ethanol (*see* **Note 2**).

3 Methods

Carry out all procedures at room temperature (r.t.) following the good practice and biosafety protocols of your institution.

This method is developed in 3 days, from the preparation of the cell culture to obtaining the results. Below, the method is described for one 96-well plate, which allows the testing of 80 molecules in a single concentration. In addition, an optional screening of a small library of compounds is described.

3.1 Day 1: Cell Culture Preparation

Confluent Vero Cell monolayers grown in 60 mm plates at 37°C with 5% CO_2 in culture medium are used (*see* **Note 3**).

1. For obtaining the cell suspension, remove the cell-culture supernatant, rinse with 1 mL of DPBS at r.t. to remove excess culture medium, then wash with 1 mL 0.1% trypsin-EDTA. Add 0.5 mL 0.1% trypsin-EDTA and incubate for 1–3 min at 37°C until cells are detached and rounded as observed under a microscope.
2. Add 5 mL of culture medium to inactivate the trypsin-EDTA and gently pipette the cells up and down until a homogeneous suspension is obtained.
3. Count the cells in a Neubauer chamber (*see* **Note 4**): It is expected that a confluent Vero cell monolayer in a 60 mm dish contains approximately 5×10^6 cells.
4. Prepare Vero cell culture in a 96-well plate: Prepare a cell suspension of 1.5×10^5 cells/mL in a sterile 15 mL tube by adding 1.5×10^6 cells in a final volume of 10 mL of culture medium. Gently pipette the cells up and down until a homogeneous suspension is obtained.
5. Add 100 μL of cell suspension per well in a 96-well cell culture plate (15,000 cell/well).
6. Incubate for 24 h at 37°C with 5% CO_2.

3.2 Day 2: Compound Preparation

Usually, potent antiviral agents should be active at low μM or nM concentrations.

If high-throughput screening (HTS) assays are to be done, compounds may be provided pre-spotted in DMSO solution in a source 96-well plate, and 10 μM final concentration is tested as a unique concentration during primary screening (Option 1). However, a small internal library of compounds can also be tested. In this case, compounds may be provided lyophilized in a vial (Option 2).

Option 1: HTS from a pre-spotted plate, one concentration.

The primary screening process is pivotal in eliminating nonspecific and cytotoxic compounds and aiming to retain potent viral inhibitors from the compound library. During this stage, the entire library is screened at a fixed concentration, typically at 10 μM using a single cell line. To test compounds at a final concentration of 10 μM, 0.1 μL of 10 mM solution in DMSO of each compound should be provided in a source 96-well plate. An example of a source plate layout is depicted in Fig. 2. If the plate was stored at −20°C, compound's solutions should be warmed at 37°C and then briefly centrifuged at 500–1000 × *g*.

Option 2: Screening of a small library of compounds, three concentrations in triplicate.

1. Prepare a 5 mM solution of compound in an adequate solvent (e.g., DMSO) (*see* **Note 5**).
2. For each compound to be tested, prepare one tube with 490 μL of infection medium and two tubes with 320 μL of infection medium–DMSO.
3. Add 10 μL of 5 mM solution to 490 μL of infection medium, homogenize with the vortex.
4. Perform two serial 1:5 dilutions by transferring 80 μL of the first dilution to the next tube with 320 μL infection medium–DMSO. Homogenize with the vortex each time. Finally, in the tubes, each compound will be in a 2× concentration with 2% DMSO.

3.3 Day 2: Infection and Treatment

Negative (infected cells treated with vehicle, 0%–1% DMSO) and mock-infected controls must be included in each plate. In addition, it is recommended to include positive controls (antiviral reference, e.g., 10 μM chloroquine).

1. Observe Vero cell monolayers prepared the day before under the microscope. Monolayers should be in a confluence of 90%–100%.
2. Prepare a solution of 5 mL of CHIKV-ZsGreen of 1.6–2.0 × 10^4 UFF/mL in infection medium. Do this procedure in an ice bath.

	1	2	3	4	5	6	7	8	9	10	11	12
A	S	S	C1	C9	C17	C25	C33	C41	C49	C57	C65	C73
B	S	S	C2	C10	C18	C26	C34	C42	C50	C58	C66	C74
C	S	S	C3	C11	C19	C27	C35	C43	C51	C59	C67	C75
D	S	S	C4	C12	C20	C28	C36	C44	C52	C60	C68	C76
E	S	S	C5	C13	C21	C29	C37	C45	C53	C61	C69	C77
F	S	S	C6	C14	C22	C30	C38	C46	C54	C62	C70	C78
G	S	S	C7	C15	C23	C31	C39	C47	C55	C63	C71	C79
H	S	S	C8	C16	C24	C32	C40	C48	C56	C64	C72	C80

Fig. 2 Example of a source plate layout. Source plates contain test compound solutions in the corresponding solvent (S). If compounds are to be tested at 10 μ M (final concentration), they can be pre-spotted in a 10 mM solution (0.1 μL) (Columns 3–12; Rows A–H). Columns 1 and 2 are pre-spotted with 0.1 μL of S and are used for mock-infected cell controls (MIC) and untreated infected cell controls (negative controls, NC)

3. Discard the culture medium from the 96-well plate containing a confluent Vero cell monolayer prepared on Day 1.
4. Wash with 100 μL of DPBS/well and discard.
5. Add 50 μL of infection medium to mock-infected control wells.
6. Add 50 μL of virus solution (prepared in Step 1) to the rest of the wells.
7. Add the compounds or infection medium according to the assay to be performed (Option 1 or 2):

   Option 1: HTS from a pre-spotted plate, single concentration. In this assay, infected cells will be treated with a final concentration of 10 μM of each compound containing 0.1% DMSO. Add 50 μL of infection medium per well in the source 96-well plate, mix gently by pipetting, and transfer to the assay plate. An example of a test plate layout is shown in Fig. 3.

   Option 2: Screening of a small library of compounds, three concentrations in triplicate. In this assay, infected cells are treated with final concentrations of 50, 10, and 2 μM of each compound. Each well in the assay will contain 1% DMSO. Add 50 μL of infection medium containing DMSO to the mock-infected and negative control wells, and 50 μL of 2× compound solutions (prepared in advance) in triplicate to the corresponding wells. An example of a test plate layout is shown in Fig. 4.
8. Incubate for 18–20 h at 37°C with 5% CO_2.

3.4 Day 3: Fixation

1. Observe the cell monolayers under the microscope. Record all observations regarding cytotoxicity. Viral cytopathic effect is not expected within 18–20 h postinfection. By comparing negative controls (infected monolayers treated with 1% DMSO vehicle) with compound-treated infected monolayers after staining with crystal violet, markedly cytotoxic compounds can

	1	2	3	4	5	6	7	8	9	10	11	12
A	MIC	NC	C1	C9	C17	C25	C33	C41	C49	C57	C65	C73
B	MIC	NC	C2	C10	C18	C26	C34	C42	C50	C58	C66	C74
C	MIC	NC	C3	C11	C19	C27	C35	C43	C51	C59	C67	C75
D	MIC	NC	C4	C12	C20	C28	C36	C44	C52	C60	C68	C76
E	MIC	NC	C5	C13	C21	C29	C37	C45	C53	C61	C69	C77
F	MIC	NC	C6	C14	C22	C30	C38	C46	C54	C62	C70	C78
G	MIC	NC	C7	C15	C23	C31	C39	C47	C55	C63	C71	C79
H	MIC	NC	C8	C16	C24	C32	C40	C48	C56	C64	C72	C80

Fig. 3 Example of a test plate layout for primary screening from a pre-spotted plate, one concentration. Column 1 is used for mock-infected cell controls (MIC), and column 2 for untreated infected cell controls (negative controls, NC). Alternatively, wells 1D–H can be used for positive control, where a known active compound (i.e., 10 μM chloroquine) is used to treat infected cells

	1	2	3	4	5	6	7	8	9	10	11	12
A	$C1_{50}$	$C1_{10}$	$C1_{2}$	$C3_{50}$	$C3_{10}$	$C3_{2}$	$C5_{50}$	$C5_{10}$	$C5_{2}$	$C7_{50}$	$C7_{10}$	$C7_{2}$
B	$C1_{50}$	$C1_{10}$	$C1_{2}$	$C3_{50}$	$C3_{10}$	$C3_{2}$	$C5_{50}$	$C5_{10}$	$C5_{2}$	$C7_{50}$	$C7_{10}$	$C7_{2}$
C	$C1_{50}$	$C1_{10}$	$C1_{2}$	$C3_{50}$	$C3_{10}$	$C3_{2}$	$C5_{50}$	$C5_{10}$	$C5_{2}$	$C7_{50}$	$C7_{10}$	$C7_{2}$
D	$C2_{50}$	$C2_{10}$	$C2_{2}$	$C4_{50}$	$C4_{10}$	$C4_{2}$	$C6_{50}$	$C6_{10}$	$C6_{2}$	$C8_{50}$	$C8_{10}$	$C8_{2}$
E	$C2_{50}$	$C2_{10}$	$C2_{2}$	$C4_{50}$	$C4_{10}$	$C4_{2}$	$C6_{50}$	$C6_{10}$	$C6_{2}$	$C8_{50}$	$C8_{10}$	$C8_{2}$
F	$C2_{50}$	$C2_{10}$	$C2_{2}$	$C4_{50}$	$C4_{10}$	$C4_{2}$	$C6_{50}$	$C6_{10}$	$C6_{2}$	$C8_{50}$	$C8_{10}$	$C8_{2}$
G	NC	NC	NC	NC	NC	NC	NC	NC	NC	NC	NC	NC
H	MIC	MIC	MIC	MIC	MIC	MIC	MIC	MIC	MIC	MIC	MIC	MIC

Fig. 4 Example of a test plate layout for primary screening of a small library of compounds provided lyophilized in a vial (three concentrations in triplicate). Row H is used for mock-infected cell controls (MIC), and row G for untreated infected cell controls (negative controls, NC). Alternatively, wells H10–12 can be used for positive control, where a known active compound (i.e., 10 μM chloroquine) is used to treat infected cells. The subscript numbers indicate the concentration of the compound (C1–C8) in μM

be detected and excluded from the antiviral activity analysis (Fig. 1).

2. Discard the culture medium. Wash each well with 100 μL PBS and discard. Add 50 μL of 4% PFA per well.
3. Incubate for 1 h at 4°C
4. Wash each well with 100 μL PBS and discard. Add 30 μL of PBS per well, avoiding bubble formation.

3.5 Day 3: Quantification of Viral Fluorescent Foci

FluoroSpot instruments allow quantitation of viral foci through imaging the wells and numbering fluorescent spots with spot-recognition algorithms [5]. Depending on available equipment, follow the manufacturer's instructions for use.

1. Scan: Set the scanning mode for a 96-well plate with the excitation and emission wavelengths appropriate for detecting the reporter gene used; for ZsGreen, 480 and 520 nm, respectively. Adjust focus to clearly distinguish viral fluorescent foci. Preset zoom and exposure can be used.
2. Quantification: Set the foci quantification to ensure detection and quantification in a way that the maximum number of fluorescent foci is counted in negative controls without unspecific counting in mock-infected controls (*see* **Note 6**).

3.6 Day 3: Staining with Crystal Violet for Cytotoxicity Estimation

1. Discard PBS from previously scanned plates.
2. Add 70 μL of 0.1% crystal violet in 20% ethanol per well.
3. Incubate for 15–30 min at r.t. until confluent monolayers are tinged with a clear violet hue. Depending on the quality of the reagent, the required incubation time may be longer.
4. Discard and dispose of the crystal violet solution according to your institution's regulations.

5. Let the plates dry with the lid open.
6. Measure absorbance at 585 nm and scan the stained cell monolayers with a spectrophotometer (*see* **Note 7**).

3.7 Data Analysis: Calculation of the Reduction in the Number of Viral Foci (%)

Antiviral activity is detected as a decrease in the number of infectious foci compared to the negative controls. Although the threshold for defining a hit may vary depending on the investigator's criteria, compounds that reduce the number of viral foci by more than 80% (without cytotoxicity effects) in a primary screening can be considered hits. The antiviral activity of these primary hits should then be confirmed in subsequent dose–response assays.

1. Calculate the average number of foci counted in mock-infected controls.
2. Subtract this average value calculated in the mock-infected controls from all infected wells (negative controls [NC], and treated infected wells [T]).
3. Calculate the average number of foci in NC.
4. Calculate the viral foci reduction (%) for each treatment replicate (T) according to the following formula: $100 - (\text{n° FFU}_{T} \times 100/\text{n° FFU}_{NC})$.

3.8 Data Analysis: Estimation of Compound Cytotoxicity by Crystal Violet Staining

Since at 18–20 h postinfection the viral cytopathic effect is not yet evident, this rapid screening method allows for examining antiviral activity and estimating the impact of compounds on cell survival and growth inhibition in the same plate. To this end, crystal violet dye, which binds to proteins and DNA, is used. The OD585 obtained is proportional to the number of cells attached in the well; thus, cytotoxic compounds are expected to reduce this OD.

1. Set the average OD585 of mock-infected and negative controls to 100% (C).
2. Determine the percentage of cells that are viable (attached) after each treatment (T) with the following formula: Viable cells $(\%) = \text{OD585}_{T} \times 100/\text{OD585}_{C}$.

4 Notes

1. For 1 L of 4% PFA, add 800 mL of 1× PBS to a glass beaker on a stir plate in a ventilated hood. Heat while stirring to approximately 60°C. Take care that the solution does not boil. Add 40 g of paraformaldehyde powder to the heated PBS solution. The powder will not immediately dissolve into solution. Slowly raise the pH by adding 1 N NaOH dropwise from a pipette until

the solution clears. Once the paraformaldehyde is dissolved, the solution should be cooled and filtered. Adjust the volume of the solution to 1 L with 1× PBS. Recheck the pH and adjust it to approximately 6.9 with small amounts of diluted HCl. The solution can be aliquoted and frozen or stored at 2–8°C for up to 1 month.

2. Prepare a solution of 0.5 L of 20% ethanol in deionized H_2O in a glass beaker protected from light (e.g., by covering it with aluminum foil) on a stir plate. Add 0.05 g of crystal violet powder and stir until dissolved. Filter through filter paper to remove possible crystals and store it in containers protected from light.

3. Vero cells are maintained in culture medium (see Sect. 2) at 37°C with 5% CO_2 and passaged at a 1:6 ratio twice a week. Cells at 80%–100% confluency are required for this procedure. For reproducibility, the cells should be maintained for up to 20 passages (P2–P22). New cells (P0) stored in liquid nitrogen should then be thawed.

4. Take 30 μL of the cell suspension and load it onto the hemocytometer. If using a glass hemocytometer, gently fill the chamber under the coverslip, allowing the cell suspension to be drawn out by capillary action. Using a microscope, focus on the grid lines of the hemocytometer with a 10× objective. With a hand tally counter, count the cells in one set of 16 squares. When counting, employ a system in which cells are only counted when they are set within a square or on the right-hand or bottom boundary line. Move the hemocytometer to the next set of 16 corner squares and carry on counting until all four sets of 16 corners are counted. Calculate the average cell count from the four sets of 16 corner squares and multiply by 10,000. The resulting value represents the number of viable cells/mL in the original cell suspension [6].

5. Organic compounds are usually soluble in DMSO at r.t. However, there are some cases in which a turbid solution is obtained. Heating the compound solution in DMSO in a water bath at 37–45°C could be useful to solubilize the compound.

6. Example of counting setting for ELISpot equipment with the InmunoSpot 7.0.30.2 software (ImmunoSpot C.T.L, Germany): Enter the Fluorospot Scan window and scan. Enter the Basic Count window and load the previously UV-scanned plate. Set the counting parameters in: Fine-tune Parameters, Advanced. Set parameters of (i) Sensitivity: 180; (ii) Background Balance: 50; (iii) Diffuse Processing: Normal; (iv) Minimum Spot Size: 0.001 mm^2; (v) Maximum Spot Size: 0.010 mm^2; (vi) Spot Separation: 1; (vii) Counted Area: 100%; and (viii) Fiber removal: active. Before using the counting data, perform

QC visualization to rapidly detect the erroneous inclusion of artifacts in the quantification (e.g., fibers). Correct if necessary.

7. Crystal violet staining can be measured by scanning the stained well surface or taking multiple reads per well in a dry plate. For example, in a Spark Tekan spectrophotometer, the measurement settings can be as follows: Measurement wavelength: 585 nm; Flashes: 1; Multiple reads per well (user-defined: Pattern Circle, Size 10×10, Border 450 μM). Alternatively, crystal violet can be eluted with 200 μL of methanol per well, and after incubation for 20 min at r.t. on a bench rocker, the optical density can be measured. Ideally, wells without cells should be included on each plate to define the background of the staining method.

References

1. Chikungunya Vaccine Information for Healthcare Providers|Chikungunya Virus|CDC, (2026). https://www.cdc.gov/chikungunya/hcp/vaccines/index.html
2. Noval MG, Rodriguez-Rodriguez BA, Rangel MV et al. (2019) Evolution-driven attenuation of alphaviruses highlights key glycoprotein determinants regulating viral infectivity and dissemination. Cell Rep 28:460
3. Battini L, Thannickal SA, Cibello MT et al (2024) Evolution of antiviral resistance captures a transient interdomain functional interaction between chikungunya virus envelope glycoproteins. bioRxiv Prepr Serv Biol
4. Castro EF, Álvarez DE (2024) New highly selective antivirals for chikungunya virus identified from the screening of a drug-like compound library. Curr Microbiol 81(10):343
5. FluoroSpot Pre-Screening with Thermo Scientific Microplate Readers|Thermo Fisher Scientific—AR, https://www.thermofisher.com/ar/es/home/life-science/cell-analysis/cell-analysis-learning-center/cell-analysis-resource-library/cell-analysis-application-notes/fluorospot-pre-screening-thermo-scientific-plate-readers.html
6. Counting cells using a hemocytometer|Abcam, https://www.abcam.com/en-us/technical-resources/protocols/counting-cells-using-a-haemocytometer?srsltid=AfmBOoox-gD1j6EiZa84S4z0ZcHe1KiYfEo6fJoiCZRnh6uQ1wUHRsZh

Chapter 16

In Silico Drug Design of Chikungunya Antivirals: Using Consensus Methods for Virtual Screening

Leandro Battini

Abstract

Structure-based virtual screening is an attractive strategy in drug design. It enables the selection of potential binders to target biomolecules with known molecular structures, providing a rational framework for studying their mechanisms of action and guiding further optimization. However, accurately predicting the free energy of binding remains an extremely challenging task, and the available tools only provide approximate estimates. In this context, one strategy to improve the predictive power of virtual screening and increase the probability of identifying true binders is to employ multiple docking algorithms and combine their results using a consensus method. In this chapter, we will describe a step-by-step methodology to perform this task. We will conduct a virtual screening using AutoDock Vina, AutoDock4, and LeDock, and then combine their results employing the Exponential Consensus Ranking algorithm. Finally, we will discuss possible extensions of this strategy, such as incorporating Molecular Dynamics simulations as an additional filter in the virtual screening or integrating the novel Artificial Intelligence–based methodologies into the consensus framework.

Keywords Chikungunya, Envelope glycoproteins, Virtual screening, AutoDock Vina, LeDock, AutoDock4, Consensus scoring

1 Introduction

Over the past century, different drug design strategies have been developed, driven by scientific and technical advances in different areas of human knowledge. Progress in experimental methods for protein structure determination, such as X-ray crystallography, NMR, and most recently cryogenic electron microscopy (cryoEM), together with the pivotal decision of sharing the experimental structures in public databases and the continuous increase in computing power, has enabled the in silico, structure-based design of potential drug candidates. Broadly, these strategies address two interrelated problems with the main objective of prioritizing a small number of compounds for experimental evaluation. First, they try to predict the structure of the protein–drug complex, in which the drug is typically a small organic

Claudia Filomatori and Eliana F. Castro (eds.), *Chikungunya Virus*, Methods in Molecular Biology, vol. 3067,
https://doi.org/10.1007/978-1-0716-5496-5_16,

molecule. Second, they attempt to estimate the binding affinity of this complex. Docking calculations, a specific approach for in silico drug design, solve both problems in parallel, generating multiple conformations of the small organic compound for a given protein target, estimating the affinity of each conformation, and selecting the one with the highest affinity. When applied systematically to a compound database, this procedure allows for ranking the molecules according to their predicted binding affinity and identifying the most promising candidates.

Accurate prediction of binding free energies is extremely challenging and computationally prohibitive for large compound libraries. For this reason, docking programs for virtual screening were designed to provide a fast and reasonable estimate, enabling compound ranking and the selection of the most promising candidates, at the expense of precision. In this sense, docking algorithms can be conceptualized as weak predictors of binding affinity. A well-established theoretical framework suggests that combining multiple weak predictors can yield a model with stronger predictive power [1], as exemplified by Random Forests and other ensemble methods in machine learning. Applied to virtual screening, this implies that combining the results from multiple docking programs can increase the probability of selecting true binders in the top-ranked compounds. This is the principle behind consensus strategies in drug design. Common approaches include Rank by Vote, Rank by Rank, Rank by Score, and the Exponential Consensus Ranking (ECR) method described in this chapter [2, 3]. An important characteristic of ECR is that it reduces the impact of poor predictions, which is particularly useful given that docking algorithm performance varies across targets, mitigating the risk of relying on a poorly performing program for a specific receptor.

In a virtual screening workflow for drug design, the aim is to identify small organic molecules that can inhibit or modulate the activity of an essential target for the pathology of concern. In this context, the first step is to identify an appropriate target, which not only needs to be essential but also must have a high-resolution experimental structure or a high-confidence prediction available. Although there are exceptions, given their diversity and functionality, in most drug design efforts, the targeted macromolecule is a protein. In the case of chikungunya, with a relatively small and simple genome, the main protein targets for drug design are limited and well established. Reported virtual screening campaigns have targeted nsP1 [4], required for RNA capping; nsP2 [5, 6], with helicase and protease activities essential for polyprotein processing; the nsP3 macrodomain [7], which exhibits ADP-ribose-1″-phosphate phosphatase activity potentially linked to the response to host antiviral mechanisms; as well as the capsid [8] and envelope glycoproteins [9], which mediate viral particle assembly and entry into the host cell. Experimental structures are available for all these proteins (Fig. 1).

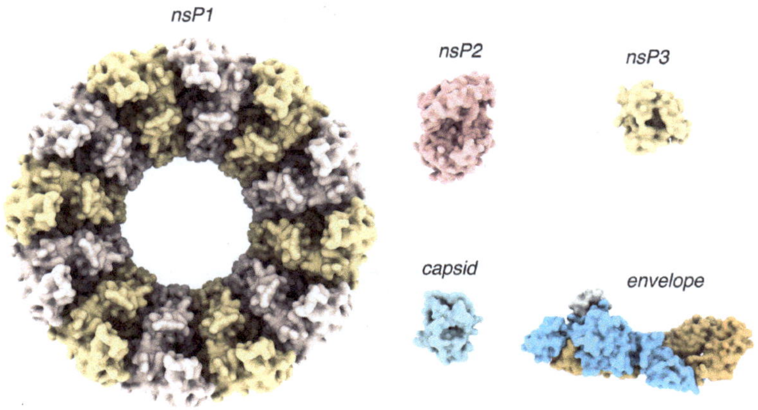

Fig. 1 Structure-based virtual screening targets in chikungunya virus. Representation of the protein structure of the different targets employed in virtual screening campaigns against chikungunya. The protein name is above each structure with the PDB ID in parentheses

Once the methodology and target have been selected, the final decision before starting a virtual screening is the selection of the compound library. There are numerous libraries available with different characteristics, such as size, degree of curation, commercial availability, chemical diversity, and pharmacological properties of the compounds. The selection of a specific library depends on multiple factors, including the available computational resources, the feasibility of synthesizing the selected compounds, particular features of the chosen target, and the ultimate objective of the virtual screening (e.g., whether the goal is to prioritize readily available compounds).

In this chapter, we will describe a detailed procedure on how to perform a virtual screening employing the consensus scoring framework, specifically the ECR, although it can be extended to other consensus metrics (general overview in Fig. 2). As an example, we will employ the crystal structure of the envelope proteins of chikungunya virus as a target and we will perform the docking calculations with three commonly used software, AutoDock Vina [10], AutoDock4 [11], and LeDock [12]. In Sect. 4, we will suggest a possible strategy to parallelize docking calculations to accelerate the virtual screening, and we will mention possible extensions to the framework described in Sect. 3. We will discuss how to take into account not only the score, but also the 3D similarity between the solutions obtained by the different docking programs using the RMSD (*Root Mean Square Deviation*) metric; how can Molecular Dynamics calculations be included as an additional filter to select the most promising compounds; and how the recently developed AI methodologies can be integrated to this framework to exploit the possible synergy between classical and AI-based strategies.

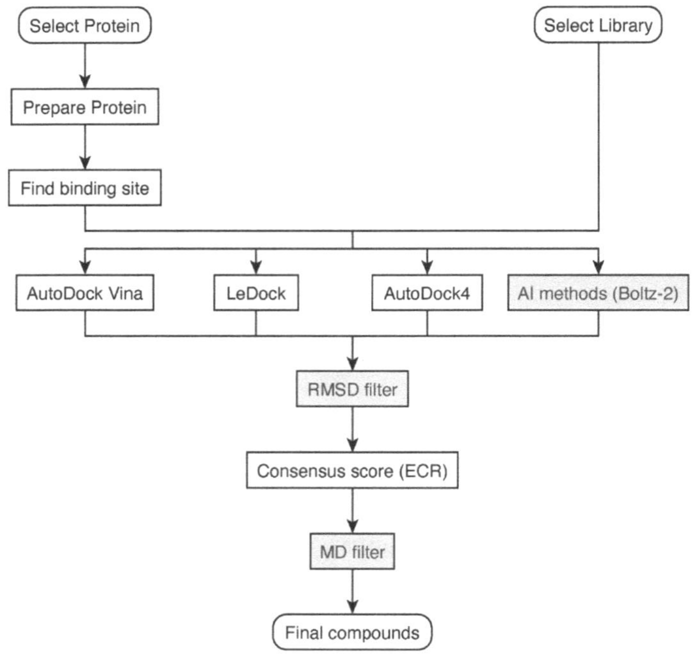

Fig. 2 Proposed virtual screening workflow under the consensus scoring framework. The possible variants suggested in Sect. 4 are represented in gray

2 Materials

To perform a virtual screening campaign, it is necessary to identify an adequate target, relevant to the biological problem being addressed, have its experimental structure or a high-confidence prediction of it, and select a suitable compound library for the objective of the drug design campaign. Additionally, it is mandatory to have sufficient computing power to perform the calculations and the necessary software to prepare the compound library and run the docking simulations. In this chapter, we will use as an example the crystal structure of chikungunya envelope proteins (PDB 3N42), the ChemBridge compound database (ChemBridge CORE Library Stock 2017_17), and the AutoDock Vina [10], AutoDock4 [11], and LeDock [12] docking software. Additionally, we will employ UCSF Chimera, AutoDockTools, and OpenBabel [13] to prepare the protein and the compound library for docking calculations, and FPocket [14], SiteHound [15], and P2Rank [16] to search for suitable pockets for drug design in the protein of interest (see Sect. 3.3). The installation of these software is system dependent, but relevant documentation can be found at the following sites:

- AutoDock Vina: https://github.com/ccsb-scripps/AutoDock-Vina; https://vina.scripps.edu/

- AutoDock4: https://autodock.scripps.edu/download-autodock4/
- LeDock: https://www.lephar.com/software
- UCSF Chimera: https://www.cgl.ucsf.edu/chimera/
- ChimeraX: https://www.rbvi.ucsf.edu/chimerax/
- AutoDockTools: https://autodocksuite.scripps.edu/adt/
- OpenBabel: https://github.com/openbabel/openbabel; https://openbabel.org/
- FPocket: https://github.com/Discngine/fpocket
- SiteHound: http://scbx.mssm.edu/sitehound/
- P2Rank: https://github.com/rdk/p2rank

3 Methods

3.1 Compound Library Preparation

Compound libraries for drug design are typically distributed as large SDF files, with each entry corresponding to a different molecule in the library. These files specify the compound name and molecular topology (atoms and connectivity) and may also include precalculated properties such as molecular formula, molecular weight, and flexibility. An initial step in a virtual screening campaign is to prepare the compound library to generate suitable inputs for docking programs. Although the exact procedure depends on the docking software used, it generally involves splitting each SDF entry into individual files, generating a reasonable 3D conformer, protonating the molecule at the desired pH, and assigning partial charges to every atom. Commonly used software for this purpose includes KNIME, an open-source data analytics platform, RDKit, or, as used in this case, OpenBabel. Outlined below is a workflow for preparing an initial **library.sdf** for docking using OpenBabel in the command line.

1. Using the flag -r, remove all but the largest contiguous fragment (strip salts) for each entry.

```
obabel -isdf library.sdf -osdf -O library_r.sdf -r
```

2. Split the SDF file into individual MOL files in the directory **mol_dir/**. Each MOL file will be assigned a consecutive number, i.e., compound1.mol, compound2.mol, etc. The flag -e is to continue the conversion after errors.

```
obabel -isdf library_r.sdf -omol -O mol_dir/
compound.mol -e -m
```

3. For each compound in **mol_dir/**, generate a MOL2 file in **mol2_dir/**, protonate the molecule at pH 7.4, assign Gasteiger partial charges to every atom, and generate a 3D conformer.

```
for file in mol_dir/*.mol
do
lig=$(basename $file.mol)
obabel -imol mol_dir/$lig.mol -omol2 -O mol2_
dir/$lig.mol2 \
-p 7.4 --partialcharge gasteiger --gen3D
done
```

These MOL2 files can be used as an input for LeDock (see Sect. 3.5). To generate the inputs for AutoDock Vina and AutoDock4, an additional step is necessary to convert the MOL2 files to PDBQT.

```
for file in mol2_dir/*.mol2
do
lig=$(basename $file.mol2)
obabel -imol2 mol2_dir/$lig.mol2 -opdbqt -O pdbqt_
dir/$lig.pdbqt
done
```

3.2 Protein Preparation

The protein receptor file must also be properly prepared for docking calculations. Experimental structures may contain missing residues or side chains that need to be modeled. In addition, solvent molecules and ions should be removed unless they play a role in the binding site, and nonstandard residues should be converted to the corresponding standard amino acids (e.g., selenomethionine to methionine). Both experimental structures and predictions generated by AI methods such as AlphaFold [17] must be protonated at the desired pH, and, for AutoDock4, partial charges need to be assigned to all protein atoms.

A convenient workflow to perform these procedures is provided by the Dock Prep tool, available in both UCSF Chimera (**Tools → Surface/Binding Analysis → Dock Prep**) and ChimeraX (**Tools → Structure Editing → Dock Prep**). For AutoDock Vina and AutoDock4, the prepared structure can be saved as a PDB file **receptor.pdb** and then converted to the required PDBQT file using AutoDockTools (ADT). In the case of LeDock, the structure needs to be passed through LePro, which outputs the **pro.pdb** file with the correct format to be used by LeDock.

```
lepro_linux_x86 receptor.pdb
```

This procedure is largely automated, so it is recommended to visually inspect the output to ensure correctness.

3.3 Pocket Identification

Once the target has been selected and prepared for in silico studies, the final step before docking is the selection of a suitable binding pocket in the protein surface. Selected pockets or cavities must have the adequate geometric and physicochemical characteristics that favor the binding of small, organic, drug-like molecules. This step is necessary to reduce the complexity of classical docking calculations, restricting the conformational search of the compound to a specific region in the protein surface. Several methods have been developed for this purpose, and, as previously discussed for docking simulations, a robust strategy is to search for druggable pockets employing different software based on different theoretical and methodological backgrounds and prioritize the pockets identified by all or most of the programs. In this chapter, we will employ three different software based on different conceptual frameworks: FPocket [14], a geometric algorithm based on Voronoi tessellation; SiteHound [15], an energy-based method; and P2Rank [16], a machine learning approach. The three software are easy-to-use command-line tools that can be run in the following way:

Fpocket> fpocket -f receptor.pdb

SiteHound> auto.py -i receptor.pdb -p CMET

P2Rank> prank predict -f receptor.pdb

For each program, various advanced options are available in their respective GitHub repositories or help messages, which can be used to tailor the results to each specific scenario.

The output of FPocket is a directory containing several files, the most relevant being **receptor_info.txt**, which provides an ordered list of identified cavities with a predicted druggability score, and **receptor_out.pdb**, which contains the original PDB structure annotated with the identified pockets. These pockets are listed under the HETATM records as groups of points with residue name STP and the corresponding residue number specified in **receptor_info.txt**. Based on these two files, it is possible to select the cavities with the highest probability of binding drug-like molecules (i.e., highest druggability score) and visualize them in molecular graphics software such as PyMOL, UCSF Chimera, ChimeraX, or VMD.

For SiteHound, the two most important output files are **receptor_CMET_summary.dat** and **receptor_clusters.pdb**, which correspond to the FPocket outputs described above. The **receptor_CMET_summary.dat** file lists all predicted binding sites ranked by their total interaction energy (TIE, the lower the better), while **receptor_clusters.pdb** is a PDB file containing the cluster of points associated with each predicted binding site. In this file, each cluster is represented as a residue with the name SXX and a residue number XX, where XX denotes the binding site index specified in **receptor_CMET_summary.dat**. As with FPocket, the information provided in these two files can be used to select the predicted binding sites with the lowest TIE and visualize them in molecular graphics software.

For P2Rank, the main output files are **receptor.pdb_predictions.csv**, which contains an ordered list of predicted pockets ranked by their probability of binding drug-like molecules, and **receptor.pdb_points.pdb**, which contains points grouped into the different predicted pockets. In this file, the points belonging to each predicted pocket are clustered into residues with the residue name STP and the rank as the residue number.

To integrate the results, it is useful to load all output PDB files into a molecular graphics software and determine which pockets predicted by the different algorithms correspond to the same physical cavity (Fig. 3). This enables the identification of binding pockets consistently supported by different prediction metrics and facilitates the selection of the most promising candidate for virtual screening. It is important to note that prior biological knowledge of the protein of interest is also important to guide the selection of the binding site; while this is often straightforward for enzymes with known active sites, it can be more challenging for other types of protein targets. In the case of the chikungunya envelope glycoproteins, the pocket located behind the fusion loop in E1 is a suitable target. An organic molecule binding this cavity may stabilize the pre-fusion conformation of the envelope proteins and therefore inhibit membrane fusion and viral entry [9].

Once a pocket has been selected, it is necessary to define an appropriate docking box to guide the docking calculations. A convenient way to achieve this is through the AutoDock Vina plugin in UCSF Chimera (**Tools → Surface/Binding Analysis → AutoDock Vina**). This tool allows the visualization of the docking box and definition of its center and dimensions (Fig. 3). The goal is to center the box on the selected binding site and set its dimensions so that it encompasses the entire cavity while allowing free rotation of

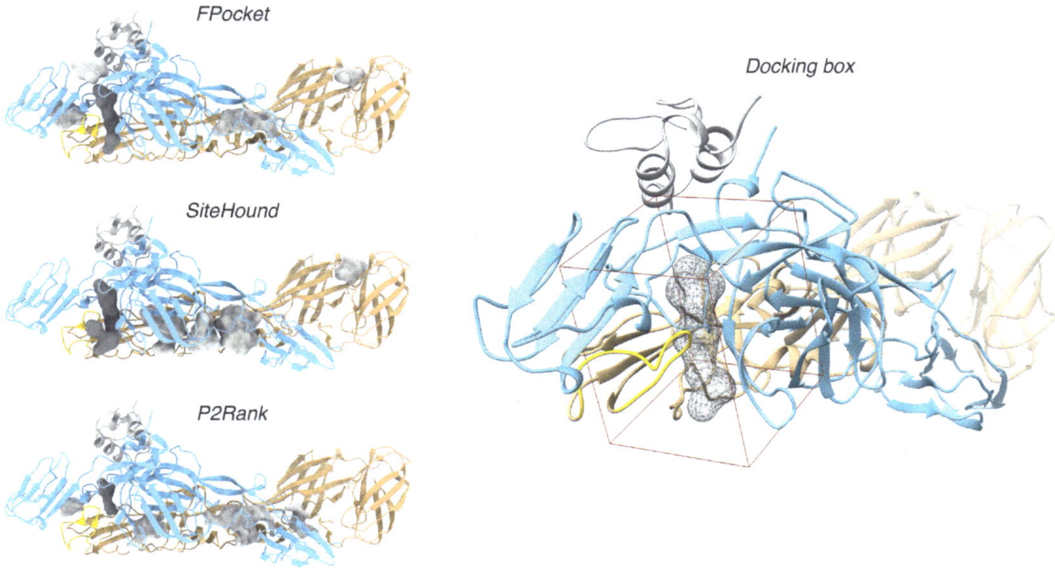

Fig. 3 Detected binding pockets in chikungunya envelope glycoproteins. The top five ranking cavities detected by FPocket, SiteHound, and P2Rank are represented as gray surfaces. The right panel shows the representation of the suggested docking box visualized with UCSF Chimera

drug-like molecules within it. Formal guidelines for box definition are provided in the work of Feinstein and Brylinski [18].

3.4 Running Docking with AutoDock Vina

Running a docking calculation with AutoDock Vina is straightforward. The inputs are a PDBQT file for the ligand, a PDBQT file for the protein receptor, and a configuration file specifying the name of the receptor file, the coordinates and dimensions of the docking box (in ångström), the exhaustiveness, and the number of poses given in the output. An example of a configuration file is the following:

receptor = receptor.pdbqt
center_x = -39.591
center_y = -32.944
center_z = -24.378
size_x = 27.5
size_y = 22.5
size_z = 22.5
exhaustiveness = 8
num_modes = 3

To run multiple dockings, one way is to define a **ligand_list*i*.list** file with the names of different compounds listed in the format *compound1.pdbqt*, *compound2.pdbqt*, and so forth, and run the simulations with the following script. The suffix *i* in the **ligand_list** file name is useful to define multiple files with different compounds and run parallel processes (*see* **Note 1**).

```
#!/bin/bash
# Script name: vina_docking.sh
# Usage:./vina_docking.sh <ligand_list_index> <
num_cpus>
# Assumes: run from the general directory and
directory vina/ exists
# Arguments
i=$1 # index of the ligand_list file (e.g., ligand_
list1.list)
N=$2 # number of CPUs per docking
# Input ligand list
ligand_list="ligand_list${i}.list"
# Loop over ligands
while read -r file; do
lig=$(basename "$file".pdbqt)
vina --ligand "pdbqt_dir/${lig}.pdbqt" \
--out "vina/${lig}.pdbqt" \
--config conf.txt \
--cpu "$N"
done < "$ligand_list"
```

Using this script, we can launch the docking calculations on the compounds specified in **ligand_list1.list** using eight CPUs with the following line.

./vina_docking.sh 1 8

The output of AutoDock Vina consists of PDBQT files, stored in our case in the **vina/** directory. These files contain the 3D conformations of the best-ranked poses (lowest scores) together with their associated scores. In each PDBQT file, every pose is enclosed within MODEL entries, and the score is reported in the *REMARK VINA RESULT:* line. To extract the scores for all poses of each compound, one approach is the one described in Sect. 3.7.

3.5 Running Docking with LeDock

To run the docking simulations with LeDock, the needed input files are the receptor file in PDB format, the ligand files in MOL2 format, and two text files: **Dock.in** file, which specifies all docking parameters, and **ligand_list.list** file, which contains the list of compounds to be docked. An example of a **Dock.in** file would be the following:

```
Receptor
pro.pdb
RMSD
1.0
Binding pocket
-53.377 -27.805
-44.194 -21.694
-35.628 -13.128
Number of binding poses
50
Ligands list
ligand_list.list
END
```

In this case, to launch the docking calculation over the compounds specified in **ligand_list.list** file, it is only necessary to run the following line:

```
ledock_linux_x86 Dock.in
```

The output files of LeDock are DOK files, which are essentially PDB files containing the different predicted poses as separate molecules, with additional information given in the REMARK entries for each pose. Among this information is the score, specified in the *REMARK Cluster* line.

3.6 Running Docking with AutoDock4

Launching a virtual screening with AutoDock4 is somewhat more complex than the previously described procedures for AutoDock Vina and LeDock. It requires the execution of three sequential steps, which are explained in more detail in the AutoDock4 user guide [19].

1. First, it is necessary to precompute the interaction energy grids between the protein receptor and all ligand atom types, which will be used later by AutoDock4 during the docking simulations. To perform this task, open the **receptor.pdbqt** file with the graphical user interface of AutoDockTools and generate a GPF file (*Grid Parameter File*) following the "Grid" tab in the main menu. This means setting the atom or map types

(*Directly* option), defining the grid box (docking box defined in Sect. 3.3, note that in this case the dimensions are not in ångström but in points, dependent on the spacing parameter), and saving the generated **receptor.gpf**. For this step, it is necessary to know all the atom types present in the compound library, which can be done using the following script.

```
#!/usr/bin/env python3
# Script name: get_atom_types.py
# Usage: python get_atom_types.py
import glob
files = glob.glob("pdbqt_dir/*.pdbqt")
atom_types = set()
for file in files:
   with open(file, "r") as f:
      for line in f:
         if line.startswith("ATOM"):
             line = line.split()
             atom_type = line[-1]
             atom_types.add(atom_type)
# save atom types to a file
with open("atom_types.txt", "w") as f:
   for atype in sorted(atom_types):
      f.write(f"{atype}\n")
```

2. Using the generated GPF file, compute the interaction energy grids with AutoGrid4:
   autogrid4 -p receptor.gpf -l receptor.glg
   This line will output the **receptor.glg** file (*Grid Log File*) and all the necessary MAP files, which are the energy grids for each ligand atom type.
3. The next step is to generate the DPF file (*Docking Parameter File*) for each compound in the library. If we create an **autodock/** directory and copy inside it the **receptor.pdbqt** and all MAP files, we can obtain the DPF files with the following script:

```
#!/bin/csh
# Script name: get_dpf.csh
# Usage:./get_dpf.csh
# Assumes: run from the general directory,
directory autodock/
# exists, and has withinit receptor.pdbqt and all
map files
set mgltools = path
# path to /MGLToolsPckgs/AutoDockTools/Utilities24/
cd autodock/
foreach file (../pdbqt_dir/*.pdbqt)
set name = 'basename $file.pdbqt'
ln -s../pdbqt_dir/"$file"
```

```
pythonsh $mgltools/prepare_dpf42.py -l 'basename
$file' \
-r receptor.pdbqt \
-p ga_num_evals=5000000 \
-p ga_pop_size=150 \
-p ga_run=100 \
-p rmstol=2.0
end
```

4. To run the previous script, it might be necessary to add the pythonsh full path. The final step is to run the docking calculations with AutoDock4, which can be done using the same strategy as for AutoDock Vina:

```
#!/bin/bash
# Script name: autodock_docking.sh
# Usage:./autodock_docking.sh <ligand_list_index>
# Assumes: run from the autodock directory
# Arguments
i=$1 # index of the ligand_list file (e.g., ligand_
list1.list)
# Input ligand list
ligand_list="ligand_list${i}.list"
# Loop over ligands
while read -r file; do
lig=$(basename "$file".pdbqt)
autodock4 -p ${lig}_receptor.dpf -l ${lig}.dlg
done < "$ligand_list"
```

The output of AutoDock4 consists of DLG files (*Docking Log Files*), which contain the final pose for each independent run of the genetic algorithm, cluster of poses, their scores, and the associated statistics. Despite containing a greater amount of useful information, to obtain the score of the lowest-energy conformation of each cluster, one possible approach is the one described in the following section.

3.7 Consensus Method: ECR

After running all the simulations, the next step is to extract the scores for each pose of every compound across all docking software. If we are only interested in the score and not in the 3D conformation similarity, one approach is to run the following script.

```
#!/usr/bin/env python3
# Script name: get_score.py
# Usage: python get_score.py
# Assumes: run from the general directory
import glob
import os
import pandas as pd
def get_score_ligand(file, program):
   ligand_score_list = []
   pose = 1
```

```
   name = os.path.basename(file).split('.')[0]
   with open(file,'r') as lig:
     for line in lig:
       line_with_score = False
       if program == 'vina' and 'REMARK VINA
       RESULT:' in line:
          line = line.split()
          score = float(line[3])
          line_with_score = True
elif program == 'ledock'
   line = line.split()
   score = float(line[7])
   line_with_score = True
elif (program == 'autodock'
      and 'Energy of Binding' in line
      and 'DOCKED:' not in line):
   line = line.split()
   score = float(line[7])
   line_with_score = True
if line_with_score:
   ligand_score_dicc = {"ligand": name
             "pose_" + program: pose
             "score_" + program: score}
   ligand_score_list.append(ligand_score_dicc)
   pose += 1
 ligand_score_df = pd.DataFrame(ligand_score_list)
 return ligand_score_df
def main(program):
   if program == 'vina':
      files = glob.glob("vina/*.pdbqt")
   elif program == 'ledock':
      files = glob.glob("ledock
   elif program == 'autodock':
      files = glob.glob("autodock/*.dlg")
   df_list = []
   for file in files:
      ligand_score_df = get_score_ligand(file,
      program)
      df_list.append(ligand_score_df)
   scores_df = pd.concat(df_list, ignore_index=
   True)
   output_name = program + "_scores.csv"
   scores_df.to_csv(output_name, index=False)
   return
main('vina')
main('ledock'
main('autodock')
```

The output of this script is a CSV file for each docking program, containing the compound name, the pose number, and the corresponding score for every ligand.

Once the scores have been obtained, ranking the compounds according to the ECR or another consensus metric is relatively simple.

As a preliminary remark, before computing the ECR, we will filter the results to retain only the best pose for each compound (i.e., pose 1). While this decision is natural in this case, it is not a trivial task, and there is no clear framework for handling different poses of different compounds across various docking software, because it is not straightforward to distinguish between statistically significant score differences and equivalent scores. If, as in this case, we consider only the best score, this limitation is of no consequence; however, if we were evaluating the 3D similarity between the solutions obtained by different software, it would be relevant (*see* **Note 2**).

To compute the ECR, we will use the following equation:

$$\mathrm{ECR}(i) = \frac{1}{\sigma}\sum_{j}\exp\left(-\frac{r_j^i}{\sigma}\right)$$

Where r^i_j is the ranking of compound *i* according to program *j*, and σ is the expected value of the exponential distribution for a given threshold. For example, if we set the threshold at 5%, σ corresponds to the ranking of the compound located in the fifth percentile of the ordered list. Regarding σ, a second consideration to take into account is that the scores given by the different software do not have the same number of decimal places. Specifically, the scores given by AutoDock Vina only have one decimal place, resulting in multiple compounds having the same score. In this case, we can give the same ranking to all compounds with equal score, but by doing this, the ranking of the compound located at a given percentile, σ, will be program dependent.

Starting from the **scores.csv** files obtained for the three docking software, we can get the ECR in the following way:

```
#!/usr/bin/env python3
# Script name: ECR.py
# Usage: python ECR.py
import numpy as np
import pandas as pd
vina_df = pd.read_csv("vina_scores.csv")
ledock_df = pd.read_csv("ledock_scores.csv")
ad4_df = pd.read_csv("autodock_scores.csv")
# retain only pose 1
vina_df = vina_df[vina_df["pose_vina"] == 1]
ledock
ad4_df = ad4_df[ad4_df["pose_autodock"] == 1]
# merge data frames
df = pd.merge(vina_df,ledock_df,on="ligand",how=
"inner")
df = pd.merge(df,ad4_df,on="ligand",how="inner")
# setting the sigma parameter
threshold = 5
sigma = int(df.shape[0]*threshold/100)
```

```
# get the ranking and sigma for every program
df["rank_vina"] = df.score_vina.rank(axis=0,
method='min')
df = df.sort_values(by="rank_vina", axis=0,
ascending=True)
sig_vina = df.iloc[sigma]["rank_vina"].copy()
df["rank_ledock"] = df.score_ledock
df = df.sort_values(by="rank_ledock", axis=0,
ascending=True)
sig_ledock = df.iloc[sigma]["rank_ledock"].copy()
df["rank_autodock"] = df.score_autodock.rank(axis=
0, method='min')
df = df.sort_values(by="rank_autodock", axis=0,
ascending=True)
sig_ad4 = df.iloc[sigma]["rank_autodock"].copy()
# compute the exponential term for every program
df["exp_vina"] = np.exp(-df["rank_vina"]/sig_vina)/
sig_vina
df["exp_ledock"]= np.exp(-df["rank_ledock
df["exp_autodock"] = np.exp(-df["rank_autodock"]/
sig_ad4)/sig_ad4
# get the exponential sum: ECR
df["ECR"] = (df["exp_vina"]+df["exp_ledock"]+
df["exp_autodock"])
# sort by ECR and save. Higher ECR is better
df["rank_ECR"]=df["ECR"].rank(axis=0,method=
'min',ascending=False)
df = df.sort_values(by="rank_ECR", axis=0,
ascending=True)
df.to_csv("ECR.csv", index=False)
```

As a final result, we obtained the consensus ECR for all compounds, based on the docking scores from AutoDock Vina, LeDock, and AutoDock4. This ranking enables the selection of the most promising candidates for experimental validation or further screening. In this regard, Molecular Dynamics simulations can be used as an additional filter in the virtual screening (*see* **Note 3**), and AI-based methodologies can be added to the consensus scoring framework detailed in this chapter (*see* **Note 4**).

Although experimental evaluation is beyond the scope of this chapter, ideally, the selected method should directly assess the binding between the compounds and the protein receptor. Common experimental approaches for this purpose include ligand-binding measurements using SPR (Surface Plasmon Resonance) or ITC (Isothermal Titration Calorimetry). Additionally, biochemical assays are often available to measure protein activity and evaluate potential inhibition by selected ligands. For chikungunya envelope glycoproteins, experimental assessment of the protein activity is more challenging, but there are published protocols for the measurement of the fusogenic activity of these proteins. Examples include membrane fusion assays between fluorescently labeled virus-like particles (VLPs) and liposomes [20], fusion-from-without (FFWO) assays [21], and cell–cell fusion assays [22]. Another approach is to evaluate the inhibitory activity of compounds using lentivirus-based pseudoviruses, pseudotyped with chikungunya envelope proteins [23]. Compounds that selectively inhibit infection by these pseudoviruses, but not by control pseudoviruses (e.g., VSV-G pseudotyped), are likely targeting the chikungunya envelope proteins. Finally, inhibition of wild-type chikungunya infection can also be assessed, although in this case it is more difficult to determine whether the inhibitors specifically target the receptor used in the virtual screening.

4 Notes

1. **Parallel implementation**
   The parallel implementation to accelerate the virtual screening is dependent on the system used to run the calculations. For example, it is important to consider that AutoDock Vina has a GPU-accelerated implementation (https://github.com/ccsb-scripps/AutoDock-GPU). However, a simple way to parallelize the run on a local machine with the code presented in this chapter is to launch several processes in the background. To do this, we have to define ***M*ligand_list*i*.list** files, with *i* from 1 to *M* and different compounds listed in each file, which could be done in the following way.

```
split -l L all_ligands.txt ligand_list --numeric-
suffixes=1 --additional-suffix=.list
```

   Where *L* is the number of compounds in each file. To launch the parallel runs, we can use the following script, exemplified for AutoDock Vina:

```
#!/bin/bash
# Script name: vina_parallel.sh
# Usage:./vina_parallel.sh <num_parallel_process>
<num_cpus>
# Assumes: run from the general directory
M=$1 # number of parallel process and ligand_list
files
N=$2 # number of CPUs per docking
for i in $(seq -w 1 $M)
do./vina_docking.sh $i $N &
#./ledock_linux_x86 Dock$i.in &
#./autodock_docking.sh $i &
done
wait
```

Using this script, with the following line, we are launching four parallel AutodDock Vina runs, each one using eight CPUs. ./vina_parallel.sh 04 8
To employ the same strategy with LeDock and AutoDock4, we just need to modify the ./vina_docking.sh $i $N & line with the corresponding one for each software. For LeDock and AutoDock4, it is not necessary to give the parameter N.
Finally, an example to launch a virtual screening with AutoDock Vina in an HPC cluster using SLURM would be the following:

```
#!/bin/bash
#SBATCH --job-name=vina_docking
#SBATCH --output=vina_%A_%a.out
#SBATCH --error=vina_%A_%a.err
#SBATCH --array=1-10
#SBATCH --cpus-per-task=4
#SBATCH --mem=8G
#SBATCH --time=12:00:00
module load autodock-vina
# Variables
LIST="ligand_list${SLURM_ARRAY_TASK_ID}.txt"
N=$SLURM_CPUS_PER_TASK
mkdir -p vina
while read file; do
lig=$(basename "$file".pdbqt)
vina --ligand "pdbqt_dir/${lig}.pdbqt" \
--out "vina/${lig}.pdbqt" \
--config conf.txt \
--cpu "$N"
done < "$LIST"
```

2. **Ligand RMSD as a filter**
In the procedure described in Sect. 3, and particularly in the ECR computation detailed in Sect. 3.7, we considered only the docking score to rank the compounds. While this represents a valid approach, one could also take into account whether the solutions identified by the different docking programs exhibit similar 3D conformations, considering only those poses recognized by all programs. As previously discussed, in such a scenario, it becomes important to consider the multiple poses provided by each software. For example, it is possible that AutoDock Vina identifies two poses with nearly indistinguishable scores. In this case, although pose 1 could differ from the best poses identified by LeDock and AutoDock4, pose 2 may, in fact, correspond to the solution found by the latter programs.
To consider the 3D conformation similarity between poses identified by different programs, it is necessary to compute the RMSD (*Root Mean Square Deviation*) between the different

ligand poses. RMSD values lower than 2 ångström could be considered as a threshold to define two poses equivalent or not. In this calculation, it is important not to fit the poses, but simply compute the RMSD between the solutions identified by each program. This computation could be done using the OpenBabel Python bindings *Pybel* calcrmsd function or the PyMol Python API rms_cur function, previously converting each structure to PDB format. As discussed in Sect. 3.7, in this case, handling the different poses is not trivial. One approach is, for every compound, compute the RMSD between all poses given by the different software, retain only those solutions found by all programs, and consider each solution as an independent compound for the ECR calculation described in Sect. 3.

3. **Molecular Dynamics as a filter**
   Molecular Dynamics (MD) simulations can be used at two stages in a virtual screening campaign, either at the beginning, running simulations of the selected protein receptor to explore its conformational space and search for new binding pockets not present in the low-energy initial structures (experimental structures or high-confidence predictions); or as an additional filter in compound selection. This latter case is the one we will address in this note. In this context, the value of MD simulations, a much more rigorous methodology in comparison with docking algorithms, in virtual screening, is to determine the stability of the binding conformation predicted by docking programs. After fitting the trajectory using the protein receptor, one could compute the ligand RMSD against the first frame as a function of time and evaluate if the compound remains stable in the initial binding pose. As mentioned in **Note 2**, a suitable threshold could be 2 ångström. It is important to notice that the value of this filter is to discard those compounds that move away from the initial pose, i.e., have an RMSD higher than 2 ångström during the simulation. Besides the RMSD, it is also possible to measure the stability of important interactions such as hydrogen bonds between the ligand and receptor. One software that allows this type of calculation is VMD.

4. **Including AI methods**
   Since the publication of the AlphaFold2 paper [17], computational biology has undergone a profound transformation. Numerous AI-based algorithms have been developed to address biological problems that were previously tackled by classical methods. Drug design is no exception, and several models are now available to predict the 3D conformation and binding affinity of protein–ligand complexes. One of the most prominent examples is Boltz-2 [24]. In this type of model, it is sufficient to

provide the protein sequence and the ligand SMILES, although it is possible to input the protein structure as a template. As for docking programs, the output consists of the predicted structure of the complex and an affinity score. While these AI-based approaches are outperforming classical methods in retrospective benchmarks [24], only the prospective results of virtual screening campaigns across diverse protein targets will allow a rigorous assessment of their value and improvement. In this context, consensus frameworks remain a valid strategy, as they can provide more reliable results and, importantly, mitigate the risk of selecting suboptimal software for a given receptor. It is worth noting that results from classical docking algorithms and AI-based methods can be readily integrated into consensus metrics based on compound ranking, potentially enhancing the overall outcome by combining orthogonal methodological approaches.

References

1. Freund Y, Schapire RE (1997) A decision-theoretic generalization of on-line learning and an application to boosting. J Comput Syst Sci 55:119–139
2. Wang R, Wang S (2001) How does consensus scoring work for virtual library screening? An idealized computer experiment. J Chem Inf Comput Sci 41:1422–1426
3. Palacio-Rodríguez K, Lans I, Cavasotto CN et al (2019) Exponential consensus ranking improves the outcome in docking and receptor ensemble docking. Sci Rep 9: 5142
4. Bhutkar M, Saha A, Tomar S (2024) Viral methyltransferase inhibitors: berbamine, venetoclax, and ponatinib as efficacious antivirals against chikungunya virus. Arch Biochem Biophys 759:110111
5. Nguyen PTV, Yu H, Keller PA (2015) Identification of chikungunya virus nsP2 protease inhibitors using structure-base approaches. J Mol Graph Model 57:1–8
6. Rabelo VW, Sanchez-Nuñez ML, Corrêa-Amorim LS et al (2024) In silico drug repurposing uncovered the antiviral potential of the antiparasitic drug oxibendazole against the chikungunya virus. ACS Omega 9:27632–27642
7. Nguyen PTV, Yu H, Keller PA (2014) Discovery of in silico hits targeting the nsP3 macro domain of chikungunya virus. J Mol Model 20:2216
8. Sharma R, Kesari P, Kumar P et al (2018) Structure-function insights into chikungunya virus capsid protein: small molecules targeting capsid hydrophobic pocket. Virology 515:223–234
9. Battini L, Fidalgo DM, Álvarez DE et al (2021) Discovery of a potent and selective chikungunya virus envelope protein inhibitor through computer-aided drug design. ACS Infect Dis 7:1503–1518
10. Trott O, Olson AJ (2010) AutoDock Vina: improving the speed and accuracy of docking with a new scoring function, efficient optimization, and multithreading. J Comput Chem 31:455–461
11. Morris GM, Huey R, Lindstrom W et al (2009) AutoDock4 and AutoDockTools4: automated docking with selective receptor flexibility. J Comput Chem 30:2785–2791
12. Zhang N, Zhao H (2016) Enriching screening libraries with bioactive fragment space. Bioorg Med Chem Lett 26:3594–3597
13. O'Boyle NM, Banck M, James CA et al (2011) Open Babel: an open chemical toolbox. J Cheminform 3:33
14. Le Guilloux V, Schmidtke P, Tuffery P (2009) Fpocket: an open source platform for ligand pocket detection. BMC Bioinform 10:168
15. Hernandez M, Ghersi D, Sanchez R (2009) SITEHOUND-web: a server for ligand binding site identification in protein structures. Nucleic Acids Res 37:W413–W416
16. Krivák R, Hoksza D (2018) P2Rank: machine learning based tool for rapid and accurate prediction of ligand binding sites from protein structure. J Cheminformatics 10:39
17. Jumper J, Evans R, Pritzel A et al (2021) Highly accurate protein structure prediction with AlphaFold. Nature 596:583–589
18. Feinstein WP, Brylinski M (2015) Calculating an optimal box size for ligand docking and virtual screening against experimental and predicted binding pockets. J Cheminform 7:18

19. Morris GM, Goodsell DS, Pique ME et al (2014) User Guide AutoDock Version 4.2, Updated for version 4.2.6. La Jolla, CA: The Scripps Research Institute
20. Van Duijl-Richter MKS, Blijleven JS, Van Oijen AM et al (2015) Chikungunya virus fusion properties elucidated by single-particle and bulk approaches. J Gen Virol 96:2122–2132
21. Noval MG, Rodriguez-Rodriguez BA, Rangel MV et al (2019) Evolution-driven attenuation of alphaviruses highlights key glycoprotein determinants regulating viral infectivity and dissemination. Cell Rep 28:460–471.e5
22. Kuo S-C, Chen Y-J, Wang Y-M et al (2012) Cell-based analysis of chikungunya virus E1 protein in membrane fusion. J Biomed Sci 19:44
23. Battini L, Thannickal SA, Cibello MT et al (2024) Evolution of antiviral resistance captures a transient interdomain functional interaction between chikungunya virus envelope glycoproteins, http://biorxiv.org/lookup/doi/; https://doi.org/10.1101/2024.11.11.623010
24. Passaro S, Corso G, Wohlwend J et al (2025) Boltz2: Towards accurate and efficient binding affinity prediction. BioRxiv.

Check for updates

Chapter 17

Surface Plasmon Resonance-Based Analysis of Ligand Binding Kinetics in Chikungunya Virus Drug Discovery

Ruchi Rani, Anushka Upadhyay and Shailly Tomar

Abstract

Chikungunya virus (CHIKV) is an emerging arbovirus lacking widely effective antiviral therapies, highlighting the need to characterize molecular interactions essential for its replication and pathogenesis. Surface plasmon resonance (SPR) is a sensitive, label-free technique that enables real-time analysis of biomolecular interactions by detecting refractive index changes at a sensor surface. SPR provides precise kinetic and affinity measurements for protein–protein, protein–RNA, and protein–ligand interactions using minimal sample quantities and supporting high-throughput screening. In CHIKV research, it has been applied to evaluate antiviral compounds, antibodies, receptor binding, and viral protein complexes. This chapter outlines key methodological considerations, including ligand preparation, immobilization strategies, assay optimization, and kinetic analysis.

Keywords SPR, Ligand binding, Chikungunya, Drug discovery

1 Introduction

Chikungunya is a mosquito-borne disease caused by chikungunya virus (CHIKV), a member of the Togaviridae family of RNA viruses. The transmission of CHIKV occurs through female mosquitoes, mainly *Aedes aegypti* and *Aedes albopictus* [1]. The uninfected mosquito acquires the virus while feeding on an infected human host. After entry, the virus replicates inside the epithelia of the mosquito's midgut, spreads into the hemocoel, thus infecting the secondary tissues, such as the salivary gland, where it continues to replicate for several days. The infected mosquito then transmits the virus to a human host by feeding. After transmission, the primary infection spreads in fibroblasts, macrophages, and muscle cells [2, 3]. The virus enters the host cells by endocytosis, followed by fusion and RNA release due to the acidic environment. The positive-sense RNA translates into nonstructural proteins (nsPs) to form a replication complex. While the negative-sense RNA is synthesized to serve as a template

Claudia Filomatori and Eliana F. Castro (eds.), *Chikungunya Virus*, Methods in Molecular Biology, vol. 3067,
https://doi.org/10.1007/978-1-0716-5496-5_17,

to produce genomic and subgenomic RNA, to yield structural proteins. The nucleocapsids (N) assemble, followed by integration of mature virions that bud out of the plasma membrane to cause further infection (Fig. 1) [4]. Understanding the virus life cycle, along with virus–host molecular interplay, including protein–RNA and protein–protein interactions (PPI), is essential in determining viral replication efficiency and modulation of host immune responses in favor of the virus. These interactions form the foundation of virus survival and immune evasion. The CHIKV replication complex assembly, highlighting virus–host protein interaction, is shown in Fig. 2.

Clinically, CHIKV infection is characterized by fever and severe joint pain along with other symptoms such as headache, nausea, swelling, etc. Because these symptoms closely resemble those of dengue and Zika infections, misdiagnosis is often common. Currently, there are no specific antiviral treatments for chikungunya,

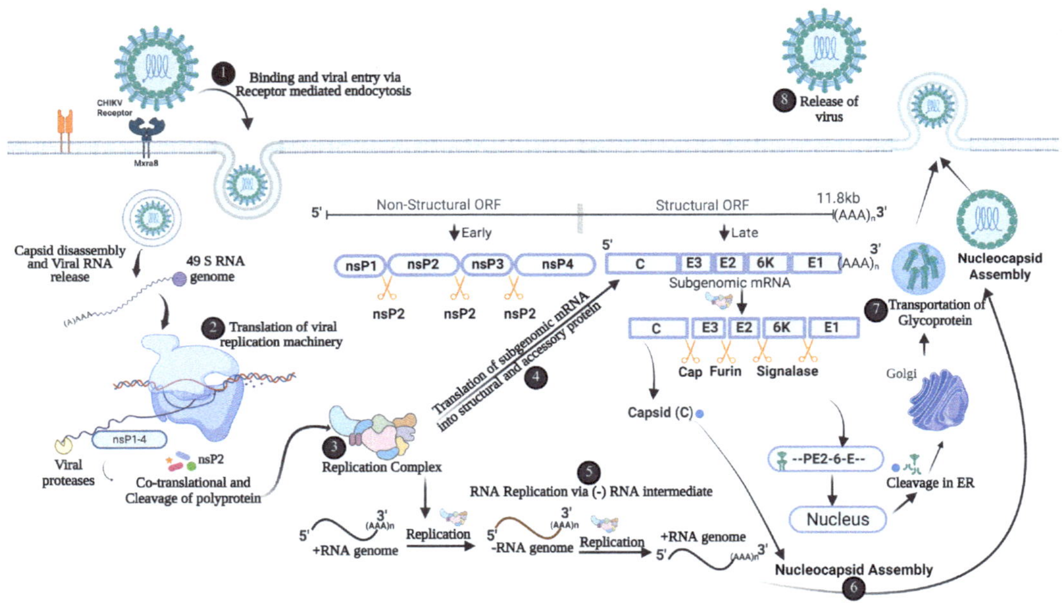

Fig.1 Schematic representation of the CHIKV life cycle. CHIKV enters the host cell via receptor-mediated endocytosis. The low pH environment of endosomes causes the disassembly and releases the viral genome into the host cytoplasm. The nonstructural ORF RNA is translated into nsPs, which are processed to form the replication–transcription complexes. These complexes use the (+)strand genomic RNA (gRNA) as a template to synthesize a (−)strand RNA intermediate. The newly synthesized (−)strand RNA then serves as a template to produce new (+)strand gRNA, which becomes the genome of the new virus particle. Subgenomic mRNAs (sg-mRNAs) generated during transcription are translated into structural proteins (C, E1/2/3, and 6K), which are further processed into mature forms required for gRNA encapsidation and nucleocapsid assembly. Viral enveloped proteins (E1, E2, E3, and 6K) enter the endoplasmic reticulum (ER) for processing and modification, and then transit through the Golgi apparatus in vesicles to reach the cell surface. At the plasma membrane, these glycoproteins are incorporated into assembling nucleocapsids, and mature virions are subsequently released from the cell. nsPs nonstructural proteins, ORF open reading frame, (−)ssRNA negative-sense single-stranded RNA, C capsid protein, E1/2/3 envelope proteins. Figure created using BioRender

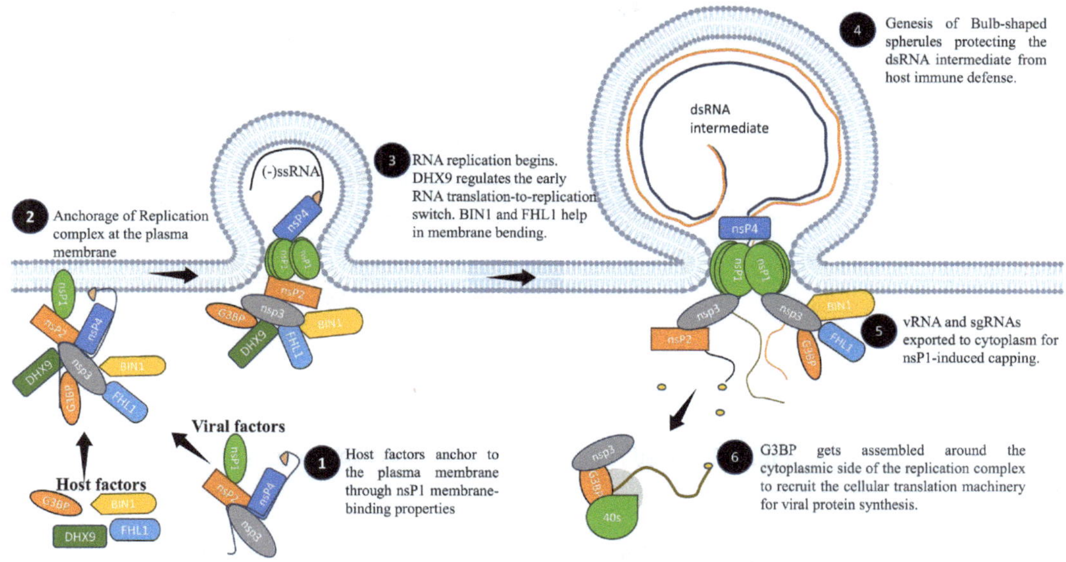

Fig.2 CHIKV replication complex assembly model for virus–host replication interaction [5]. dsRNA (double-stranded RNA), nsPs (nonstructural proteins), ssRNA (single-stranded RNA), vRC (viral replication complex)

and the disease management mainly relies on antipyretics and analgesics. Additionally, two vaccines have been approved by regulatory bodies—IXCHIQ (Valneva, VLA1553), which is a live-attenuated vaccine, and VIMKUNYA (Bavarian Nordic, VLP vaccine), a virus like particle-based vaccine. However, IXCHIQ was suspended in August 2025 due to reports of adverse complications [6]. To develop antivirals and vaccines against CHIKV, it is crucial to characterize the molecular interactions that sustain its life cycle, which requires sensitive biophysical techniques, such as surface plasmon resonance (SPR).

SPR is a label-free optical sensing method for real-time monitoring of molecular interactions in terms of a change in refractive index. The principle of SPR relies on total internal reflection (TIR), where light travels from a more optically dense medium to a less optically dense medium and gets reflected. In simpler terms, the polarised light is incident on an electrically conducting sensor (usually made of gold), and an energy transfer occurs, generating a wave of oscillating electrons on the metal surface called plasmons. The energy absorbed by these plasmons causes a significant dip in the intensity of reflected light recorded in the form of a resonance angle. This resonance angle is highly sensitive to the medium adjacent to the metal surface. When the ligand is immobilized on the sensor, and then it is exposed to an analyte (in solution), the resonance angle shift occurs in proportion to the bound mass, and the detector measures this shift to plot the real-time sensorgram (Fig. 3) [7]. However, there are several other methods for measuring biomolecular interactions, such as biolayer

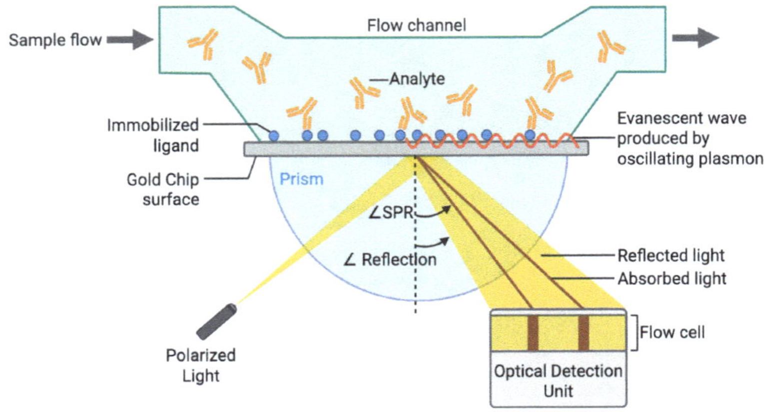

Fig.3 Schematic representation of the principle of surface plasmon resonance

interferometry (BLI), isothermal titration calorimetry (ITC), and microscale thermophoresis (MST). Compared to BLI, SPR provides approximately 100-fold higher sensitivity of detection and resolution. Additionally, SPR is a multifaceted and robust technique that offers label-free detection, whereas MST requires internal labeling of molecules and can only provide kinetic information. Unlike ITC, which requires a high quantity of sample and is not suitable for high-throughput screening assays, SPR can generate kinetic and affinity data using samples in picograms and allows multiple analytes to be tested simultaneously [8].

SPR has proved to be an important utility for studying protein–protein, protein–nucleic acid, and protein–ligand interactions in virology. For instance, SPR was employed to screen antiviral molecules and RNA oligonucleotides against the nsP4 protein (RNA-dependent RNA polymerase) of CHIKV and Sindbis virus for binding studies [9]. Another study by Nehul and colleagues on antivirals targeting the CHIKV capsid utilized SPR for binding kinetics of selected antiviral molecules [10]. Additionally, SPR has been used to characterize the interaction between the CHIKV E2 antigen and neutralizing monoclonal antibody [11]. Furthermore, numerous studies on other viruses also employed SPR, highlighting its growing relevance in identifying and characterizing potential antiviral candidates across diverse viral pathogens [12–17].

1.1 Application of SPR in CHIKV Research

The life cycle of CHIKV begins with the interaction of the E1–E2 glycoprotein complex with host-cell receptor Mxr8, mediating viral entry for infection [18]. Further, several other factors of the ESCRT protein complex facilitate CHIKV replication and post-translational steps in infection [19]. Likewise, nsP3 and capsid protein of CHIKV interact with G3BP1/2 to block antiviral stress response and facilitate replication, respectively [20–22]. These virus–host protein interactions involved in the CHIKV life cycle can be explored as potential therapeutic targets (Table 1) [23, 24]. Similar strategies have been applied to other viruses such as SARS-CoV-2, where inhibitory peptides were screened against the interaction of ACE2 and RBD protein using SPR [25].

Similarly, interactions between viral components, including proteins and vRNA, are significantly important for replication and survival of CHIKV in host cells. One of the important PPIs involved domains of the envelope protein (E1–E2) to form the spike of CHIKV [53, 54]. CRISPR-Cas9-based screening identified YBX-1 as a vRNA-binding factor in CHIKV, which augments the binding of nsP3 to vRNA for replication [55]. Moreover, host factors such as DDX39 and DDX56, which are evolutionarily conserved, bind to the secondary structure of vRNA to impose antiviral activity to control alphavirus infection [56, 57]. Studying such interactions involving host antiviral factors gives insight into host defense mechanisms, which can be valuable in understanding host–virus relationships.

The virus interactome or viral proteins are favorable antiviral targets, and SPR-based antiviral screenings have emerged as a high-throughput, semiautomated, and time-saving platform to evaluate them. Examples include screening of antiviral molecules against nsP4 of Alphavirus, validating inhibitors for nsP2 protease of CHIKV, as well as characterizing antiviral binding to the 2-E protein, RdRp, and Mpro of SARS-CoV-2 [9, 13, 58–62]. Furthermore, researcher shows the usage of SPR in the evaluation of RNA aptamers with E2DA (domain A of Envelope 2) of CHIKV as potential attachment inhibitors [48]. The binding and kinetics studies for antiviral molecules against the CHIKV capsid protein were performed using SPR [63]. The macrocycle candidates against the E2 protein of CHIKV were tested using SPR for their binding [49].

SPR has been a gold standard for quantification of antigen–antibody interactions [64–66]. The mouse-derived mAbs against recombinant E2 protein of CHIKV showing anti-CHIKV activity were analyzed using SPR for their binding kinetics [11, 67]. As reported, Fab molecules against CHIKV E1–E2 heterodimer were tested for binding kinetics in the sub-nanomolar range [46]. Characterization of nanobodies designed against the p62–E1 heterodimer of CHIKV was tested for binding kinetics studies [50]. CHIKV nsP2-targeting nanobodies derived from phage display have been reported to be tested with SPR for evaluating their binding affinities [30]. Moreover, a combinational monoclonal antibody therapy highlights the sensitivity of SPR within the nanomolar range kinetics [47].

Researchers have been using SPR for testing VLPs for their efficient binding with the receptor and for mutagenesis studies. For instance, CHIKV VLPs were tested for their binding with the human MXRA8 receptor and its mutants to investigate key interaction residues [68]. Another study used amine coupling for binding CHIKV mAb to CHIKV VLP and validated their binding to recombinant Mxr8 [69]. Serum-derived antibodies were tested

Table 1
The list of CHIKV therapeutic targets and the corresponding therapeutic approaches

Target (CHIKV/ host protein)	Function	Type of molecules	Reference
nsP1	vRNA capping and methylation	Berbamine hydrochloride; Venetoclax; Ponatinib	[26]
		Pyrantel pamoate; garcinolic acid; lobaric acid	[27]
		MADTP-314	[28]
		4-Hydroxytamoxifen; tamoxifen; clomifene	[29]
nsP2	Helicase, polyprotein cleavage, transcriptional shut-off	Nb-A9 (nanobody)	[30]
		Simeprevir; Voxilaprevir	[31]
		P1 (VMPWDEWLTKRKPELP)	[32]
		Withaferin A	[33]
		Pantinin-1	[34]
nsP4	Viral replication	MMs03131094; MMs03927237	[35]
		LabMol-309	[36]
		4′-Fluorouridine	[37]
		Piperine; 2-Thiouridine; Pyrazinamide; Chlorogenic acid	[9]
		Doxycycline; Ribavirin	[38]
		Favipiravir (T-705)	[39]
		Sofosbuvir	[40]
Capsid	Nucleocapsid core formation for vRNA protection	Efavirenz	[10]
		P1,P4-Di (adenosine-5′) tetraphosphate; Paromomycin sulphate (PSU	[41]
CHIKV E1/E2 glycoproteins	Virus attachment and entry	Suramin; Suramin conjugates	[42]
		CHK-263 (mAb)	[43]
		IM-CKV063; CHK-166	[44]
		CHE19	[45]
		C34; C37 (Fab)	[46]
		CHK-102 + CHK-152; CHK-166 + CHK-152	[47]
CHIKV E2	Virus attachment and entry	CHK-124 (mAb)	[43]
		RNA aptamers	[48]
		Macrocyclic peptides	[49]
		CHK-152	[11]

Table 1 (continued)

Target (CHIKV/ host protein)	Function	Type of molecules	Reference
Host factor interactions targeted:			
CHIKV E1–p62	Protects the fusion loop of the immature virus	N033-Fc; N053-Fc (nanobody)	[50]
nsP3–G3BP1	Inhibits host antiviral stress granule formation	Naltrindole; L-798106; WIN_62577; Fluspirilene; SB_222200; Dihydroergotamine; GSK1838705A	[23]
nsP3–PI3 kinase	Important for CHIKV replication and pathogenesis	Miltefosine	[51]
nsP2–Hsp90	CHIKV replication	Radicicol	[52]

against the receptor-binding domain of SARS-CoV-2 induced from chimeric VLP candidates for SARS-CoV-2 and CHIKV using SPR [70]. Researchers have extended the use of SPR for developing a fluorescence-assisted biosensor for virus detection using a VLP-based platform [71]. Researchers are extending the use of SPR into a virus detection device by integrating the machine with an AI-based multi-layered MDM (metal–dielectric–metal) configuration to surpass the traditional and time-consuming detection methods [72].

SPR has fundamentally revolutionized viral research through label-free, real-time analysis of molecular interactions with high resolution. In CHIKV studies, it facilitates detailed mapping of virus–host protein–protein interactions, which might serve as potential antiviral targets. SPR enables studies of protein–RNA interaction, high-throughput drug screening with kinetic profile, and therapeutic antibody development with quantitative measurement of antigenicity and affinity maturation in greater detail. The binding parameters help to predict lead molecules for further in vitro testing. As CHIKV remains a continuous threat to human health, integrating SPR with structural biology, bioinformatics, and cell-based assays forms an effective strategy for CHIKV research and therapeutic development.

2 Materials (*see* Note 1)

2.1 Running Buffers

1. 10× phosphate-buffered saline (PBS), with surfactant (PBS-P+), pH 7.4.
2. HEPES-buffered saline (HBS-N), HBS with EDTA and surfactant (HBS-EP+), and HBS with surfactant only (HBS-P+).

2.2 Protein Elution Buffers

1. 80–500 mM imidazole for Ni-NTA/IMAC (His-tagged proteins).

2. 10–20 mM reduced glutathione in running/binding buffer for Glutathione Sepharose (GST-tagged proteins).
3. 10 mM maltose in running/binding buffer for maltose-binding protein (MBP).
4. 2.5 mM desthiobiotin in running/binding buffer for Strep-tagged proteins.

2.3 Analyte Dilution Buffer

Same composition as running buffer.

2.4 Coupling Kits

1. Amine Coupling Kit for CM5 chip immobilization, including 0.4 M 1-ethyl-3-(3-dimethylaminopropyl)carbodiimide (EDC), 0.1 M N-hydroxysuccinimide (NHS), and 1 M ethanolamine-HCl, pH 8.5.
2. Thiol Coupling Kit for covalent immobilization of molecules and/or surface thiol couplings.

2.5 Capture Kits

1. NTA Reagent Kit for Sensor Chip NTA to capture His-tagged protein.
2. Biotin CAPture Kit for reversible capture of biotinylated molecules.
3. GST Capture Kit to capture GST-tagged protein.
4. Mouse or Human Antibody Capture Kit for capturing of mouse or human antibodies.
5. Human Fab Capture Kit for capturing human Fab antibody fragments.

2.6 Immobilization Buffer

10 mM sodium acetate buffers, pH 4.0–5.5. Adjust pH depending on the protein's isoelectric point.

2.7 Regeneration Buffers, Select Based on Ligand Stability (see Note 2)

1. Low-pH buffer: 10 mM glycine-HCl, pH 1.5–3.0.
2. High-pH buffer: 50 mM NaOH.
3. High-salt buffer: 1 M NaCl in running buffer.
4. 0.05%–0.5% SDS in water.
5. Ni-NTA Sensor Chip Regeneration Solution: 350 mM EDTA, pH 8.0.

2.8 Recombinant Proteins/Antibody Purification

1. His/GST-tagged or any other tagged proteins.
2. Affinity chromatography using appropriate columns.
3. SDS-PAGE.
4. Nanodrop or Bradford assay.

2.9 Analytes

1. Commercially available or synthesized small molecules, purity: ≥95% (HPLC/LC-MS verified). Dissolve in nuclease-free

water or DMSO, depending on solubility. Store at −20°C or −80°C, depending on compound stability.

2. Nucleic acids: Short oligonucleotides, cloned DNA fragments, or in vitro-transcribed CHIKV RNA fragments. Purity: ≥95%, free of protein and nuclease contaminants. Store at −80°C.
3. Small proteins, purity: ≥90% verified by SDS-PAGE before starting the experiment. Store at −80°C and avoid repeated freeze–thaw cycles.

2.10 Consumables

1. Low-binding RNase/DNase-free microcentrifuge tubes.
2. 1 and 10 mL disposable syringes.
3. PVDF or PES 0.22 μm syringe filters.
4. Pipettes and low-retention, sterile filter tips.
5. Vials compatible with SPR autosampler.
6. Gloves.

2.11 Sensor Chips (see Note 3)

1. Ni-NTA Sensor Chip for capture of His-tagged viral or host proteins.
2. CM5 Sensor Chip for covalent amine coupling of proteins.
3. Streptavidin (SA) Sensor Chip for capturing biotinylated ligands.
4. Protein A Sensor Chip for oriented immobilization of Fc-tagged antibodies or receptors.

2.12 Instruments

1. Surface Plasmon Resonance System (e.g., Biacore T200 for high-sensitivity kinetic studies or OpenSPR for smaller-scale interaction studies).
2. Auxiliary laboratory equipment: microcentrifuge (≥12,000 × *g*), vortex mixer, refrigerated centrifuge for protein purification, degassing unit or vacuum pump, pH meter (±0.01 accuracy).

2.13 Software

1. Instrument control and data collection: Biacore T200 control software version 2.0 or higher, or OpenSPR OneView software.
2. Kinetic data analysis and model fitting: BIAevaluation software version 3.2 or TraceDrawer software for fitting to 1:1 Langmuir, heterogeneous, or bivalent bound analyte models.
3. Graphing and visualization (e.g., GraphPad Prism for plotting and statistical analysis, and Origin software for advanced graphing and data presentation).

3 Methods

3.1 Ligand Preparation: Expression and Purification of CHIKV Proteins/Antigen (see Note 4)

1. Express recombinant CHIKV proteins (e.g., E1, E2, capsid protein, nsP1–nsP4 nonstructural proteins, etc.) in suitable systems such as *E. coli*, mammalian cells, and insect cells, depending on the folding and post-translational modification requirements (*see* **Note 5**).
2. Purify using affinity chromatography corresponding to the protein tag (e.g., Ni-NTA for His-tagged proteins, Glutathione Sepharose for GST-tagged proteins) (*see* **Note 6**).
3. Characterize the protein/antigen purity and determine the concentration using a spectrophotometric method or NanoDrop (*see* **Notes 7 and 8**).

3.2 Analyte Preparation (see Note 9)

1. Small molecules: Thaw the stock solution dissolved in nuclease-free water or DMSO and dilute in running buffer immediately before SPR injection. Keep the final DMSO concentration $\leq$1% to prevent refractive index artifacts.
2. Nucleic acids: Thaw RNA (with RNase inhibitors) or DNA fragments on ice, centrifuge to remove particulates, and dilute in running buffer.
3. Small proteins used as an analyte for PPI studies. Thaw on ice, centrifuge or filter to remove precipitates, and degas them to avoid bubble formation.

3.3 Sensor Chip Preparation (see Note 10)

Choose the immobilization strategy depending on the ligand type, desired orientation, tag, and experimental requirements. Capture His-tagged proteins on Ni-NTA chips for reversible binding. Capture biotinylated proteins or biotinylated nucleic acids on streptavidin-coated chips for high-affinity and irreversible immobilization. Covalently immobilize proteins with primary amines via amine coupling on CM5 or similar carboxylated sensor chips (typically used when the protein lacks an affinity tag) (*see* **Note 10**).

3.4 Sensor Chip Installation and Equilibration (see Note 11)

1. Prime the system using the running buffer to make sure that all system tubing is properly equilibrated with the running buffer.
2. Install the newly opened sensor chip into the SPR instrument according to the manufacturer's instructions.
3. Prime the chip (three times) with a running buffer until a stable baseline is achieved.

3.5 pH Scouting for Amine Coupling or Direct Immobilization (see Notes 12 and 13)

1. Perform pH scouting to determine the optimal pH at which the protein is stable and efficiently immobilized.
2. Open Biacore T200 control software, then go to: File → Open → New wizard template → Immobilization pH scouting wizard → New.

3. Specify the buffers used for the experiments and their corresponding pH values in the wizard. Enter the protein/analyte injection time (typically 90–120 s), dissociation times (usually 60–90 s), and flow rate (10 μL/min). Also specify the regeneration solution, commonly used 50 mM NaOH.
4. Dilute the protein in different pH immobilization buffers (pH 4.0, 4.5, 5.0, and 5.5) (*see* **Note 12**).

3.6 Surface Activation for Covalent Amine Coupling on CM5 Chips (see Note 14)

1. Activate carboxyl groups on the chip surface using a mixture of NHS and EDC.
2. Open Biacore T200 control software, then go to: File → Open → New wizard template → Immobilization → New.
3. Select the sensor surface chip (CM5), the flow path (flow cell 2, as flow cell 1 acts as a reference), and the amine coupling method.
4. The wizard template prompts you to choose either a target immobilization level in RUs or to set a contact time and flow rate for your ligand (*see* **Note 12**).
5. If we choose a target immobilization level, the system will first inject 10 μL of the ligand at a flow rate of 5 μL/min. The system initially estimates the preconcentration rate based on the level reached and the slope of the sensorgram. It then completes the immobilization by adjusting the injection according to this preconcentration calculation to achieve the desired target level.
6. After setting the wizard template, prepare the reagents provided in the amine coupling kit (thaw EDC, NHS, and ethanolamine solutions).
7. Prepare the ligand in an appropriate immobilization buffer selected based on the above pH scouting results.
8. Eject the tray and place all the vials in their respective position as specified in the wizard template, then run the immobilization program.
9. For the manual method, open the Biacore T200 software and select Manual Run. Set the flow rate to 10 μL/min, choose flow cell 2, and click start.
10. Prepare and place the vials containing 10–50 μg/mL protein, the activation mixture (120 μL of 0.4 M EDC and 120 μL of 0.1 M NHS), and 1 M ethanolamine pH 8.5 (120 μL).
11. After a straight baseline (120 s), inject EDC + NHS mixture over the surface for 6–10 min for activation and then inject the ligand, and once the immobilization target is reached, inject ethanolamine for 5 min to block the remaining active groups of the surface chip.

3.7 Ligand Capturing for Affinity-Based Sensor Chips

1. For His-tag, charge Ni-NTA chip with 50 µM $NiCl_2$ and inject His-tagged protein at 50–200 nM until the desired immobilization is reached.
2. For biotinylated ligands, inject ligand onto a streptavidin-coated chip at a low flow rate (10–20 µL/min) until the desired immobilization response is achieved.
3. For Fc-tagged protein/antibody, inject onto Protein A surface for oriented capture.

3.8 Blocking/Quenching

1. For covalent surfaces, deactivate with 1 M ethanolamine (pH 8.5).
2. For capture surfaces, wash with the running buffer to remove the unbound ligand.

3.9 Baseline Stabilization

Flow the running buffer until a stable baseline is achieved before analyte injection.

3.10 Running Binding Assay

1. Inject analyte solutions at various concentrations (typically 1–1000 nM) over the immobilized ligand surface, depending on the affinity of the ligand and analyte (*see* **Notes 9, 15, and 16**).
2. Maintain a constant flow rate of 25 µL/min to ensure optimal mass transport and reproducible binding.
3. Association phase: Allow analyte binding for 120–180 s to monitor the association kinetics.
4. Dissociation phase: Continue buffer flow without analyte for 300–600 s.
5. Surface regeneration: For other sensor chips, it is suggested to test different regeneration solutions for the optimal buffer. For some ligand–analyte interactions, the running buffer itself can serve as the regeneration solution. For strong interactions, 10 mM Glycine-HCl (pH 1.5–3) is typically recommended. A list of regeneration buffers is provided in Sect. 2.7 under "Regeneration buffer." For the Ni-NTA chip, after each binding cycle, regenerate the sensor surface with 350 mM EDTA to remove bound analyte or His-tagged protein, followed by recharging with 50 µM $NiCl_2$ for subsequent runs.
6. Replicates and concentration range: Perform binding assays at a minimum of five analyte concentrations to enable accurate global kinetic curve fitting of k_a, k_d, and K_D values. Always perform the experiment in three independent replicates.

3.11 Data Processing and Affinity/Kinetic Analysis (see Note 17)

1. Reference subtraction: Subtract the response from the reference channel (without immobilized ligand) and the blank injection from all sensorgrams to remove bulk refractive index and nonspecific binding effects.

2. Baseline correction: Normalize all sensorgrams to establish a stable baseline before curve fitting.
3. Model fitting: Analyze the processed binding curves using a 1:1 Langmuir binding model to derive kinetic parameters: association rate constant (k_a), dissociation rate constant (k_d), and equilibrium dissociation constant (K_D). When the data deviate from 1:1 binding, evaluate fitting with heterogeneous ligand or bivalent analyte models to confirm binding behavior.
4. Perform kinetic fitting and residual analysis using BIAevaluation, Tracedrawer, or any other software packages.
5. Data visualization: Export fitted data and kinetic parameters to GraphPad Prism or Origin software for graphical representation and statistical analysis.

4 Notes

1. All reagents should be of molecular grade. Prepare all buffers using ultrapure water (Milli-Q) and filter through a 0.22 μm filter. Degas all buffers before use to avoid air bubbles during the SPR assay.
2. Always test regeneration conditions on a control surface before applying them to the experimental chip.
3. Select the chip based on the nature of the ligand and immobilization strategy.
4. Prepare the ligand prior to the start of the experiment. Always thaw the ligand on ice, followed by centrifugation for 5 min at 10,000 × *g* to remove aggregates, and optionally filter protein samples through 0.22 μm syringe filters.
5. Recombinant antigens can be produced in different systems depending on the complexity and post-translational requirements of the antigen. Mammalian cells (HEK293T, ExpiCHO, Expi293): Provide proper folding and post-translational modification, such as glycosylation, for viral glycoproteins. Insect cells (Sf9/Baculovirus): Suitable for high-level expression of complex proteins. Bacterial systems (*E. coli*): Useful for non-glycosylated proteins or small protein domains, and for obtaining high concentrations of protein.
6. Purification involves affinity chromatography, polishing steps such as size-exclusion chromatography to isolate a properly folded pure fraction of antigen, and buffer exchange and concentration into a physiologically compatible buffer. Purified antigen should always be quantified by standard protein assays, and small aliquots (0.1–2 mg/mL) of proteins should

be stored in appropriate buffers at −80°C to avoid repeated freeze–thaw cycles as well as to maintain stability.

7. Confirmatory analyses ensure that the antigen is suitable for downstream interaction studies, including SDS-PAGE to assess the purity and size of the antigen, Western blotting using specific-tag or CHIKV-specific antibodies, functional ELISA (where relevant) to confirm antigenicity, and analytical size-exclusion chromatography to assess aggregation.
8. Tagging consideration. Protein tags facilitate immobilization on SPR sensor chips or improve solubility. The tag should be positioned to avoid interference with binding sites. Protein tag cleavage can be performed to remove the tag if necessary for functional studies. Common tags included: His-tag, GST tag, biotinylation, and Fc-fusion.
9. Prepare a series of serial dilutions of the analyte, ranging in concentration above and below the expected K_D range, to obtain precise kinetic parameters if the analyte is known. However, if it is unknown, always perform preliminary binding (affinity) experiments before conducting new kinetic studies to determine the appropriate concentration range for subsequent kinetic analysis. If the analyte is sensitive, e.g., RNA as an analyte or small molecules as an analyte, perform all preparation steps on ice to prevent degradation. Gently mix the serial dilution properly and briefly centrifuge the samples to remove air bubbles before use.
10. Keep one channel as a reference (no protein immobilization) for baseline subtraction. Avoid over-immobilization to prevent mass transport limitations. Maintain consistent temperature during immobilization and experiments (typically 25°C). Ensure ligand and analyte buffers are compatible to avoid refractive index artifacts.
11. Perform all experiments at 25°C unless otherwise specified. Ensure buffer compatibility and remove all air bubbles before injections.
12. If the isoelectric point (pI) of the protein/antigen is nearly 5.5, use the pH between 4.0 and 5.0. If the ligand is acidic in nature, it is preferable to use covalent immobilization or non-covalent immobilization, as very low pH may negatively affect the protein/antigen.
13. The selection of optimal pH for immobilization depends on the sensorgram. The best selection sensorgram shows rapid binding of the ligand to the sensor surface and complete dissociation after the injection ends. If two or more pH conditions produce a similar pattern of sensorgram, always prefer the highest pH buffer for immobilization to minimize potential damage caused by a lower pH.

14. The immobilization is ideal when the active form of the ligand is homogeneously oriented on the chip surface. Importantly, a high concentration of ligand on the surface should be avoided, as it causes mass transport effects. Hence, low to moderate levels of ligand immobilization are generally recommended for kinetic measurements. To determine the appropriate immobilization level, one must calculate a theoretical Rmax. Rmax represents the theoretical maximum analyte binding response on the surface of the chip, recorded in terms of response Units (RUs). It is calculated using the Eq. (1):

$$R_{\max} = \frac{\text{Molecular weight of analyte}}{\text{Molecular weight of ligand}} \times \text{Amount of ligand in RU } (R_i) \times \text{Stoichiometry of reaction } (S) \quad (1)$$

For instance, to analyze the interaction between a 45 kDa ligand and a 15 kDa analyte, with one analyte binding site per ligand ($S=1$), ~600 RUs of ligand are immobilized on the chip surface to achieve an $R_{\max}$ of 200 RU.

$$R_i = \frac{45 \text{ kDa}}{15 \text{ kDa}} \times 200 \times 1 = 600 \text{ RUs}$$

15. Before going for kinetic analysis, it is recommended to first conduct a binding analysis using a concentration range of 0.1–50 μM. The results from this binding analysis help to understand the ligand–analyte interaction affinity to set the optimal concentration range for the kinetic study. For example, if the binding analysis shows that the RU response is optimal between 1 and 10 μM, then the kinetics experiment should focus on this range using a twofold dilution series to generate an appropriate sensorgram.
16. If the binding analysis produces a very high RU response within this range, it is better to use the nanomolar range, which is commonly employed for antigen–antibody interaction studies. Some PPIs are very strong, and in such cases, the nanomolar range is particularly suitable for kinetic experiments such as the interaction of the ACE2 receptor and the RBD of SARS-CoV-2.
17. For detailed information, the Getting Started BIAcore T200 and the instrument handbook are provided on the Cytiva website [73, 74].

Acknowledgments

The author acknowledges the constant support of the core facilities at the Department of Biosciences and Bioengineering, Indian Institute of Technology Roorkee, in the completion of this work. This work was supported by the Scheme for Transformational and Advanced Research in Sciences (STARS), Ministry of Education (MoE), under Project Reference No. STARS2/2023-0209.

Conflict of Interest The authors declare no competing interests.

References

1. Chikungunya. Available online: https://www.who.int/health-topics/chikungunya#tab=tab_1
2. Tsetsarkin KA, Vanlandingham DL, McGee CE, Higgs S (2007) A single mutation in chikungunya virus affects vector specificity and epidemic potential. PLoS Pathog 3:1895–1906. https://doi.org/10.1371/JOURNAL.PPAT.0030201
3. Vazeille M, Moutailler S, Coudrier D, Rousseaux C, Khun H, Huerre M, Thiria J, Dehecq JS, Fontenille D, Schuffenecker I et al (2007) Two chikungunya isolates from the outbreak of La Reunion (Indian Ocean) exhibit different patterns of infection in the mosquito, Aedes Albopictus. PLoS One 2:e1168. https://doi.org/10.1371/JOURNAL.PONE.0001168
4. Solignat M, Gay B, Higgs S, Briant L, Devaux C (2009) Replication cycle of chikungunya: a re-emerging arbovirus. Virology 393:183–197. https://doi.org/10.1016/J.VIROL.2009.07.024
5. Kril V, Aïqui-Reboul-Paviet O, Briant L, Amara A (2021) New insights into chikungunya virus infection and pathogenesis. Annu Rev Virol 8(1):327–347
6. Chikungunya Vaccines|Chikungunya Virus|CDC. Available online: https://www.cdc.gov/chikungunya/vaccines/index.html
7. Homola J, Yee SS, Gauglitz G (1999) Surface plasmon resonance sensors: review. Sens Actuators B Chem 54:3–15. https://doi.org/10.1016/S0925-4005(98)00321-9
8. Exploring Optimal Interaction Techniques|SPR vs ITC vs MST vs BLI. Available online: https://nicoyalife.com/blog/spr-vs-itc-vs-mst-vs-bli/
9. Pareek A, Kumar R, Mudgal R, Neetu N, Sharma M, Kumar P, Tomar S (2022) Alphavirus antivirals targeting RNA-dependent RNA polymerase domain of NsP4 divulged using surface plasmon resonance. FEBS J 289:4901–4924. https://doi.org/10.1111/FEBS.16397
10. Nehul S, Rani R, Walia P, Panda PK, Singh S, Chattopadhyay S, Tomar S (2025) Repurposing efavirenz, the HIV antiretroviral drug for chikungunya virus infection. ACS Infect Dis 11(4):963–976
11. Moore CM, Paul MJ, Pinneh E, Shanmugaraj B, Ashall J, Ramalingam S, Hewson R, Diamond MS, Fox JM, Ma JKC (2025) Characterisation of chikungunya virus neutralising monoclonal antibodies expressed in tobacco plants. J Biotechnol 406:53–59. https://doi.org/10.1016/J.JBIOTEC.2025.06.014
12. Si L, Meng K, Tian Z, Sun J, Li H, Zhang Z, Soloveva V, Li H, Fu G, Xia Q et al (2018) Triterpenoids manipulate a broad range of virus-host fusion via wrapping the HR2 domain prevalent in viral envelopes. Sci Adv 4:eaau8408. https://doi.org/10.1126/SCIADV.AAU8408
13. Zhang H, Shi S, Sun L, Li S, Zhang Y, Li Z, Hou J, Li P, Shen J, Cheng X et al (2025) Screening and discovery of an antiviral candidate inhibiting the SARS-CoV-2 envelope (2-E) channel. Curr Res Microb Sci 9:100409. https://doi.org/10.1016/J.CRMICR.2025.100409
14. Cairns TM, Ditto NT, Atanasiu D, Lou H, Brooks BD, Saw WT, Eisenberg RJ, Cohen GH (2019) Surface plasmon resonance reveals direct binding of herpes simplex virus glycoproteins GH/GL to GD and locates a GH/GL binding site on GD. J Virol 93:e00289-19. https://doi.org/10.1128/JVI.00289-19
15. Yoo H, Shin J, Sim J, Cho H, Hong S (2020) Reusable surface plasmon resonance biosensor chip for the detection of H1N1 influenza virus. Biosens Bioelectron 168:112561. https://doi.org/10.1016/J.BIOS.2020.112561
16. Barth H, Schnober EK, Zhang F, Linhardt RJ, Depla E, Boson B, Cosset F-L, Patel AH, Blum HE, Baumert TF (2006) Viral and cellular determinants of the hepatitis C virus envelope-heparan sulfate interaction. J Virol

80:10579–10590. https://doi.org/10.1128/JVI.00941-06

17. Genova-Kalou P, Dyankov G, Marinov R, Mankov V, Belina E, Kisov H, Strijkova-Kenderova V, Kantardjiev T (2021) SPR-based kinetic analysis of the early stages of infection in cells infected with human coronavirus and treated with hydroxychloroquine. Biosensors (Basel) 11:251. https://doi.org/10.3390/BIOS11080251
18. Zhang R, Kim AS, Fox JM, Nair S, Basore K, Klimstra WB, Rimkunas R, Fong RH, Lin H, Poddar S et al (2018) Mxra8 is a receptor for multiple arthritogenic alphaviruses. Nature 557:570–574. https://doi.org/10.1038/S41586-018-0121-3
19. Torii S, Orba Y, Sasaki M, Tabata K, Wada Y, Carr M, Hobson-Peters J, Hall RA, Takada A, Fukuhara T et al (2020) Host ESCRT factors are recruited during chikungunya virus infection and are required for the intracellular viral replication cycle. J Biol Chem 295:7941. https://doi.org/10.1074/JBC.RA119.012303
20. Liu Y, Wang J, Han Y, Xia X, Zeng R, Fan X, Zhang B, Wang K, Lei J (2025) Cryo-EM reveals a double oligomeric ring scaffold of the CHIKV NsP3 peptide in complex with the NTF2L domain of Host G3BP1. mBio 16:e0396724. https://doi.org/10.1128/MBIO.03967-24
21. Fros JJ, Domeradzka NE, Baggen J, Geertsema C, Flipse J, Vlak JM, Pijlman GP (2012) Chikungunya virus NsP3 blocks stress granule assembly by recruitment of G3BP into cytoplasmic foci. J Virol 86:10873–10879. https://doi.org/10.1128/JVI.01506-12
22. Yin P, Jian X, Liu Y, Liu Y, Lv L, Cui H, Zhang L (2023) Elucidating cellular interactome of chikungunya virus identifies host dependency factors. Virol Sin 38:497–507. https://doi.org/10.1016/J.VIRS.2023.05.007
23. Mahajan S, Kumar R, Singh A, Dhaka P, Pareek A, Kumar P, Tomar S (2025) Targeting the host protein G3BP1 for the discovery of novel antiviral inhibitors against chikungunya virus. Virology 608:110551. https://doi.org/10.1016/J.VIROL.2025.110551
24. Yang Z, Johnson BA, Meliopoulos VA, Ju X, Zhang P, Hughes MP, Wu J, Koreski KP, Clary JE, Chang TC et al (2024) Interaction between host G3BP and viral nucleocapsid protein regulates SARS-CoV-2 replication and pathogenicity. Cell Rep 43:113965. https://doi.org/10.1016/J.CELREP.2024.113965
25. Broderick K, Moutaoufik MT, Saccon T, Malty R, Amin S, Phanse S, Joseph TP, Zilocchi M, Hosseinnia A, Istace Z et al (2025) Human protein interaction networks of ancestral and variant SARS-CoV-2 in organ-specific cells and bodily fluids. Nat Commun 16: 5784. https://doi.org/10.1038/s41467-025-60949-1
26. Bhutkar M, Saha A, Tomar S (2024) Viral methyltransferase inhibitors: berbamine, venetoclax, and ponatinib as efficacious antivirals against chikungunya virus. Arch Biochem Biophys 759:110111
27. Feibelman KM, Fuller BP, Li L, LaBarbera DV, Geiss BJ (2018) Identification of small molecule inhibitors of the Chikungunya virus nsP1 RNA capping enzyme. Antivir Res 154:124–131
28. Delang L, Li C, Tas A, Quérat G, Albulescu IC, De Burghgraeve T, Guerrero NS, Gigante A, Piorkowski G, Decroly E, Jochmans D (2016) The viral capping enzyme nsP1: a novel target for the inhibition of chikungunya virus infection. Sci Rep 6(1):31819
29. Mudgal R, Bharadwaj C, Dubey A, Choudhary S, Nagarajan P, Aggarwal M, Ratra Y, Basak S, Tomar S (2022) Selective estrogen receptor modulators limit alphavirus infection by targeting the viral capping enzyme nsP1. Antimicrob Agents Chemother 66(3):e01943–21
30. Deng Q, Guo Z, Hu H, Li Q, Zhang Y, Wang J, Liao C, Guo C, Li X, Chen Z et al (2022) Inhibition of chikungunya virus early replication by intracellular nanoantibodies targeting NsP2 epitope rich region. Antiviral Res 208:105446. https://doi.org/10.1016/J.ANTIVIRAL.2022.105446
31. Kalam N, Ali R, Balasubramaniam VR (2025) Exploring the potential of direct-acting antivirals against chikungunya virus through structure-based drug repositioning and molecular dynamic simulations. Comput Biol Med 189:109989
32. Mastalipour M, Gering I, Coronado MA, González JE, Willbold D, Eberle RJ (2025) Novel peptide inhibitor for the chikungunya virus nsP2 protease: identification and characterization. Curr Res Microb Sci 8:100376
33. Sharma KB, Subramani C, Ganesh K, Sharma A, Basu B, Balyan S, Sharma G, Ka S, Deb A, Srivastava M, Chugh S (2024) Withaferin A inhibits chikungunya virus nsP2 protease and shows antiviral activity in the cell culture and mouse model of virus infection. Plos Pathogens 20(12):e1012816
34. Mastalipour M, Coronado MA, Hernández González JE, Willbold D, Eberle RJ (2026) Inhibition of Chikungunya virus nsP2 protease in vitro by scorpion venom peptide pantinin-1. PloS One. 21(4):e0346930. https://doi.org/10.1371/journal.pone.0346930
35. Singh H, Mudgal R, Narwal M, Kaur R, Singh VA, Malik A, Chaudhary M, Tomar S (2018) Chikungunya virus inhibition by peptidomimetic inhibitors targeting virus-specific cysteine protease. Biochimie. 149:51–61

36. Freire MC, Basso LG, Mendes LF, Mesquita NC, Mottin M, Fernandes RS, Policastro LR, Godoy AS, Santos IA, Ruiz UE, Caruso IP (2022) Characterization of the RNA-dependent RNA polymerase from chikungunya virus and discovery of a novel ligand as a potential drug candidate. Sci Rep.22 12(1):10601
37. Yin P, May NA, Lello LS, Fayed A, Parks MG, Drobish AM, Wang S, Andrews M, Sticher Z, Kolykhalov AA, Natchus MG (2024) 4′ -Fluorouridine inhibits alphavirus replication and infection in vitro and in vivo. Mbio.12 15(6):e00420–24
38. Rothan HA, Bahrani H, Mohamed Z, Teoh TC, Shankar EM, Rahman NA, Yusof R (2015) A combination of doxycycline and ribavirin alleviated chikungunya infection. PloS one. 13 10(5):e0126360
39. Franco EJ, Rodriquez JL, Pomeroy JJ, Hanrahan KC, Brown AN (2018) The effectiveness of antiviral agents with broad-spectrum activity against chikungunya virus varies between host cell lines. Antivir Chem Chemother 26:2040206618807580
40. Ferreira AC, Reis PA, de Freitas CS, Sacramento CQ, Villas Bôas Hoelz L, Bastos MM, Mattos M, Rocha N, Gomes de Azevedo Quintanilha I, da Silva Gouveia Pedrosa C, Rocha Quintino Souza L (2019) Beyond members of the Flaviviridae family, sofosbuvir also inhibits chikungunya virus replication. Antimicrob Agents Chemother 63(2):10–128.
41. Fatma B, Kumar R, Singh VA, Nehul S, Sharma R, Kesari P, Kuhn RJ, Tomar S (2020) Alphavirus capsid protease inhibitors as potential antiviral agents for chikungunya infection. Antivir Res 179:104808
42. Ho YJ, Wang YM, Lu JW, Wu TY, Lin LI, Kuo SC, Lin CC (2015) Suramin inhibits chikungunya virus entry and transmission. PLoS One 10(7):e0133511
43. Zhou QF, Fox JM, Earnest JT, Ng TS, Kim AS, Fibriansah G, Kostyuchenko VA, Shi J, Shu B, Diamond MS, Lok SM (2020) Structural basis of chikungunya virus inhibition by monoclonal antibodies. Proc Natl Acad Sci. 3 117(44):27637–27645
44. Quiroz JA, Malonis RJ, Thackray LB, Cohen CA, Pallesen J, Jangra RK, Brown RS, Hofmann D, Holtsberg FW, Shulenin S, Nyakatura EK (2019) Human monoclonal antibodies against chikungunya virus target multiple distinct epitopes in the E1 and E2 glycoproteins. PLoS Pathog. 7 15(11):e1008061
45. Tumkosit U, Siripanyaphinyo U, Takeda N, Tsuji M, Maeda Y, Ruchusatsawat K, Shioda T, Mizushima H, Chetanachan P, Wongjaroen P, Matsuura Y (2020) Anti-chikungunya virus monoclonal antibody that inhibits viral fusion and release. J Virol.15 94(19):10–128
46. Han X, Ji C, Tian S, Wang F, Cao GP, Li D, Duan X, Tong Z, Qi J, Wang Q et al (2025) Neutralizing antibodies against chikungunya virus and structural elucidation of their mechanism of action. Nat Commun 16:9682. https://doi.org/10.1038/s41467-025-64687-2
47. Pal P, Dowd KA, Brien JD, Edeling MA, Gorlatov S, Johnson S, Lee I, Akahata W, Nabel GJ, Richter MKS et al (2013) Development of a highly protective combination monoclonal antibody therapy against chikungunya virus. PLoS Pathog 9:e1003312. https://doi.org/10.1371/JOURNAL.PPAT.1003312
48. Goto K, Amano R, Ichinose A, Michishita A, Hamada M, Nakamura Y, Takahashi M (2025) Generation of RNA aptamers against chikungunya virus E2 envelope protein. J Virol 99:e0209524. https://doi.org/10.1128/JVI.02095-24
49. Coppock MB, Hurley M, Jones C, Erickson D, Stratis-Cullum DN (2019) A novel discovery, maturation, and assay integration approach for the development of ruggedized multi-valent capture receptors exemplified against the chikungunya virus E2 protein. Sens Bios Res 22:100248. https://doi.org/10.1016/J.SBSR.2018.100248
50. Song L, Fu G, Li J, Chen Z, Fu L, Yu C, Qiang L, Li J, Fang T, Yuan H et al (2025) Highly potent neutralizing nanobodies acting against chikungunya virus infection via inhibiting multiple stages of the viral life cycle. Int J Mol Sci 26:3982. https://doi.org/10.3390/IJMS26093982
51. Sharma A, Bhomia M, Yeh TJ, Singh J, Maheshwari RK (2018) Miltefosine inhibits chikungunya virus replication in human primary dermal fibroblasts. F1000Research 7:9
52. Nam S, Ga YJ, Lee JY, Hwang WY, Jung E, Shin JS, Chen W, Choi G, Zhou B, Yeh JY, Go YY (2021) Radicicol inhibits chikungunya virus replication by targeting nonstructural protein 2. Antimicrob Agents Chemother 65(7):10–128
53. Voss JE, Vaney MC, Duquerroy S, Vonrhein C, Girard-Blanc C, Crublet E, Thompson A, Bricogne G, Rey FA (2010) Glycoprotein organization of chikungunya virus particles revealed by X-ray crystallography. Nature 468:709–712. https://doi.org/10.1038/nature09555
54. Gibbons DL, Vaney MC, Roussel A, Vigouroux A, Reilly B, Lepault J, Kielian M, Rey FA (2004) Conformational change and protein–protein interactions of the fusion protein of Semliki Forest virus. Nature 427:320–325. https://doi.org/10.1038/nature02239
55. Li Z-Q, Zhao L-X, Wang S-Y, Hu C-Y, Wang Y-Y, Yang Y (2024) YBX1 is required for assembly of viral replication complexes of chikungunya virus and replication of multiple

alphaviruses. J Virol 99:e02015–24. https://doi.org/10.1128/JVI.02015-24
56. Taschuk F, Tapescu I, Moy RH, Cherry S (2020) Ddx56 binds to chikungunya virus RNA to control infection. mBio 11:1–16. https://doi.org/10.1128/MBIO.02623-20
57. Thompson MG, Horner SM (2023) The evil DExH/D: chikungunya virus runs but cannot hide from DDX39A. Mol Cell 83:3948–3949. https://doi.org/10.1016/J.MOLCEL.2023.10.032
58. Tripathi PK, Soni A, Singh Yadav SP, Kumar A, Gaurav N, Raghavendhar S, Sharma P, Sunil S, Ashish, Jayaram B et al (2020) Evaluation of novobiocin and telmisartan for anti-CHIKV activity. Virology 548:250–260. https://doi.org/10.1016/J.VIROL.2020.05.010
59. Tripathi PK, Singh J, Gaurav N, Garg DK, Patel AK (2020) In-silico and biophysical investigation of biomolecular interaction between naringin and NsP2 of the chikungunya virus. Int J Biol Macromol 160:1061–1066. https://doi.org/10.1016/J.IJBIOMAC.2020.05.165
60. Rani R, Nehul S, Choudhary S, Upadhyay A, Sharma GK, Kumar P, Tomar S (2023) Revealing and evaluation of antivirals targeting multiple druggable sites of RdRp complex in SARS-CoV-2. bioRxiv, 2023.07.24.550324. https://doi.org/10.1101/2023.07.24.550324
61. Li Y, Zhang J, Duan Z, Wang N, Sun X, Zhang Y, Fu L, Liu K, Yang Y, Pan S et al (2021) High-throughput screening and evaluation of repurposed drugs targeting the SARS-CoV-2 main protease. Signal Transduct Target Ther 6:356. https://doi.org/10.1038/s41392-021-00763-5
62. Singh A, Jangid K, Nehul S, Dhaka P, Rani R, Pareek A, Sharma GK, Kumar P, Tomar S (2025) Structural and mechanistic insights into the main protease (Mpro) dimer interface destabilization inhibitor: unveiling new therapeutic avenues against SARS-CoV-2. Biochemistry 64:1589–1605. https://doi.org/10.1021/ACS.BIOCHEM.4C00535
63. Sharma R, Kesari P, Kumar P, Tomar S (2018) Structure-function insights into chikungunya virus capsid protein: small molecules targeting capsid hydrophobic pocket. Virology 515:223–234. https://doi.org/10.1016/J.VIROL.2017.12.020
64. Khalenkov AM, Norton MG, Scott DE (2023) Method for screening influenza neutralizing antibodies in crude human plasma and its derivatives using SPR. Heliyon 9:e15651. https://doi.org/10.1016/j.heliyon.2023.e15651
65. Khurana S, Chung KY, Coyle EM, Meijer A, Golding H (2016) Antigenic fingerprinting of antibody response in humans following exposure to highly pathogenic H7N7 avian influenza virus: evidence for anti-PA-X antibodies. J Virol 90:9383–9393. https://doi.org/10.1128/JVI.01408-16
66. Khurana S, Hahn M, Coyle EM, King LR, Lin TL, Treanor J, Sant A, Golding H (2019) Repeat vaccination reduces antibody affinity maturation across different influenza vaccine platforms in humans. Nat Commun 10:3338. https://doi.org/10.1038/s41467-019-11296-5
67. Scott Smith AA, Silva LA, Fox JM, Diamond MS, Dermody TS, Crowe JE, Smith SA, Flyak AI, Kose N et al (2015) Isolation and characterization of broad and ultrapotent human monoclonal antibodies with therapeutic activity against chikungunya virus. Cell Host Microbe 18(1):86-95. https://doi.org/10.1016/j.chom.2015.06.009
68. Song H, Zhao Z, Chai Y, Qi J, Gao F, Gao G F (2019) Molecular basis of arthritogenic alphavirus receptor MXRA8 binding to chikungunya virus envelope. Cell 177:1714–1724.e12. https://doi.org/10.1016/j.cell.2019.04.008
69. Basore K, Kim AS, Nelson CA, Zhang R, Smith BK, Uranga C, Vang L, Cheng M, Gross ML, Smith J et al (2019) Cryo-EM structure of chikungunya virus in complex with the Mxra8 receptor. Cell 177:1725. https://doi.org/10.1016/J.CELL.2019.04.006
70. Singh V, Nehul S, Kumar CS, Banerjee M, Kumar P, Sharma G, Tomar S (2023) Chimeric chikungunya virus-like particles with surface exposed SARS-CoV-2 RBD elicits potent immunogenic responses in mice. bioRxiv, 2023.01.29.526074. https://doi.org/10.1101/2023.01.29.526074
71. Ashiba H, Sugiyama Y, Wang X, Shirato H, Higo-Moriguchi K, Taniguchi K, Ohki Y, Fujimaki M (2017) Detection of norovirus virus-like particles using a surface plasmon resonance-assisted fluoroimmunosensor optimized for quantum dot fluorescent labels. Biosens Bioelectron 93:260–266. https://doi.org/10.1016/J.BIOS.2016.08.099
72. Rumi RB, Paul AK, Alyami SA, Moni MA (2023) Multi-disease detection using a prism-based surface plasmon resonance sensor: a TMM and FEM approach. IEEE Trans NanoBiosci 23(1):51–62
73. Cytiva: Biacore T200 Software Handbook (Version 3.2.1). Cytiva (2024)
74. GE Healthcare: Biacore T200 Getting Started (Edition AB 28-9840-98). GE Healthcare (2013)

Check for updates

Chapter 18

Analysis of Cell Populations from Arbovirus-Infected Tissues by Flow Cytometry

Jenniffer Ramos Martins, Viviane Lima Batista, Celso M. Queiroz-Junior, Mauro Martins Teixeira and Vivian Vasconcelos Costa

Abstract

Arboviruses still represent a major challenge to public health in many parts of the world. Despite shared epidemiological characteristics, each arboviral disorder displays distinct cellular tropism and activation patterns, progressing to immunopathological outcomes that remain not fully elucidated. A deeper understanding of these processes requires methodologies capable of characterizing the often-complex cellular responses. In this context, flow cytometry emerges as a valuable tool, enabling the characterization of heterogeneous cell populations, the assessment of activation and proliferation markers, the quantification of inflammatory mediators, and the detection of cell death pathways. By integrating these parameters, flow cytometry can enhance our understanding of arbovirus pathogenesis, clarify their cellular tropism, and understand immune responses to infection. This methodological article outlines the main steps for applying the technique to different target tissues infected by chikungunya, dengue, yellow fever, and Zika viruses, emphasizing the most effective strategies.

Keywords Flow cytometry, Arbovirus, Chikungunya, Dengue, Zika, Yellow fever

1 Introduction

Arboviruses are viruses transmitted to humans through the bites of infected female *Aedes aegypti* mosquitoes, during which viral transmission occurs via the transfer of blood [1]. This mode of transmission exerts a profound impact on population health, particularly in tropical and subtropical regions where environmental conditions favor both vector proliferation and disease spread [2]. Among the arboviruses of greatest public health importance are Chikungunya virus (CHIKV), belonging to the family *Togaviridae* [3] and genus *Alphavirus*, primarily employing the glycoproteins E1 and E2 to interact with host receptors such as Mxra8 [4]. Following infection, the virus replicates in fibroblasts, keratinocytes, macrophages, and

The authors Jenniffer Ramos Martins and Viviane Lima Batista contributed equally to this work.

Claudia Filomatori and Eliana F. Castro (eds.), *Chikungunya Virus*, Methods in Molecular Biology, vol. 3067,
https://doi.org/10.1007/978-1-0716-5496-5_18,

subsequently disseminates to musculoskeletal (myocytes) and synovial tissues [5, 6].

Infection induces a strong immune response characterized by cellular activation and the production of inflammatory cytokines and chemokines during the acute phase (IL-1Rα, IL-6, IL-7, IL-8, IL-12, IL-15, IFNα, CCL2, CCL4, CXCL10, and IL-16). This phase lasts 2–6 days and manifests with fever, headache, myalgia, arthralgia, and nausea [7, 8]. In the chronic phase, the disease is associated with a specific cytokine profile (IL-1β, IL-6, IL-17, TNF, GM-CSF, RANKL, and CCL2), which contributes to extracellular matrix degradation, leading to bone resorption, erosion, and joint damage in patients with chronic chikungunya arthritis [9]. This immune activation can also drive skeletal muscle atrophy [10], resulting in persistent arthralgia, highlighting the virus's tropism for both musculoskeletal and immune system cells [11].

Other clinically relevant arboviruses within the *Flaviviridae* family, including dengue virus (DENV), Zika virus (ZIKV), and yellow fever virus (YFV), elicit a robust host immune response following cell entry mediated by the envelope (E) protein [1, 12, 13]. Viral RNA is detected by monocytes, macrophages, and dendritic cells through pattern recognition receptors, leading to the induction of type I interferons and a wide range of proinflammatory cytokines and chemokines. Excessive activation of these pathways may drive systemic inflammation, tissue injury, and multiorgan dysfunction, as exemplified in severe dengue [14]. Despite their distinct cellular tropism, these viruses share common mechanisms of innate immune recognition and inflammatory activation that are central to disease pathogenesis [15–17].

Despite these similarities, these arboviruses induce virus-specific immune response signatures, highlighting the need for advanced techniques to investigate such responses. Among the various biomedical techniques currently employed, flow cytometry stands out in this context for enabling the identification of individual cell populations, the evaluation of surface proteins, intracellular structures, and cellular pathophysiological status [18]. The principle of flow cytometry relies on laser excitation of fluorochromes conjugated to antibodies bound to cell surface or intracellular targets. Labeled cells are carried in a continuous hydrodynamic flow and pass through one or more laser beams, which excite the fluorochromes and generate wavelength-dependent light emission [19, 20].

Forward scatter (FSC) provides information on cell size, while side scatter (SSC) reflects cellular granularity [21]. The emitted light and fluorescence signals are detected, converted into digitized electrical pulses, and processed by specialized software, enabling the identification and characterization of heterogeneous cell populations [22]. Flow cytometry is a powerful approach for elucidating viral tropism and advancing the understanding of immunopathological pathways underlying severe clinical outcomes. Its broad applicability to diverse sample types further reinforces its value as a key technique in the study of viral infections, particularly arboviruses [23]. In this chapter, we describe the application of a myeloid panel for flow cytometric immunophenotyping to investigate cellular recruitment in murine models of arbovirus infection.

2 Materials

2.1 Processing of Tissues

1. RPMI-1640 medium.
2. Fetal bovine serum (FBS).
3. Serological pipettes or Pasteur pipettes.
4. 2 mL microtubes.
5. 0.5 mL EDTA-K2 blood collection tubes.
6. 50 mL centrifuge tubes.
7. 15 mL Falcon tubes.
8. ACK lysing buffer: 150 mM NH_4Cl, 10 mM $KHCO_3$, 0.1 mM EDTA.
9. Micropipettes.
10. Sterile tips.
11. Refrigerated centrifuge.
12. 10 mL syringe.
13. Collagenase type IV: 3 mg/mL diluted in RPMI-1640.
14. Collagenase type III: 1 mg/mL diluted in RPMI-1640.
15. PBS 1×: 137 mM NaCl, 2.7 mM KCl, 10 mM Na_2HPO_4, 2 mM KH_2PO_4 pH 7.4.
16. 40 or 70 μm nylon cell strainer.
17. 96-well V-bottom plate.
18. Formaldehyde 37% (analytical grade).
19. Stainless steel scalpel blade No. 21.
20. Percoll.
21. Digestion solution (see Sect. 2.3.1).
22. Water bath.

2.2 Cell Suspension Counting

1. 0.4% w/v Trypan blue.
2. Micropipettes.
3. Sterile tips.
4. Microtubes.

5. Neubauer chamber and microscope, or an automated cell counter.

2.3 Solutions

1. Digestion solution: 1 mg/mL collagenase type I and 1 mg/mL DNase, in RPMI-1640 medium supplemented with 10% fetal bovine serum (FBS). Mix thoroughly to resuspend the enzymes. Always prepare this solution fresh immediately before use.
2. FACS Buffer: 2% FBS in PBS 1×.
3. Blocking Buffer: 10% FBS in PBS 1×.
4. 2% Formaldehyde: To prepare 50 mL, add 2.70 mL of 37% formaldehyde stock solution and adjust the final volume with PBS 1× pH 7.2.
5. Fixation and Permeabilization Buffer: Use a commercial Intracellular Fixation and Permeabilization Buffer Set (*see* **Note 1**).

3 Methods

3.1 Chikungunya Virus Infection

For tissue collection and flow cytometry analysis of CHIKV infection, a previously described murine model of CHIKV-induced arthralgia in immunocompetent mice is recommended. In this model, animals develop persistent hypernociception, joint inflammation, and tissue damage, providing a relevant system for investigating the pathogenesis of CHIKV infection. This model allows standardized tissue sampling and cellular analysis using flow cytometry (see [24]).

Paw (Fig. 1A)

1. Remove the right ankle joint paw, mince the entire paw with scissors, and transfer the fragments into a 2 mL microtube containing 500 μL of RPMI medium supplemented with collagenase type III (1 mg/mL) (*see* **Note 2**).
2. Incubate the tube at 37°C in a water bath for 1 h, gently agitating every 15 min.
3. Add 500 μL of fetal bovine serum (FBS) to stop the collagenase activity.
4. Pass the digested suspension through a 70 μm cell strainer. Use the plunger of a 10 mL syringe to gently grind the tissue debris and rinse the strainer with 10 mL of RPMI medium supplemented with 10% FBS to recover the cells.
5. Centrifuge the cell suspension at 400 × *g* for 10 min at 4°C.
6. Carefully discard the supernatant (*see* **Note 3**).

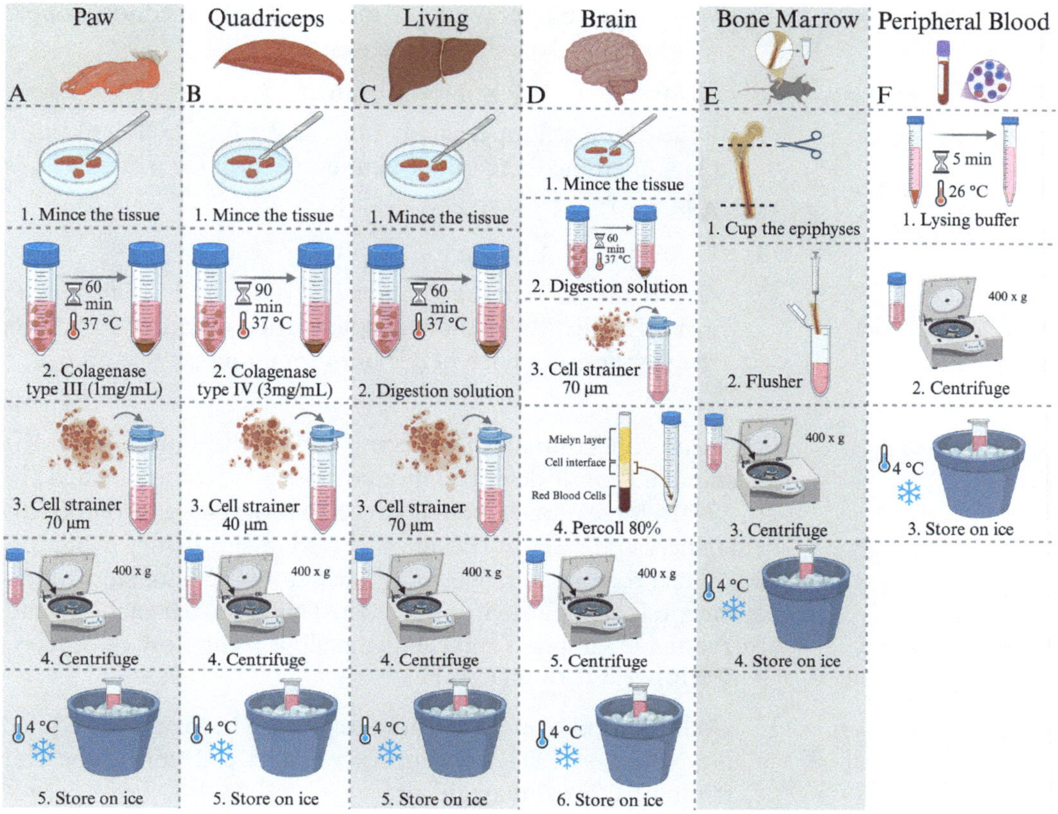

Fig. 1 The main steps in processing different tissues for flow cytometry. (Created with BioRender.com. (see [26]) Available at: https://BioRender.com/dwaj989)

7. Resuspend the pellet in 1 mL of RPMI medium supplemented with 10% FBS and keep the cell suspension on ice (4°C) until staining.

Quadriceps (Fig. 1B)

1. Dissect the hind limb muscles of the mice, identify, and remove the quadriceps muscles (*see* details in [25]).
2. After removing the quadriceps, cut the tissue into several pieces using sterile scissors.
3. Transfer the tissue pieces to a tube containing RPMI medium supplemented with DNase (1 mg/mL) and collagenase type IV (3 mg/mL) (*see* **Note 2**).
4. Incubate the tube at 37°C for 1 h and 30 min in a water bath, gently agitating every 15 min.
5. Add 500 µL of fetal bovine serum (FBS) to stop the collagenase activity.
6. Pass the suspension through a 40 µm cell strainer, using the plunger of a 10 mL syringe to gently grind the tissue debris.

7. Rinse the strainer with 10 mL of RPMI medium supplemented with 10% FBS to recover the cells.
8. Centrifuge at 400 × *g* for 10 min at 22°C.
9. Discard the cell supernatant and add 9 mL of ACK lysis buffer (1×), then incubate under gentle agitation for 5 min.
10. Add 1 mL of PBS 10× to stop the lysis.
11. Centrifuge at 400 × *g* for 10 min at 4°C.
12. Gently discard the supernatant and resuspend the final pellet in 1 mL of RPMI medium supplemented with 10% FBS. Keep the suspension on ice (4°C) until staining (see Sect. 3.6.1)

3.2 Dengue Virus Infection

For tissue collection and flow cytometry analysis in dengue virus infection, previously described murine models are recommended [14, 26, 27]. In these models, animals develop progressive thrombocytopenia, increased vascular leakage, elevated inflammatory mediators, and cellular recruitment, providing a relevant system for studying the pathogenesis of dengue virus infection. This approach enables standardized tissue sampling and cellular analysis using flow cytometry.

Peripheral Blood (Fig. 1F)

1. Collect blood by cardiac puncture using a 0.45 mm (26G) needle bathed in EDTA and transfer it to a collection tube containing EDTA anticoagulant.
2. Transfer the blood from the EDTA tube to a 50 mL Falcon tube containing ACK lysing buffer (1×), using a ratio of 10 mL buffer per 1 mL of blood. Incubate at room temperature (20°C–26°C) with agitation for 5 min.
3. Add 1 mL of PBS 10× to stop the lysis.
4. Centrifuge at 400 × *g* for 10 min at 4°C.
5. Discard the supernatant (*see* **Note 3**).
6. Resuspend the pellet in 1 mL of RPMI medium supplemented with 10% fetal bovine serum and keep the suspension on ice until staining. Keep the suspension on ice (4°C) until staining (see Sect. 3.6).

Bone Marrow (Fig. 1E)

1. Make a midline incision along the lower abdomen and extend it along both hind limbs to expose the leg bones (*see* details [28]).
2. Remove excess muscle tissue to fully expose the femur and tibia.
3. Dislocate the femur from the pubic bone and carefully remove it from the hind limb.
4. Remove remaining muscles and tendons from the bones.

5. Separate the femur from the tibia.
6. Cut the bone ends to expose the bone marrow (Fig. 1).
7. Fill a 1 mL syringe with 500 μL of cold RPMI-1640 medium and attach a 21G needle.
8. Flush the bone marrow into a 2 mL microtube (Fig. 1).
9. Gently resuspend bone marrow aggregates by pipetting several times using a serological pipette.
10. Inspect the bones: they should appear completely white. If red staining remains, additional marrow is present in the canal.
11. Centrifuge the cell suspension at 400 × *g* for 10 min at 4°C.
12. Carefully discard the supernatant (*see* **Note 3**).
13. Resuspend the pellet in 1 mL of RPMI-1640 supplemented with 10% FBS.
14. Keep the cell suspension on ice (4°C) until further staining (see Sect. 3.6).

3.3 Yellow Fever Infection

For tissue collection and flow cytometry analysis of yellow fever virus infection, previously described murine models are recommended [29, 30]. In these models, AG129 mice, which lack type I and II interferon receptors, are used. Following infection, animals develop hepatic injury, leukocytosis, and clinical signs consistent with human disease pathogenesis.

Liver processing (Fig. 1C) (*see* **Note 4**)

1. Administer a lethal dose of anesthetic to the mouse.
2. Working quickly, place the mice on a styrofoam support that allows blood to drain (*see* details in [31, 32]).
3. Lift the abdominal skin with forceps and, using scissors, cut the anterior abdominal wall. Extend the incision up to the diaphragm.
4. Take care not to cut the liver and make a small hole in the diaphragm.
5. Continue the incision laterally on both sides, right and left, until the heart is exposed.
6. Insert a 21G needle into the left ventricle of the heart and hold it firmly with forceps.
7. Slowly perfuse PBS 1× through the left ventricle until the liver is desvascularized (*see* details in [31, 32]).
8. After perfusion, carefully collect the liver (*see* details in [32]).
9. Transfer the fragments into a 2 mL round-bottom tube containing 1 mL of digestion solution (1 mg/mL Collagenase Type III,1 mg/mL DNAse in RPMI—see Sect. 2.2.3)

10. Incubate at 37°C in a bain-marie for 60 min.
11. Pass the suspension through a 70 μm cell strainer into a 50 mL tube, rinsing the filter (using a syringe) with 10 mL PBS 10×.
12. Centrifuge at 400 × *g* for 5 min at 4°C and discard the supernatant.
13. Resuspend the pellet in 9 mL of ACK lysing buffer and incubate for 5 min at room temperature (20°C–26°C).
14. Add 1 mL of PBS 10× to stop lysis, centrifuge again at 400 × *g* for 5 min at 4°C, and discard the supernatant.
15. Resuspend the pellet in 1 mL of FACS buffer.
16. Keep the cell suspension on ice (4°C) until further staining (see Sect.3.6).

3.4 Zika Virus Infection

For tissue collection and flow cytometry analysis, previously standardized murine models are recommended [33]. In this model, mice deficient in the type I interferon receptor (IFN-α/βR$^{-/-}$) show intense brain inflammation, observed at the peak of ZIKV infection.

Brain (Fig. 1D)

1. Perfuse the animal with PBS 1x prior to brain collection (see Sect. 3.3 Liver processing, items 1–7).
2. After perfusion, collect the brain by removing the mouse's head with large scissors, cutting through the neck.
3. Make a shallow cut in the auditory canal using fine dissection scissors to access the cranial cavity. Extend the cut transversely along the entire auditory canal. Then, make a longitudinal cut and remove tissue up to the region near the eyes (*see* details in [31]).
4. Chop the brain with scissors and transfer it to a microtube containing 1 mL of digestion solution (1 mg/mL collagenase + 1 mg/mL DNase in RPMI—see Sect. 2.3.1).
5. Incubate the tube at 37°C in a water bath for 60 min.
6. Dissociate the digested tissue using a syringe fitted with a 1.2 mm internal diameter needle until no solid residue remains.
7. Pass the suspension through a 70 μm cell strainer into a new 50 mL tube, rinsing the filter with 10 mL PBS.
8. Add 5 mL of 80% Percoll to a 50 mL round-bottom polypropylene tube.
9. Carefully layer 8 mL of the brain cell suspension on top of the 80% Percoll using a Pasteur pipette.
10. Centrifuge at 400 × *g* for 20 min at room temperature (20°C–26°C) with low acceleration (setting 1) and no brake.

11. Collect the intermediate interface layer (*see* Fig. 1D) and transfer it to a new 15 mL tube containing 10 mL PBS.
12. Centrifuge at 400 × *g* for 5 min at 4°C.
13. Discard the supernatant and resuspend the pellet in 2 mL of FACS buffer.
14. Keep the cell suspension on ice until staining (see Sect. 3.6).

3.5 Preparation of the Antibody Mix for Staining

Extracellular Staining. Prepare the antibody mix for each sample prior to starting the experiment. The selection of the panel should be based on the experimental question and the tissue of interest. The flowchart provides an example for evaluating the profile of myeloid cells in mouse peripheral blood, bone marrow, paw, quadriceps, and liver (*see* **Note 5**). Prepare the antibody mixture according to Table 1, using a final volume of 30 μL per sample, consisting of 1.65 μL of antibody and 28.35 μL of FACS buffer.

Intracellular staining. Prepare the antibody mixture according to Table 2, using a final volume of 40 μL per sample, consisting of 0.88 μL of antibody and 39.12 μL of FACS buffer.

3.6 Flow Cytometry Staining Protocol

3.6.1 Cell Counting

1. Take 10 μL of the cell suspension from each tissue indicated in Sect. 3.5 and mix with 0.4% Trypan blue at a 1:10 ratio in a microtube.
2. Allow the mixture to stand for 1–2 min at room temperature to stain dead cells.
3. Place a clean coverslip over a Neubauer counting chamber.
4. Load 10 μL of the cell-Trypan blue mixture into the chamber, avoiding bubble formation.

Table 1
Preparation of the antibody mix for extracellular staining

Fluorophore	Marker	Company	Clone	Final dilution	Volume (μL)
Percp-Cy5.5	CD45	BioLegend	Ly-5.2	1/200	0.15
Super Bright 600	CD11b	ThermoFisher	M1/70	1/200	0.15
Super Bright 645	CD11c	ThermoFisher	N418	1/200	0.15
FITC	Ly6G	eBioscience	1A8-Ly6g	1/200	0.15
APC	F4/80	ThermoFisher	BM8	1/200	0.15
APC-eFluor 780	MHC II	ThermoFisher	M5/114.15.2	1/50	1.66
PE	CD206	ThermoFisher	MR6F3)	1/100	0.3

Table 2
Preparation of the antibody mix for intracellular staining

Fluorophore	Marker	Company	Clone	Final dilution	Volume (μL)
Alexa fluor 700	IL10	BioLegend	M5/114.15.2	1/100	0.4
Texas Red	iNos	ThermoFisher	M5/114.15.2	1/500	0.08
Pacific blue	TNF	BioLegend	MP6-XT22	1/100	0.4

5. Let the cells settle for 1–2 min.
6. Count viable (unstained) and non-viable (blue-stained) cells under a light microscope (10× or 20× objective) in the four large corner squares.
7. Calculate the total cell concentration using:

Cell concentration (cells/mL) = Average cell counts per square × dilution factor × 10^4.

3.6.2 Cell Preparation

1. Transfer 1×10^6 cells from the tissue suspension into a 96-well V-bottom plate.
2. Centrifuge at 400 × *g* for 5 min at 4°C.
3. Discard the supernatant.
4. Add Brefeldin A (0.2 μL/well) diluted in RPMI-1640 medium supplemented with 10% FBS. For 50 wells, add 10 μL Brefeldin A to 500 μL complete medium and distribute 10 μL/well (*see* **Note 6**).
5. Incubate for 3–4 h at 37°C, protected from light.

3.6.3 Viability Staining

1. Centrifuge the plate (400 × *g*, 10 min, 4°C) and discard the supernatant.
2. Wash cells with 100 μL PBS 1×.
3. Add 50 μL of Live/Dead dye (1:1000 dilution in PBS 1×).
4. Incubate for 20 min at room temperature, protected from light.
5. Wash twice by adding 100 μL of PBS 1× (400 × *g*, 5 min, 4°C).

3.6.4 Blocking

1. Resuspend cells in 100 μL blocking buffer (PBS 1× + 10% FBS).
2. Incubate for 15 min at 4°C.
3. Wash once with 100 μL FACS buffer (PBS 1× + 2% FBS).
4. Centrifuge at 400 × *g* for 5 min at 4°C and discard the supernatant.

3.6.5 Extracellular Staining

1. Resuspend cells in 30 µL extracellular antibody mix.
2. Incubate for 20 min at 4°C, protected from light.
3. Centrifuge at 400 × *g* for 5 min at 4°C and discard the supernatant.
4. Wash with 100 µL FACS buffer and centrifuge (400 × *g*, 5 min, 4°C).
5. Repeat the wash step once more.

3.6.6 Fixation and Permeabilization with a Commercial Kit

1. Add 100 µL Fix/Perm solution (25% Fixation/Permeabilization Concentrate + Diluent, eBioscience).
2. Incubate for 35 min at 4°C, protected from light.
3. Centrifuge at 400 × *g* for 5 min at 4°C and discard the supernatant.
4. Resuspend cells in 100 µL Perm/Wash buffer.
5. Centrifuge again (400 × *g*, 5 min, 4°C) and discard the supernatant.

3.6.7 Intracellular staining (see Note 7)

1. Resuspend cells in 40 µL intracellular antibody mix (myeloid and/or lymphoid panels).
2. Incubate for 45 min at 4°C, protected from light.
3. Centrifuge at 400 × *g* for 5 min at 4°C and discard the supernatant.
4. Wash with 100 µL FACS buffer (400 × *g*, 5 min, 4°C).
5. Repeat wash once with Perm/Wash buffer (400 × *g*, 5 min, 4°C).

3.6.8 Final Fixation

If acquisition is planned within 24 h:

1. Resuspend cells in 200 µL of 2% formaldehyde.
2. Transfer cells to FACS tubes.
3. Store at 4°C, protected from light, until acquisition.

If acquisition is performed on the same day:

1. Resuspend cells in 200 µL FACS buffer.
2. Acquire samples immediately on the flow cytometer.
3. Exclude acquisition noise by retaining only events that are stable over time.

3.7 Expected Results

This protocol describes the flow cytometry steps for analyzing a myeloid panel across different tissues in murine models previously standardized in the context of arbovirus infection. The basic principle of the gating strategy is to use negative markers to exclude unwanted cell populations and positive markers to define the population of interest. Below, we provide a step-by-step guide for data analysis in FlowJo following flow cytometer acquisition. The gating analysis described below was adapted from previously published [34].

Initial Quality Control (Fig. 2A). After flow cytometer acquisition and data export to FlowJo, define a Time vs. FSC-A plot. This plot allows the exclusion of acquisition-related noise and ensures retention of stable events over time, improving the overall quality of the analysis.

1. Singlet selection (Fig. 2B). Define a plot: FSC-A vs. FSC-H (or FSC-A vs. FSC-W). The diagonal population represents single cells, while doublets and aggregates fall outside this gate and should be excluded. This step ensures that only single cells are analyzed, preventing errors in immunophenotyping caused by cell aggregates.
2. Viable cell selection (Fig. 2C). Define a plot: Live/Dead dye vs. FSC-A. Gate on viable cells, which are negative for the viability dye. This step ensures that only live cells are included in the analysis, increasing the accuracy of downstream immunophenotyping.
3. $CD45^+$ cell selection (Fig. 2D). Define a plot: CD45 vs. SSC-A. Gate on $CD45^+$ cells to identify the total leukocyte population. This step narrows the analysis to leukocytes, serving as the starting point for further immunophenotyping.

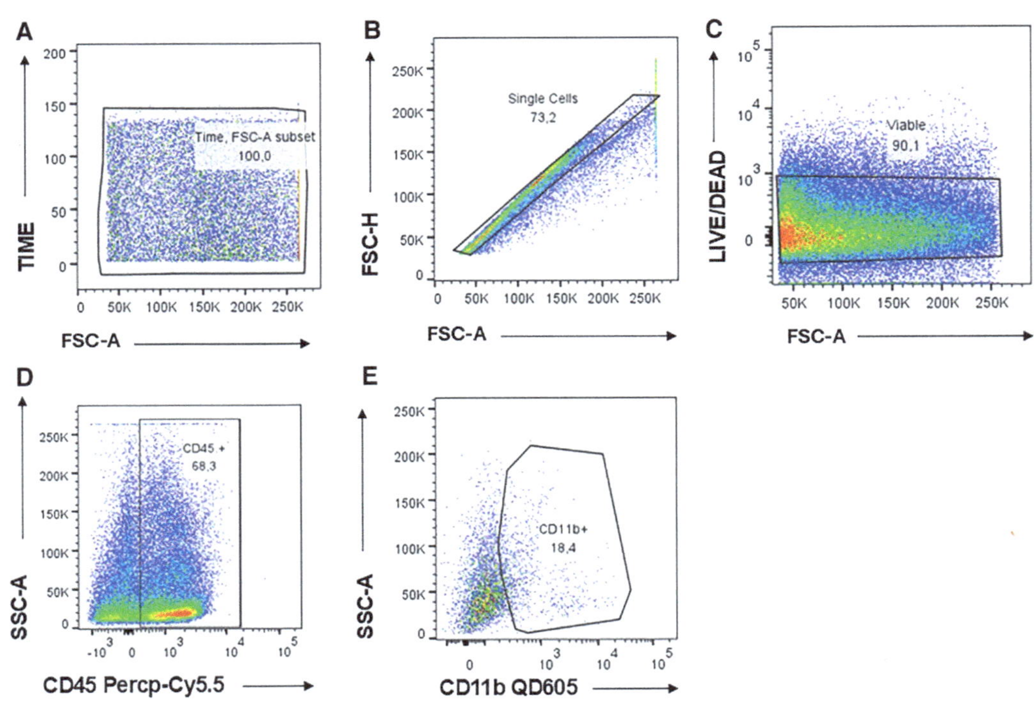

Fig. 2 Common gating strategy is used to identify cell subpopulations across different tissues. (a) Initial quality control, (b) Singlet selection, (c) Viable cell selection, (d) $CD45^+$ cell selection, and (e) Total myeloid cell gating.

4. Total myeloid cell gating (Fig. 2E). Define a plot: CD11b vs. SSC-A. Gate on CD11b⁺ cells, which correspond to the total myeloid population. This step establishes the basis for further subdivision of myeloid subsets according to the specific panel used in this model.

5. Myeloid Subpopulation Differentiation (Fig. 3). Within the CD11b⁺ gate, define subpopulations using the following markets:

Blood (peripheral blood) (Fig. 3A)**:** Neutrophils: Ly6G⁺ F4/80⁻, Monocytes: Ly6G⁻ F4/80^{low}, Macrophages: Typically, absent in peripheral blood, low frequency.

Tissues (Fig. 3B)**:**Neutrophils: Ly6G⁺F4/80⁻, Macrophages: Ly6G⁻ F4/80⁺

Dendritic cells (blood and tissues) (Fig. 3C)**:** Identified as CD11c⁺ MHC-II⁺ events.

4 Notes

1. Option 2 (in-house preparation): Fix cells with 2% formaldehyde, then permeabilize using 1% BSA, 0.1% NaN_3, 0.15% saponin in PBS 0.01 M pH 7.4.

2. Collagenase is activated at elevated temperatures. Therefore, it is essential to keep the tissue on ice immediately after collecting it to prevent premature enzymatic activation. The optimal incubation time with collagenase is approximately 1 h; longer periods may negatively affect cell staining.

3. At this step, inspect the tissue pellet for the presence of red blood cells. If present, lyse them by adding 5 mL of ACK lysing buffer and incubating the sample at 20°C–26°C for 5 min with gentle mixing. Terminate the lysis by adding 500 μL of 10×

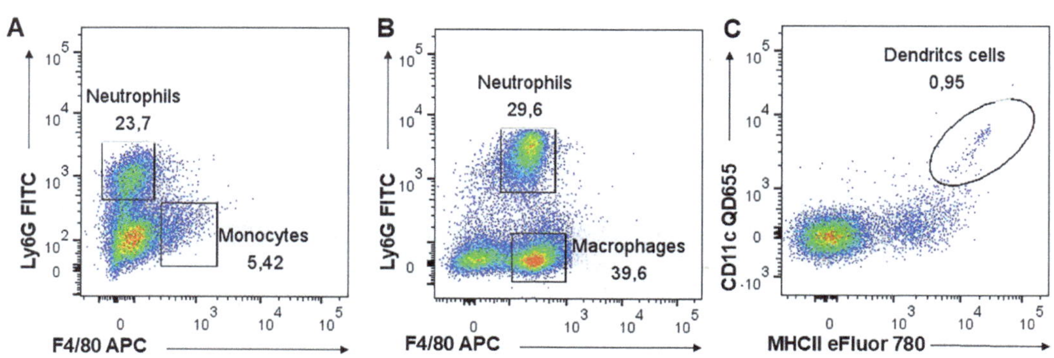

Fig. 3 Set of subpopulations in the myeloid panel. (**a**) Neutrophils and monocytes in blood; (**b**) Neutrophils and macrophages in tissues; and (**c**) Dendritic cells in all tissues

PBS. Centrifuge at 400 × *g* for 10 min at 4°C and proceed to the next step.

4. For liver and brain collection, it is necessary to perfuse the animal prior to tissue harvesting. This can be performed using a syringe with 1× PBS or a dedicated perfusion apparatus.
5. Before setting up the multicolor panel for flow cytometry, it is essential to understand the properties of the fluorophores used and to avoid inappropriate combinations of tandem dyes. It is not recommended to pair brighter tandems with weaker parental fluorophores (e.g., PE-Cy5 with PE), as spillover from the tandem may generate artifactual signals and compromise the analysis.
6. Brefeldin A is used to block vesicular transport in the Golgi complex and is essential for the evaluation of intracellular cytokines. The incubation time with brefeldin A should not exceed 4 h, as longer periods may induce neutrophil death. An incubation period of 3–4 h is recommended.
7. Option 2 (In-house Fixation)

Add 200 μL of 2% formaldehyde and incubate for 20 min at room temperature.

Centrifuge at 400 × *g* for 5 min at 4°C and discard the supernatant.

Wash with 100 μL PBS 1× and centrifuge (400 × *g*, 5 min, 4°C).

Add 100 μL permeabilization solution (1% BSA, 0.1% NaN_3, 0.15% saponin in 0.01 M PBS, pH 7.4).

Incubate for 35 min at 4°C, protected from light.

Centrifuge at 400 × *g* for 5 min at 4°C and discard the supernatant.

Resuspend in 100 μL Perm/Wash buffer and centrifuge again (400 × *g*, 5 min, 4°C).

Acknowledgments

This study was funded by the Fundação de Amparo à Pesquisa do Estado de Minas Gerais (FAPEMIG), grant numbers RED-00202–22, APQ-02281–18, APQ-04650-23, APQ-02618-23, and APQ-04983-24; and by the Coordenação de Aperfeiçoamento de Pessoal de Nível Superior (CAPES/Brazil), through the project: CAPES—Programa 9951—Programa Estratégico Emergencial de Prevenção e Combate a Surtos, Endemias, Epidemias e Pandemias, AUX 0641/2020—Process 88881.507175/2020–01. This work also received support from the National Institute of Science and Technology in Dengue and Host-Microorganism Interaction (INCT Dengue), funded by the Conselho Nacional de Desenvolvimento Científico e Tecnológico (CNPq, Brazil) [grant number

465425/2014-3], and by FAPEMIG [grant number 25036]. Additional support was provided by FINEP—Financiadora de Estudos e Projetos under MCTI/FINEP—MS/SCTIE/DGITIS/CGITS [grant number 6205283BBB28-4F9C-AA65-808FE4450542], and by the International Society of Neurochemistry–Committee for Aid and Education in Neurochemistry (ISN-CAEN) [project number 29344, ISID/ICB/DMORF/Zika Virus].

References

1. Rocklöv J, Dubrow R (2020) Climate change: an enduring challenge for vector-borne disease prevention and control. Nat Immunol 21(5):479–483. https://doi.org/10.1038/s41590-020-0648-y
2. Guzman M, Gubler D, Izquierdo A et al (2016) Dengue infection. Nat Rev Dis Primers 2:16055. https://doi.org/10.1038/nrdp.2016.55
3. Bettis AA, L'Azou Jackson M, Yoon IK et al (2022) The global epidemiology of chikungunya from 1999 to 2020: a systematic literature review to inform the development and introduction of vaccines. PLoS Negl Trop. Dis 16(1):e0010069. https://doi.org/10.1371/journal.pntd.0010069
4. Feng F, Bouma EM, Hu G et al (2023) Colocalization of chikungunya virus with its receptor MXRA8 during cell attachment, internalization, and membrane fusion. J Virol 97(5):e0155722. https://doi.org/10.1128/jvi.01557-22
5. Ozden S, Huerre M, Riviere JP et al (2007) Human muscle satellite cells as targets of Chikungunya virus infection. PLoS one 2(6):e527. https://doi.org/10.1371/journal.pone.0000527
6. Zhang R, Kim AS, Fox JM et al (2018) Mxra8 is a receptor for multiple arthritogenic alphaviruses. Nature 557(7706):570–574. https://doi.org/10.1038/s41586-018-0121-3
7. Simon F, Javelle E, Oliver M et al (2011) Infecção pelo vírus Chikungunya. Relatórios atuais de doenças infecciosas 13(3):218–228. https://doi.org/10.1007/s11908-01-0180-1
8. Babu N, Mahilkar S, Jayaram A et al (2023) Cytokine profile, neutralisation potential and viral replication dynamics in sera of chikungunya patients in India: a cross-sectional study. Lancet Reg Health Southeast Asia 19:100269. https://doi.org/10.1016/j.lansea.2023.100269
9. Chow A, Her Z, Ong EK et al (2011) Persistent arthralgia induced by chikungunya virus infection is associated with interleukin-6 and granulocyte macrophage colony-stimulating factor. J Infect Dis 203(2):149–157. https://doi.org/10.1093/infdis/jiq042
10. da Silva MOL, Figueiredo CM, Neris RLS et al (2024) Chikungunya and Mayaro viruses induce chronic skeletal muscle atrophy triggered by pro-inflammatory and oxidative response. Int J Mol Sci 25(16):8909. https://doi.org/10.3390/ijms25168909
11. Lentscher AJ, McCarthy MK, May NA et al (2020) Chikungunya virus replication in skeletal muscle cells is required for disease development. J Clin investig 130(3):1466–1478. https://doi.org/10.1172/JCI129893
12. Barros JBS, da Silva PAN, Koga RCR et al (2018) Acute Zika virus infection in an endemic area shows modest proinflammatory systemic immunoactivation and cytokine-symptom associations. Front Immunol 9:821. https://doi.org/10.3389/fimmu.2018.00821
13. Gaythorpe KA, Hamlet A, Jean K et al (2021) The global burden of yellow fever. eLife 10:e64670. https://doi.org/10.7554/eLife.64670
14. Costa VV, Sugimoto MA, Hubner J et al (2022) Targeting the Annexin A1-FPR2/ALX pathway for host-directed therapy in dengue disease. eLife 11:e73853. https://doi.org/10.7554/eLife.73853
15. Pereira LHS, Alves ADC, Lopes GFM et al (2024) Soluble isoforms of the DC-SIGN receptor can increase the dengue virus infection in immature dendritic cells. The Braz J Infect Dis 28(6):103873. https://doi.org/10.1016/j.bjid.2024.103873
16. Khasa R, Ogden SC, Wang Y et al (2025) A single mutation in the *PrM* gene of Zika virus determines AXL dependency for infection of human neural cells. J Virol 99:e01873–24. https://doi.org/10.1128/jvi.01873-24
17. Woodson SE, Holbrook MR (2011) Infection of hepatocytes with 17-D vaccine-strain yellow fever virus induces a strong pro-inflammatory host response. J Gen Virol 92(10(Pt):2262–2271. https://doi.org/10.1099/vir.0.031617-0
18. Telford WG (2024) Basic principles of flow cytometer operation: the make your own flow cytometer. Methods Mol Biol (Clifton, N.J.) 2779:11–31. https://doi.org/10.1007/978-1-0716-3738-8_2
19. Chan HE, Jilani I, Chang R, Albitar M (2007) Monitoring cell signaling pathways

by quantitative flow cytometry. Methods Mol Biol (Clifton, N.J.) 378:83–90. https://doi.org/10.1007/978-1-59745-323-3_7
20. Maecker HT, McCoy JP, Nussenblatt R (2012) Standardizing immunophenotyping for the Human Immunology Project. Nat Rev Immunol 12(3):191–200. https://doi.org/10.1038/nri3158
21. Welsh JA, Arkesteijn GJA, Bremer M et al (2023) A compendium of single extracellular vesicle flow cytometry. J Extracell Vesicles 12(2):e12299. https://doi.org/10.1002/jev2.12299
22. Austin Suthanthiraraj PP, Graves SW (2013) Fluidics. In: Current protocols in cytometry, Chapter 1, 1.2.1–1.2.14. John Wiley & Sons, Hoboken, New Jersey, USA. https://doi.org/10.1002/0471142956.cy0102s65
23. de Lima RC, da Costa Faria NR, de Carvalho AT (2025) Flow cytometry as immunoassay tool for research on yellow fever virus. Methods Mol Biol (Clifton, N.J.) 2913:1–17. https://doi.org/10.1007/978-1-0716-4458-4_1
24. de Araújo S, de Melo Costa VR, Santos FM et al (2022) Annexin A1-FPR2/ALX signaling axis regulates acute inflammation during chikungunya virus infection. Cells 11(17):2717. https://doi.org/10.3390/cells11172717
25. Benedetti A, Cera G, De Meo D et al (2021) A simple method for the isolation and *in vitro* expansion of highly pure mouse and human satellite cells. Bio-protocol 11(23):e4238. https://doi.org/10.21769/BioProtoc.4238
26. Batista VL, Martins JR, Dias ASL et al (2025) P2Y12–P-selectin mediated platelet activation drives dengue-associated thrombocytopenia. bioRxiv. https://doi.org/10.1101/2025.09.08.674864
27. Martins JR, Batista VL, Ramos LG et al (2025) Cyclodextrin-based delivery of the annexin A1 mimetic peptide Ac2-26 enhances anti-inflammatory effects and prevents dengue-induced lethality in combination with antiviral therapy. bioRxiv. https://doi.org/10.1101/2025.09.02.673789
28. Song R, Bafit M, Tullett KM et al (2024) A simple and rapid protocol for the isolation of murine bone marrow suitable for the differentiation of dendritic cells. Methods Protoc 7(2):20. https://doi.org/10.3390/mps7020020
29. LeCher JC, Costa VV, Rust LN, Bassit LC et al (2025) Combating yellow fever virus with 7-deaza-7-fluoro-2′-C-methyladenosine. Antimicrob Agents Chemother 69:e0188924. https://doi.org/10.1128/aac.01889-24
30. LeCher JC, Zandi K, Costa VV, Amblard et al (2022) Discovery of a 2′-fluoro,2′-bromouridine phosphoramidate prodrug exhibiting anti-yellow fever virus activity in culture and in mice. Microorganisms 10(11):2098. https://doi.org/10.3390/microorganisms10112098
31. Rana A, Massa PT, Chen XJ (2022) A gravity-fed transcardial perfusion method for histologic analysis of the mouse central nervous system. J Vis Exp (179):e63386. https://doi.org/10.3791/63386
32. Vanekova L, Polidarova MP, Veverka V et al (2022) Multiparametric flow cytometry-based immunophenotyping of mouse liver immune cells. Methods Protoc 5(5):70. https://doi.org/10.3390/mps5050070
33. Costa VV, Del Sarto JL, Rocha RF et al (2017) *N*-methyl-D-aspartate (NMDA) receptor blockade prevents neuronal death induced by zika virus infection. mBio 8(2):e00350–17. https://doi.org/10.1128/mBio.00350-17
34. Liu Z, Gu Y, Chakarov S et al (2019) Fate mapping via Ms4a3-expression history traces monocyte-derived cells. Cell 178(6):1509–1525.e19. https://doi.org/10.1016/j.cell.2019.08.009

Chapter 19

Chikungunya Virus Infection in Paraffin-Embedded Tissue: Analysis of Histological Alterations and Viral Detection by Immunohistochemistry

Natália G. Salomão and Kíssila Rabelo

Abstract

Histopathological analysis of tissues infected with Chikungunya virus (CHIKV), including biopsy and autopsy specimens, can provide valuable insights into the pathogenesis of atypical and fatal cases. By examining tissue architecture and cellular alterations under the microscope, it is possible to identify patterns of injury, inflammation, and cellular degeneration. These morphological findings provide evidence of how the pathogen interacts with host cells and tissues. Moreover, immunohistochemistry may be performed in tissue sections for the detection of viral antigens. Importantly, this technique allows the characterization of diverse biologically relevant targets in tissue samples, such as distinct immune cell subsets, cytokines, and additional molecular markers. This is accomplished by the use of antibodies selected for their specificity toward the antigen of interest. In this chapter, we present a concise overview of how histological analysis and immunohistochemical approaches can enhance the understanding of CHIKV-associated pathological mechanisms.

Keywords CHIKV, Human tissues, Histopathological analysis, Viral antigens, Pathogenesis

1 Introduction

Chikungunya virus (CHIKV) is an arthropod-borne virus that belongs to the *Togaviridae* family and *Alphavirus* genus. Its first isolation was in 1953 in Tanzania, East Africa, from the serum of an infected patient during a major outbreak [1]. The virus was mainly circulating in Africa and Asia during the 1970s; however, since 2004, it has spread to various regions worldwide. In the Americas, the first reports appeared in 2013, and since then, there have been significant outbreaks with a large number of infected individuals. CHIKV has expanded so extensively that it has already affected more than 110 countries across Asia, Africa, the Americas, and Europe [2]. The name "chikungunya" comes from the dialect of the "Makonde" people due to the hunched posture that infected patients acquire as a

Claudia Filomatori and Eliana F. Castro (eds.), *Chikungunya Virus*, Methods in Molecular Biology, vol. 3067,
https://doi.org/10.1007/978-1-0716-5496-5_19,

consequence of the intense arthralgia pain, the hallmark of the disease. The term can be translated as "disease that bends up the joints" [3]. The virus is transmitted mainly by *Aedes* mosquitoes, such as *Aedes aegypti* and *Aedes albopictus.* In the latter species, a mutation occurred in the envelope protein gene (E1-A226V), resulting in improvement of infection and dissemination of the virus [4]. However, vertical transmission and transmission through blood transfusion have also been reported [5, 6]. After infection, it takes 3–7 days for the onset of the symptoms, which correspond to the incubation period. The symptomatic chikungunya infection is classified into three phases: acute, post-acute, and chronic. During the acute phase (until 21 days), the most common symptoms are high fever, polyarthralgia/polyarthritis (usually accompanied by edema), myalgia, headache, photophobia, and rash. After 21 days, the individual begins in the post-acute phase, with persistence of polyarthralgia or polyarthritis [7, 8]. Meanwhile, the chronic phase occurs with more than 3 months of symptoms, which affect 40%–80% of the infected individuals [9].

Atypical and severe cases of chikungunya infection may occur, including neurological, cardiac, and renal manifestations, which are more common in individuals with preexisting medical conditions, such as obesity and diabetes, as well as in older individuals [10, 11]. A study conducted in Brazil with the "100 Million Brazilian Cohort," between 2015 and 2018 showed an increased risk of death up to 84 days after symptom onset, including deaths from cerebrovascular diseases, ischemic heart diseases, and diabetes [12]. In addition, de Lima et al. investigated 68 fatal chikungunya cases, and 42 autopsy tissues (brain, lungs, heart, and liver). All the organs showed histopathological changes, such as vascular congestion, edema, and hemorrhage, and viral antigen was detected in 41.1% by immunohistochemistry [13].

Histological and immunohistochemical techniques are essential for characterizing tissue pathology and identifying specific molecular or cellular targets, respectively, thereby providing critical insights into the pathogenesis of a disease. Hematoxylin and Eosin (H&E) staining remains the standard method for routine histopathological evaluation, in which cell nuclei are stained purple by hematoxylin affinity, and eosin, an acidic dye, stains the cytoplasm and extracellular matrix in varying shades of pink, depending on their basic protein content. The contrast generated by the differential affinity of hematoxylin and eosin allows for detailed assessment of cellular morphology, tissue architecture, inflammatory processes, and patterns of degeneration or necrosis [14].

Immunohistochemistry (IHC) is an immunoenzymatic technique based on the binding between an antibody and antigen enabling the localization of target antigens in tissue samples. The antibody–antigen interaction is visualized through an enzyme that reacts with an appropriate chromogenic substrate and then examined with a light microscope. The advantage of using IHC is to identify the protein localization on the tissue, as well as to analyze the status of the whole tissue [15, 16].

In chikungunya cases that result in death, the visceral fragments/ tissues may be collected to perform anatomopathological studies followed by viral antigen detection using immunohistochemistry [17]. In addition, since CHIKV vertical transmission occurs, it is valuable to collect the placenta to investigate the pathogenesis of the disease in this tissue, which, in general, is discarded [18–20]. Although serological and molecular techniques are the most used to diagnose chikungunya infection, histopathology and IHC are powerful alternatives focusing on organs/tissues [21].

2 Materials

2.1 Fixing Solution

10% buffered formalin pH 6.8–7.4. To prepare, add 900 mL of distilled water, 100 mL of formalin—37% formaldehyde solution, 4 g of sodium phosphate, monobasic, monohydrate, and 6.5 g of sodium phosphate, dibasic, anhydrous. Store at room temperature.

2.2 Staining Solutions

1. Hematoxylin dye. Add 900 mL of distilled water, 5 g of Harris' hematoxylin, 100 mL of 100% (v/v) ethanol, 2.5 g red mercury oxide, and 4 mL of acetic acid. Store at room temperature in an amber bottle.
2. Eosin dye. Add 1000 mL of distilled water, 20 g of yellowish eosin, and 4 mL of acetic acid. Store at room temperature in an amber bottle.

2.3 Antigen Retrieval Buffer

10 mM sodium citrate, 0.05% Tween 20, pH 6.0. Weigh 2.94 g of tri-sodium citrate (dihydrate) and add to 1 L of distilled water. Mix to dissolve. Adjust pH to 6.0 with 1 N HCl. Add 0.5 mL Tween 20 and mix well. Store at 4°C.

2.4 Washing Buffer

TBS 1×—pH 7.4. Dissolve 3.02 g of Tris base and 8 g of sodium chloride in distilled water. Adjust the pH to 7.4 using 1 M HCl, then bring the final volume to 1 L with distilled water.

2.5 Histological Processing

1. Automatic tissue processor.
2. Scalpel.
3. Scissors.
4. Tweezers.
5. Tissue cassettes.
6. Tissue Embedding Center.
7. Paraffin.

8. Paraffin embedding molds.
9. Microtome.
10. Slides.
11. Water bath.
12. Slide rack.
13. Incubator.
14. Staining jars or Coplin jars.
15. Xylene.
16. 100% (v/v), 95% (v/v), 90% (v/v), and 70% (v/v) ethanol solutions.
17. Running water.
18. Hematoxylin solution.
19. Eosin solution.

2.6 Immunohistochemistry

1. Microtome.
2. Positively charged slides.
3. Microtome Blades.
4. Water Bath.
5. Incubator.
6. Distilled water.
7. Xylene.
8. 70% (v/v), 90% (v/v), 95% (v/v), and 100% (v/v) ethanol solutions.
9. Citrate-NaOH buffer pH 6.0.
10. 3% hydrogen peroxidase in methanol.
11. Tris–HCl pH 7.4 or PBS pH 7.4.
12. Protein blocker solution: 5% bovine seroalbumin (BSA).
13. Primary antibodies.
14. Secondary complement (optional).
15. Anti-goat/mouse/rabbit IgG-HRP conjugate.
16. Diaminobenzidine (DAB) chromogen.
17. Harris' hematoxylin solution.

2.7 Slide Mounting

1. Mounting medium (e.g., Entellan, Permount).
2. Pasteur pipette.
3. Glass coverslips.

2.8 Analysis

1. Optical microscope with camera.
2. Computer with software installed (e.g., Image Pro Plus software version 7.0).

3 Methods

All procedures must be carried out using appropriate personal protective equipment (PPE), such as lab coats and disposable gloves, five and collective protective equipment (laboratory fume hood). Many of the solutions used are toxic, and caution should be exercised to avoid skin and eye contact, as well as inhalation or ingestion.

3.1 Organ Collection and Histological Processing

1. Procedures to euthanize infected animals vary according to the experiment and should be consulted with the Ethics Committee. In the case of human tissues, use biopsies and necropsies according to ethical regulations.
2. Perform organ collection with the aid of a scalpel, scissors, and tweezers.
3. Fix sections of the organ in 10% buffered formalin for 24–48 h.
4. Cleave the tissue into smaller fragments using a scalpel.
5. Rinse the tissue under running water for approximately 1 h to remove excess fixative.
6. For histological processing, use an automatic tissue processor to dehydrate in graded ethanol series: 70% (v/v), 80% (v/v), 95% (v/v), and 100% (v/v). Clarify in xylene (three baths, 30 min each), and then melt paraffin at 60°C to allow complete infiltration.

3.2 Embedding of Tissue Samples in Paraffin

1. Embed the tissue in paraffin blocks using a Tissue Embedding Center.

3.3 Microtomy

1. Section the paraffin blocks in a microtome to obtain blades with cuts.
2. Gently spread the sections on the surface of the warm water bath and then collect them onto microscope slides.
3. Incubate the slides at 60°C to ensure proper adhesion of the tissue sections to the slide and to remove residual paraffin before proceeding to the following steps.

3.4 Hematoxylin and Eosin (H&E) Staining

1. Deparaffinize in xylene (three baths), and rehydrate with ethanol (100% [v/v], 95% [v/v], 90% [v/v], and 70% [v/v]) and in running water.
2. Stain with hematoxylin (~5 min; the time may vary depending on reagent; consult the manufacturer).
3. Rinse in running water for 2–5 min to remove excess dye.
4. Differentiate briefly in alcohol.
5. Immerse slides in eosin Y (~3 min—the time depends on concentration and desired intensity).

6. Dehydrate in ethanol (70% [v/v], 90% [v/v], 95% [v/v], and 100% [v/v]) and in xylene (three baths).
7. Mount the slides with mounting medium (e.g., Entellan, Permount) resin with glass coverslips and visualize under a light microscope attached to a camera.

3.5 Immunohistochemistry by Peroxidase Protocol Using a PolymerDetection System

1. Incubate the slides with sections (4-μm thick units) (*see* **Note 1**) in the lab oven at 60°C for 1 h to melt the paraffin.
2. Deparaffinize in xylene (three baths) (*see* **Note 2**) and hydrate with ethanol (100% [v/v], 95% [v/v], 90% [v/v], and 70% [v/v]) and in distilled water.
3. Perform the antigen retrieval by heating the tissue in the presence of citrate buffer pH 6.0 for 20 min (*see* **Note 3**).
4. Block for endogenous peroxidase with 3% hydrogen peroxide in methanol for 10 min and rinse in Tris–HCl pH 7.4 or PBS pH 7.4 (*see* **Note 4**).
5. To reduce nonspecific binding, incubate sections with protein blocker solution for 10 min at room temperature.
6. Incubate the sections with primary antibodies diluted as recommended by the manufacturer overnight at 4°C.
7. On the following day, incubate the sections with secondary complement for 10 min (some commercial products take this step to ensure the connection of the anti-rabbit IgG-HRP) at room temperature.
8. Incubate the sections with an anti-goat/mouse/rabbit IgG-HRP conjugate for 15 min at room temperature.
9. Rinse in Tris–HCl pH 7.4 or PBS pH 7.4.

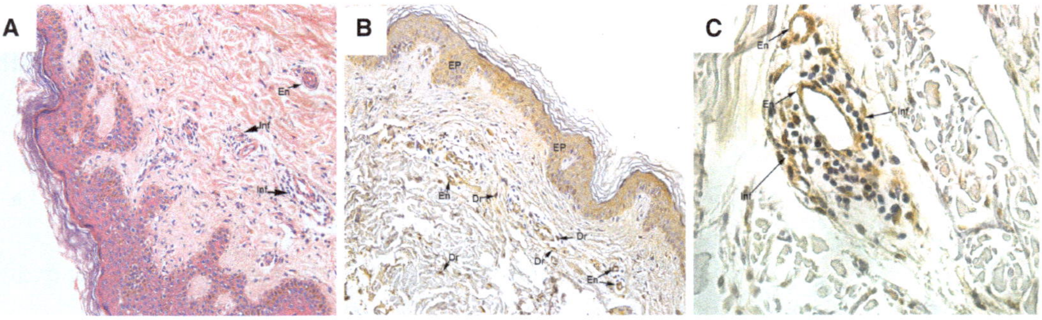

Fig. 1 Histological analysis and CHIKV antigen detection by immunohistochemistry assay: (**A**) Hematoxylin and eosin staining: Biopsy skin exhibiting areas of inflammatory infiltrate and thickening of the vascular endothelium –I Immunohistochemistry (**B/C**): the brown staining highlights the CHIKV-positive cells in (**A**) epithelial, endothelial, and dermal cells and (**B**) perivascular inflammatory infiltrate. EP: Epithelial cells, Dr: Dermal cells, En: endothelial cells, Inf: Inflammatory infiltrate

10. Reveal the reactions with diaminobenzidine (most reactions occur in 5 min or less) and rinse with distilled water.
11. Counterstain the sections with Mayer's hematoxylin for 30 s.
12. Bathe in distilled water, dehydrate in ethanol (70% [v/v], 90% [v/v], 95% [v/v], and 100% [v/v]) and in xylene (three baths) for 2 min each.
13. Mount the slides with Entellan resin with glass coverslips and visualize under a light microscope attached to a camera.
14. Take digital images using software, for example, Image Pro Plus software version 7.0 (Fig. 1).

4 Notes

1. For immunohistochemistry, 4-μm-thick units are recommended in positively charged slides or silanized slides.
2. Ensure that no residual paraffin remains so as not to hinder the detection of your antigen of interest.
3. This step can vary according to the primary antibody to be used. Here, we are describing what is most usual for our experiments; however, antigen retrieval may be performed with other reagents, such as Tris-EDTA buffer pH 9.0 at 90°C for 1 h, among others.
4. This step can vary according to the enzyme used to reveal the reaction, such as alkaline phosphatase, which is blocked with 1–5 mM levamisole, and biotin with avidin or skim milk.

References

1. Robinson MC (1955) An epidemic of virus disease in Southern Province, Tanganyika territory, in 1952–1953. I. Clinical features. Trans R Soc Trop Med Hyg 49(1):28–32
2. World Health Organization Chikungunya. https://www.who.int/health-topics/chikungunya#tab=tab_1
3. Thiberville SD, Moyen N, Dupuis-Maguiraga L et al (2013) Chikungunya fever: epidemiology, clinical syndrome, pathogenesis and therapy. Antiviral Res 99(3):345–370
4. Tsetsarkin KA, Vanlandingham DL, McGee CE et al (2007) A single mutation in Chikungunya virus affects vector specificity and epidemic potential. PLoS Pathog 3(12):e201
5. Lenglet Y, Barau G, Robillard PY et al. (2006) Chikungunya infection in pregnancy: evidence for intrauterine infection in pregnant women and vertical transmission in the parturient. Survey of the Reunion Island outbreak. J Gynecol Obstet Biol Reprod (Paris) 35:578–583
6. Appassakij H, Silpapojakul K, Promwong C et al. (2020) The potential impact of chikungunya virus outbreaks on blood transfusion. Transfus Med Rev 34:23–28
7. Silva LA, Dermody TS (2017) Chikungunya virus: epidemiology, replication, disease mechanisms, and prospective intervention strategies. J Clin Invest 127:737
8. da Cunha RV, Trinta KS (2017) Chikungunya virus: clinical aspects and treatment—a review. Mem Inst Oswaldo Cruz 112:523–531
9. Mateo L, Roure S (2019) Chronic arthritis in chikungunya virus infection. Reumatol Clin 15:113–116

10. Rajapakse S, Rodrigo C, Rajapakse A (2010) Atypical manifestations of chikungunya infection. Trans R Soc Trop Med Hyg 104(2):89–96
11. Economopoulou A, Dominguez M, Helynck B et al (2009) Atypical chikungunya virus infections: clinical manifestations, mortality and risk factors for severe disease during the 2005-2006 outbreak on Réunion. Epidemiol Infect 137(4):534–541
12. Cerqueira-Silva T, Pescarini JM, Cardim LL et al (2024) Risk of death following chikungunya virus disease in the 100 Million Brazilian Cohort, 2015-18: a matched cohort study and self-controlled case series. Lancet Infect Dis 24(5):504–513
13. de Lima STS, , de Souza WM, Cavalcante JW et al. (2021) Fatal outcome of chikungunya virus infection in Brazil. Clin Infect Dis 73:e2436–e2443
14. Feldman AT, Wolfe D (2014) Tissue processing and hematoxylin and eosin staining. Methods Mol Biol 1180:31–43
15. Kim SW, Roh J, Park CS (2016) Immunohistochemistry for pathologists: protocols, pitfalls, and tips. J Pathol Transl Med 50:411
16. Magaki S, Hojat SA, Wei B et al. (2019) An introduction to the performance of immunohistochemistry. Methods Mol Biol 1897:289
17. Sharp TM, Keating MK, Shieh W-J et al (2021) Clinical characteristics, histopathology, and tissue immunolocalization of chikungunya virus antigen in fatal cases. Clin Infect Dis 73(2):e345–e354
18. Salomão N, Brendolin M, Rabelo K et al. (2021) Spontaneous abortion and chikungunya infection: pathological findings. Viruses 13:554
19. Salomão N, Araújo L, Rabelo K et al (2022) Placental alterations in a chikungunya-virus-infected pregnant woman: a case report. Microorganisms 10:872
20. Salomão N, Rabelo K, Avvad-Portari E et al (2022) Histopathological and immunological characteristics of placentas infected with chikungunya virus. Front Microbiol 13:1055536
21. Ministério da Saúde, Secretaria de Vigilância em Saúde e Ambiente, Departamento de Vigilância das Doenças Transmissíveis (2024) Chikungunya: manejo clínico [recurso eletrônico]. Ministério da Saúde, Secretaria de Vigilância em Saúde, Departamento de Doenças Transmissíveis, Brasília

Check for updates

Chapter 20

Collection and Processing of Adult Mosquitoes (Diptera: Culicidae) for Molecular Studies of Arbovirus Infection

Griselda Ines Oria

Abstract

The processing of adult mosquitoes collected in the field for studies of natural infection with arboviruses of public health interest (genera: *Flavivirus*, *Alphavirus*, *Orthobunyavirus*) follows the assembly of homogenized mosquito pools and extraction of genetic material using commercial or manual techniques, and the detection of viral RNA by polymerase chain reaction (nested RT-PCR) [11]. This paper describes the procedures for capturing, taxonomic determination, obtaining genetic material, and molecular techniques for detecting alphaviruses.

Keywords Mosquito vectors, Molecular detection, Alphavirus, Nested RT-PCR, RNA

1 Introduction

Currently, 244 species of mosquitoes (Diptera: Culicidae) have been recorded in Argentina [1]. Those recognized as transmitters of pathogenic arboviruses are: *Aedes* (*Stegomyia*) *aegypti* (Linnaeus), *Aedes albopictus* (Skuse), species of *Aedes* subgenus *Ochlerotatus*, *Culex quinquefasciatus* (Say), and species of *Psorophora* and *Haemagogus* [2]. *Aedes aegypti* is the main vector of alphaviruses, such as Chikungunya virus (CHIKV) and flavivirus such as Dengue (DENV), Yellow fiver (YFV), and Zika (ZIKV) [3]. Alphaviruses have been identified in humans, wild birds, wild mammals, mosquitoes, and horses [4–8]. Some viruses remain in nature with complex dynamics, such as EEEV and VEEV, forming transmission networks with multiple hosts and vectors [9], while others, such as CHIKV, are considered specialists as they have a single main vector (*Ae. aegypti* and *Ae. albopictus*) and a definitive host (humans) [10]. The capture of mosquito species of epidemiological interest is the first step in arbovirus studies, and sampling strategies are selected according to the research objectives and entomological criteria. Collection methods include: light traps (CDC-CO_2) and mechanical aspiration

Claudia Filomatori and Eliana F. Castro (eds.), *Chikungunya Virus*, Methods in Molecular Biology, vol. 3067,
https://doi.org/10.1007/978-1-0716-5496-5_20,

(Fig. 1a,b). The former traps were designed at the Centers for Disease Control and Prevention (CDC-USA) as part of arbovirus surveillance programs and are based on the fact that mosquitoes are attracted to a light source. Mechanical aspirators are aspirators portables battery devices that allow adult specimens to be aspirated from their hosts, whether animals (vacuuming up mosquitoes after using a capture method) or humans, provided that this does not pose a risk of pathogen transmission to the operators. For this purpose, one person must remain motionless while another operator vacuums up the mosquitoes that are attempting to feed on the first host (Fig. 1c), or on and around farm or domesticated vertebrates (farmyard animals) in their resting or feeding places. This method provides information on the mosquito selection. Mechanical aspirators can also be used to collect resting mosquitoes (e.g., garages, house roofs, internal walls of chicken coops, drainage networks, etc.)

Homogenized mosquito pools are used for virus detection in various tissues, including nested reverse transcription-polymerase chain reaction (nested RT-PCR), viral isolation in cell cultures, ELISA for

Fig.1 Entomological aspirators for collecting adult mosquitoes resting on surfaces or foliage: (**a**) battery-powered modified CDC portable backpack; (**b**) battery operated aspirator; (**c**) collection of host-seeking mosquitoes by aspiration

antigen detection, and immunofluorescence. This chapter describes a conventional nested RT-PCR assay for the amplification and generic identification of alphaviruses [11]. To this end, viral genetic material is extracted with a manual technique (TRIzol–chloroform) [12, 13]. RNA extraction is based on the fact that these biomolecules contain negatively charged and polar phosphate groups, which strongly repel each other and allow nucleic acids to remain dissolved in aqueous solutions surrounded by a hydration layer. In the presence of ethanol, this layer is disrupted, and the phosphate groups become exposed, favoring the binding of cations such as Na^+. This interaction reduces the repulsive forces between phosphates in the nucleotide chains and allows the RNA to precipitate [14].

After precipitation, the nucleic acids are pelleted by centrifugation, and the supernatant is carefully discarded. The pellet is then washed, usually with 70% ethanol, to remove residual salts and contaminants, and subsequently dried to eliminate traces of ethanol that may inhibit downstream enzymatic reactions. Finally, the RNA is resuspended in nuclease-free water or buffer and stored at −80°C until use. Throughout this procedure, it is essential to minimize RNA degradation by avoiding repeated freeze–thaw cycles and by working with RNase-free materials and solutions.

Reverse transcription of the extracted RNA is then performed to obtain complementary DNA (cDNA), which serves as the template for subsequent amplification reactions. The cDNA is amplified using a nested PCR strategy to increase sensitivity and specificity, particularly in samples with low viral loads or a high background of host RNA. In this approach, two successive rounds of amplification are carried out using two pairs of primers: an external pair that amplifies a larger fragment and an internal pair that amplifies a specific region within the first amplicon.

Amplified products are analyzed by electrophoresis on a 2% agarose gel to confirm the expected fragment size. Positive samples can then be purified and subjected to Sanger sequencing in order to confirm viral identity and perform phylogenetic or molecular epidemiology analyses.

2 Materials

2.1 Collection of Mosquitoes Using CDC-Type Light Traps

1. Circular aluminum container.
2. An acrylic cylindrical chamber containing a motor-driven fan.
3. Lamp (light source).
4. Battery as an electrical circuit (12 V).
5. Fabric collection bag (Voile).
6. Two wire rings as a frame.

7. CO_2 provided in dry ice format (*see* **Note 1**) [15].
8. Portable polypropylene coolers.
9. Petri dishes and tweezers.
10. Refrigerated plate for taxonomic determination of the mosquito species.
11. Sterile 1.5 mL tubes.

2.2 Homogenization of the Mosquito Pools

1. Sterile mortars and pestles (*see* **Note 2**).
2. Sterile, DNase-free disposable supplies for molecular biology: 1000 μL filter tips, 1.5 mL Eppendorf tubes.
3. Variable volume micropipettes.
4. Minimal essential medium (MEM): Earle's salts, L-glutamine, and sodium bicarbonate. Sterile-filtered, 1000 μL for pools of up to 25 mosquitoes and 2000 μL for pools of up to 50 individuals.
5. Centrifuge.

2.3 RNA Extraction with TRIzol–Chloroform

1. TRIzol: phenol, guanidine isothiocyanate.
2. Chloroform.
3. Isopropyl alcohol.
4. 75% (v/v) ethanol.
5. Nuclease-free water.
6. Centrifuge.
7. Sterile 1.5 mL microtubes.
8. Thermal bath (wet or dry) between 50°C and 60°C.

2.4 Reverse Transcription Using Improm II® Reverse Transcriptase

1. 5× buffer.
2. 25 mM $MgCl_2$.
3. 10 pmol random primers.
4. 2.5 mM dNTPs (dATP, dCTP, dGTP, and dTTP).
5. RNaseOUT/RNasin ribonuclease inhibitor (optional but recommended).
6. Nuclease-free water.
7. Improm II® Reverse Transcriptase.

2.5 Conventional Nested PCR for the Detection of Generic Alphaviruses

1. Nuclease-free water.
2. 5× buffer with 25 mM $MgCl_2$.
3. 10 mM dNTP mix.
4. 10 μM ALFA +1 primer: 5′GAYGCITAYYTIGAYATGGTIGAIGG3′, where Y= C or T; I=inosine (universal base).

5. 10 μM ALFA −1 primer: 5′KYTCYTCIGTRTGYTTIG TICCIGG3′, where K= G or T; Y=C or T; R=A or G; and I=inosine (universal base).
6. 10 μM ALFA +2 primer: 5′GCRAAIARIGCIGCIGCYTY IGGICC3′, where Y= C or T; R=A or G; and I=inosine (universal base).
7. 10 μM ALFA−2 primer: 5′GIAAYTGYAAYGTIACICARATG3′, where Y=C or T; R=A or G; and I=inosine (universal base).
8. 5 U/μL Taq polymerase.

2.6 Agarose Gel Electrophoresis

1. TE buffer: 10 mM Tris-HCl, 1 mM EDTA, pH 8.0.
2. Agarose.
3. GelRed or lactophenol blue dyes.
4. 100 bp molecular weight marker.
5. 70% (v/v) glycerol or loading buffer.
6. Tank, combs, and power supply.

3 Methods

3.1 Adult Mosquito Collection (see Note 1)

1. Collect mosquitoes using CDC light traps supplemented with dry ice (as a source of CO_2), placed in varying numbers depending on the objectives of the study, with a distance of 50 m between each trap to avoid interference. Keep traps active between 6 p.m. and 9 a.m. (Fig. 2a,b).
2. Before deactivating the trap by disconnecting it from the battery, remove the cloth bag containing the collected mosquitoes (Fig. 2f,g). Handle the bag with care to avoid crushing the mosquitoes, which could make it difficult to identify specific individuals in the laboratory.
3. Label the collected bags with information about the location and date of capture, and temporarily store them in refrigerators until they are transferred to the laboratory for processing.
4. Once in the laboratory, place them in a freezer at −20°C to kill the mosquitoes, and then transfer the collected material to plastic tubes (Fig. 2h,i).
5. Identify the mosquitoes using a stereomicroscope on a refrigerated table, following the characteristics outlined in Darsie [15], Consoli and Lourenço-de-Oliveira [16], Forattini [17], and Faran and Linthicum [18].
6. Pools of mosquitoes by species, sex, date, and sampling location, in batches of 1–50 specimens, and store at −70°C.

Fig. 2 Components and assembly of a CDC light trap used in the field. (**a**) Assembled device consisting of cylindrical acrylic tube with a fan and wire mesh cover, fitted with a spotlight at the top and screwed onto an aluminum cap that protects against inclement weather; (**b**) dry ice block placed on newspaper to form a package weighing approximately 1 kg; (**c**) 12-V battery with double polarity cables (marked red and black); (**d** and **e**) complete trap in operation with the light on, the fabric unfolded, and the dry ice package releasing CO_2 as an attractant; (**f**) mesh with the catch obtained after 15 h of activity; (**g**) contents of the collection recovered; (**h**) arrangement of the specimens collected by site and date of capture in tubes and jars; (**i**) storage at −70°C

3.2 Mosquito Pool Homogenization *(Adapted from Dr. J. M. Vanella, Institute of Virology, UNC, Córdoba, Argentina)*

1. Homogenize mosquitoes pools under a biological safety cabinet in sterile mortars, using 1000 μL of MEM for pools of up to 25 mosquitoes and 2000 μL of MEM for pools of up to 50 individuals (*see* **Note 3**).
2. Centrifuge at 10,000 × *g* for 30 min and transfer the supernatants to cryovials. Store at –70°C.

*3.3 RNA Extraction Using TRIzol–Chloroform (*see *Note 4)*

1. Centrifuge mosquito homogenates for 5 min at 4°C to clarify and pellet cellular debris.
2. Take 200 μL of the homogenates and add 800 μL of TRIzol.
3. Pipet well to homogenize the solution and let stand for 5 min at room temperature.
4. Add 200 μL of chloroform to each tube. Shake vigorously for 15 s, using a vortex mixer or shaking by hand. Leave to stand at room temperature for 10 min. Opalescence can be observed in the upper phase, indicating digestion of the sample components.
5. Centrifuge for 15 min at 12,000 × *g* at 4°C.
6. Carefully aspirate the aqueous phase without disturbing the interphase and transfer it to a sterile 1.5 mL tube.
7. Add 500 μL of isopropanol (approximately 0.5 μL/μL of TRIzol) (*see* **Note 5**).
8. Incubate at room temperature for 10 min, then centrifuge at 12,000 × *g* for 10 min at 4°C. RNA will precipitate at the bottom of the tube as a white or transparent gelatinous pellet, which may not always be visible.
9. Discard the isopropanol.
10. Wash the pellet with 1 mL of 75% (v/v) ethanol.
11. Repeat the washing step *(optional; not necessary if a transparent pellet is observed)*.
12. Centrifuge at 12,000 × *g* for 5 min at 4°C
13. Discard the ethanol, ensuring that no droplets remain. Remove most of the ethanol by decanting and the remainder with a pipette, aspirating from the side opposite the pellet.
14. Allow the open tube to air-dry at room temperature for approximately 10 min.
15. Resuspend the pellet in 30 μL of nuclease-free water.
16. Use 10 μL of RNA for reverse transcription and store the remaining RNA at –70°C.

3.4 Retrotranscription Using Improm II Kit and Random Hexamer Primers (see Note 6)

1. Prepare the RNA–primer mix: Add 1 μL of random hexamer primers (resuspended at 0.5 μg/μL) to 10 μL of target RNA. Incubate the RNA–primer mixture at 70°C for 5 min, then immediately chill on ice or place the tube on a cold rack in the freezer for 5 min to shock-cool the sample (*see* **Note 7**).
2. Assemble the RT Master Mix:

RT master mix	Volume per reaction (μL)
Water	1.1
RNA/random primer mix	11
Buffer 5× (without $MgCl_2$)	4
25 mM $MgCl_2$	2.4
10 mM dNTP mix	1
5 U/μL Improm II reverse transcriptase	1
40 U/μL RNaseOUT	0.5
Final volume	10

3. Add 10 μL of Master Mix to the RNA–random primer mixture to obtain a final reaction volume of 20 μL.
4. Incubate for 10 min at 25°C, followed by 60 min at 42°C, and finally 15 min at 70°C. Store at 4°C.

3.5 Nested PCR for the Detection of Generic Alphaviruses

1. For the first round of PCR amplification, prepare the reaction mixture as follows:

Reagent	Volume per reaction (μL)
Nuclease- free water	14.35
cDNA	2.5
Buffer 5× with $MgCl_2$	5
25 mM $MgCl_2$	0.5
10 mM dNTP mix	0.5
10 μM ALFA +1	1
10 μM ALFA −1	1
5 U/μL Taq polymerase	0.15
Total	25

2. Perform PCR using the following cycling conditions: an initial denaturation at 94°C for 2 min, followed by 40 cycles of 94°C for 30 s, 52°C for 1 min, and 72°C for 1 min, with a final extension at 72°C for 7 min.
3. For the second round of amplification, assemble the reaction mixture as follows:

Reagents	Volume per reaction (μL)
Nuclease- free water	16.35 μL
First PCR product	1 μL
Buffer 5× with $MgCl_2$	5 μL
10 mM dNTP mix	0.5 μL
10 μM ALFA +2	1 μL
10 μM ALFA −2	1 μL
Taq polymerase	0.15 μL
Total volume	25 μL

4. Include negative and positive controls.
5. Perform PCR using the following cycling conditions: an initial denaturation at 94°C for 2 min, followed by 35 cycles of 94°C for 30 s, 52°C for 1 min, and 72°C for 30 s, with a final extension at 72°C for 7 min. Using the degenerate primers ALFA +1 and ALFA −1 in the first round of PCR, a 481 bp amplicon is generated, whereas nested amplification with ALFA +2 and ALFA −2 primers yields a 195 bp product (Fig. 3) (*see* **Note 8**).

3.6 Product Resolution on a 2% Agarose Gel

1. Assemble the gel tray with the combs in place.
2. Weigh the appropriate amount of agarose, transfer it to an Erlenmeyer flask, and add 1× TE gel buffer.

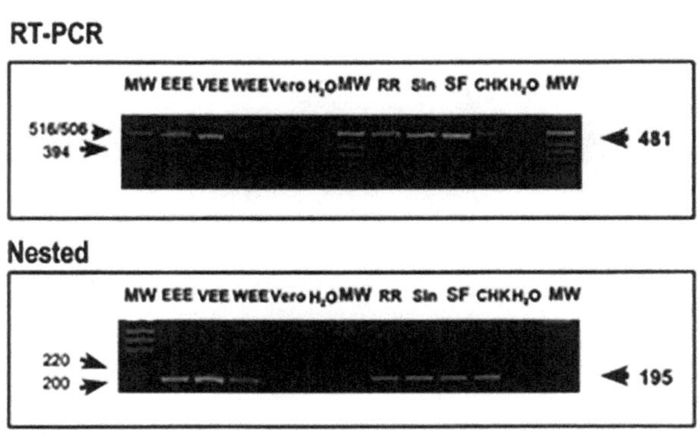

Fig.3 Alphavirus-specific amplicons derived from RT-PCR and nested RT-PCR. Agarose gels showing the 481 bp product of the RT-PCR amplification and 195 bp product of the nested reaction. Amplifications were carried out with RNA extracted from EEEV, VEEV, WEEV (West equine encephalitis Virus), RRV (Ross river virus), SINV (Sindbis virus), SF (Semliki forest virus), and CHIKV, with Vero cell culture and water used as negative controls. DNA marker was included (MW), and molecular weights are indicated

3. Heat the agarose solution in a microwave until completely dissolved.
4. Allow the molten agarose to cool to 50–60°C.
5. Add 1 μL of GelRed to 33 mL of molten agarose [19, 20].
6. Gently pour the agarose into the tray and ensure that no air bubbles are trapped in the wells.
7. Allow the gel to solidify at room temperature.
8. Remove the samples from the freezer and allow them to thaw at room temperature.
9. Once the gel has solidified, place the tray into the electrophoresis tank, ensuring correct orientation so that samples migrate from the negative (black) electrode to the positive (red) electrode.
10. Add 1× TE gel buffer to the tank until the gel is fully submerged.
11. Mix 10 μL of each sample with 5 μL of 70% (v/v) glycerol and load into the wells.
12. Load 1 μL of the molecular weight marker (Fig. 4) (*see* **Note 9**).

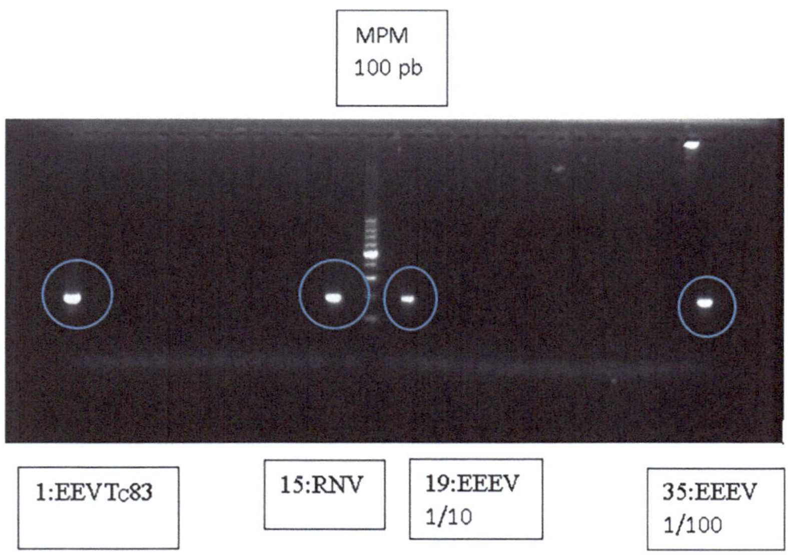

Fig.4 Electrophoretic analysis of amplified products from field-collected mosquitoes positive for generic alphaviruses. Bands of 195 bp are indicated by circles. The gel shows the expected 195 bp amplicons confirming alphavirus detection using primers amplified from EEV Tc83, RNV, and EEEV (1/10 and 1/100 dilutions). A 100 bp DNA ladder was used as a molecular weight marker. Photograph courtesy of Stechina O., IMR-UNNE-CONICET

4 Notes

1. It is important to conduct a preliminary study of the environment where the traps will be set (chicken coops, backyards, pastures, swamps, groves, etc.) to ensure an abundant catch of mosquitoes. Another factor to consider is the presence of light in the environment (residential lighting, street lighting), as the presence of other light sources reduces the attractiveness of the trap light, limiting the number of mosquitoes captured. Approximately 1 kg of dry ice is sufficient to bait the trap for a period of 16 h, which varies according to environmental conditions (temperature, relative humidity, air currents) that influence the rate of volatilization.
2. Mortars and pestles can be made of porcelain or glass, and their size may vary depending on the pool size. For small pools, small plastic pestles that fit directly into 1.5 mL tubes can be used to disrupt mosquito tissues.
3. When mosquito pools contain fewer than ten specimens per tube, mechanical homogenization can be performed using sterile plastic pestles directly inside the storage tube. Mosquito pools should be homogenized under a biosafety cabinet in sterile mortars using 1000 μL of MEM for pools of up to 25 mosquitoes and 2000 μL of MEM for pools of up to 50 individuals.
4. Mosquitoes may also be homogenized directly with TRIzol, leaving it to act for at least 5 min at room temperature. In that case, the volume of TRIzol depends on the size of the mosquito pools (<25 mosquitoes = 1 mL of TRIzol; between 25 and 50 mosquitoes = 2 mL of TRIzol).
5. The volume of isopropanol is proportional to the volume of the aqueous phase. Approximately 500 μL of isopropanol is used for aqueous phase volumes ranging between 200 and 800 μL.
6. Since the viral genome is composed of RNA, a preliminary reverse transcription step to generate complementary DNA (cDNA) is required. Either thermolabile reverse transcriptases (e.g., MMLV from Moloney murine leukemia virus or AMV from avian myeloblastic virus) or thermostable enzymes (e.g., *Thermus thermophilus* DNA polymerase, Tth Pol) may be used.
7. Aliquots of 1 μL of resuspended random hexamer primers can be stored at −20°C. When preparing the reverse transcription mix after RNA extraction, add 10 μL of target RNA to the random primer aliquot and perform the heat–cold shock step.
8. Most common problems encountered in PCR reactions include:

- Lack of amplification or poor performance due to low viral load in mosquito vectors.
- Presence of nonspecific amplification products (sub-bands).
- Primer–dimer formation.
- Mutations in the amplified product resulting from nucleotide misincorporation by the DNA polymerase.

9. Electrophoresis can be performed using different variants. When GelRed (nucleic acid gel stain, 10,000×) is used, it can be incorporated directly into the molten agarose or added during preparation of the loading buffer, followed by sample loading. Although the reference protocol (Sánchez-Seco et al. [11]) and its initial modifications used ethidium bromide as the staining agent, we use GelRed because it is less toxic to laboratory personnel and reduces environmental contamination. In our experience, GelRed also provides improved performance and greater versatility.

Acknowledgments

To Dr Adrián Díaz, Dr. Viviana Ré, and Dr. María Belén Pisano from the Arbovirus Laboratory, Dr. J. M. Vanella Institute of Virology (InVIV), Faculty of Medical Sciences, National University of Cordoba, for technical support in developing the techniques and procedures described in the protocols. Dr. Gerardo Deluca for his technical advice on the development and adaptation of these protocols, and Dr. Ornela Stechina for the setup and testing of alphavirus nested RT-PCR reactions with samples and controls. To Dr. Ailin Sotelo and Dr. Cecilia Urquijo for technical assistance and quality control. The molecular studies of arboviruses at the Institute of Regional Medicine of the UNNE were funded by PI 22L004 (Res. N° 313/2023, C.S., Secretaría General de Ciencia y Técnica, UNNE), PICTO N° 0246/2011 (FONCyT), and PICT-2018-02403.

References

1. Campos RE, Laurito M, Muttis E, Micieli MV. Culicidae (Diptera) species from Argentina and Uruguay. (Retrieved 27 July 2026). https://biodar.unlp.edu.ar/culicidae/
2. Marcondes CB, Contigiani M, Gleiser RM (2017) Emergent and reemergent arboviruses in South America and the Caribbean: why so many and why now. J Med Entomol 54(3):509–532. https://doi.org/10.1093/jme/tjw209
3. Tabachnick WJ (2016) Research contributing to improvements in controlling Florida's mosquitoes and mosquito-borne diseases. Insects 7:50. https://doi.org/10.3390/insects7040050
4. Stechina OS, Oria GI, Torres C, Diaz LA, Contigiani M, Marina S (2019) First detection of Madariaga virus in mosquitoes collected in a wild environment of Northeastern Argentina. Am J Trop Med Hyg 101(4):916–918. https://doi.org/10.4269/ajtmh.19-0475
5. Diaz LA, Diaz M, Almirón W, Contigiani M (2007) Infection by una virus (Alphavirus; Togaviridae) and risk factor analysis in black

howler monkeys (*Alouatta caraya*) from Paraguay and Argentina. Trans R Soc Trop Med Hyg 101:1039–1041

6. Sabattini MS, Monath TP, Mitchell CJ, Daffner JF, Bowen GS, Pauli R, Contigiani MS (1985) Arbovirus investigations in Argentina, 1977–1980. I. Historical aspects and description of study sites. Am J Trop Med Hyg 34:937–944
7. Monath TP, Sabattini MS, Pauli R, Daffner JF, Mitchell CJ, Bowen GS (1985) Arbovirus investigations in Argentina 1977–1980. IV. Serologic surveys and sentinel equine program. Am J Trop Med Hyg 34:966–975
8. Mitchell CJ, Darsie JrRF (1985) Mosquitoes of Argentina. Part II. Geographic distribution and bibliography (Diptera, Culicidae). Mosq Syst 17(4):279–360
9. Diaz LA, Flores FS, Quaglia A, Contigiani MS (2013) Intertwined arbovirus transmission activity: reassessing the transmission cycle paradigm. Front Physiol 3:493
10. Álvarez Escobar M del C, Torres Álvarez A, Torres Álvarez A, Semper González A, Romeo Almanza D (2018) Dengue, chikungunya, virus de zika. Determinantes sociales. Rev Méd Electrón [Internet]. 40(1). (Accessed Feb 2018). Disponible en: http://www.revmedicaelectronica.sld.cu/index.php/rme/article/view/1821/3693
11. Sanchez-Seco MP, Rosario D, Quiroz E, Guzmán MG, Tenorio A (2001) A generic nested-RT-PCR followed by sequencing for detection and identification of members of the alphavirus genus. J Virol Methods 2001(95):153–161
12. Chomczynski P, Sacchi N (1987) Single-step method of RNA isolation by acid guanidinium thiocyanate-phenol-choloroform extraction. Anal. Biochem. 162:156–159
13. Anderson MLM (2020) Nucleic acid extraction: fundamentals of sample preparation methodologies, current advancements and future endeavors. Trends Anal Chem Garland Science, p 260. ISBN 1000102246, 9781000102246
14. Service MW (1993) Mosquito ecology: field sampling methods, 2nd edn. Elsevier Science Publishers, Essex, p 988
15. Darsie Jr RF (1985) Mosquitoes of Argentina. Part I. Keys for identification of adult females and fourth stage larvae in English and Spanish (Diptera: Culicidae). Mosq Syst 17(3):153–278
16. Consoli R, Lourenço-de-Oliveira R (1994) Principias Mosquitos de Importancia Sanitâria no Brasil. Editora Fiocruz, Rio de Janeiro
17. Forattini OP (2003) Epidemiology and phylogenetic relationships of dengue viruses. Dengue Bull 27:91–94
18. Faran ME, Linthicum K (1981) A handbook of the amazonian species of Anopheles (Nyssorhynchus) (Diptera: Culicidae). Mosq Syst 13(1):1–80
19. Chomczynski P (1993) A reagent for the single-step simultaneous isolation of RNA, DNA and proteins from cell and tissue samples. BioTechniques 15:532–537
20. Kumar S, Stecher G, Li M, Knyaz C, Tamura K (2018) MEGA X: Molecular Evolutionary Genetics Analysis across computing platforms. Mol Biol Evol 35:1547

Check for updates

Chapter 21

Assessing the Vector Competence for Chikungunya Virus in Mosquitoes

Alexander W.E. Franz

Abstract

Chikungunya virus (CHIKV) is a human pathogenic, mosquito-borne virus (arbovirus), which is causing epidemic outbreaks among human populations in Africa, Asia, Europe, and South- and Central America including the Caribbean. The development of novel approaches to prevent mosquito transmission of the virus in the field requires a detailed study of CHIKV interactions with its mosquito vectors in the laboratory. In this chapter, we describe how to prepare *Aedes aegypti* mosquitoes for the infection with freshly cultivated CHIKV. We also describe quantitative and qualitative viral detection assays in mosquitoes based on plaque assays and the amplification of saliva samples in cell culture.

Keywords Chikungunya virus, Mosquito vector, *Aedes aegypti*, Infection, Bloodfeeding, Virus injection, Plaque assay, Mosquito saliva

1 Introduction

Chikungunya virus (*Togaviridae*, *Alphavirus*; CHIKV) is considered worldwide the most important and widespread alphavirus infecting humans [1, 2]. Originally detected in Tanzania, Africa in the 1950s, the virus caused sporadic outbreaks in Africa and Asia. However, in 2004–2005, a large epidemic outbreak occurred in Kenya, which spread across the southwestern Indian Ocean region to Southeast Asia [3]. In 2013, the virus was introduced to the Western Hemisphere causing severe outbreaks in the Caribbean and then in Brazil [4]. Since then, CHIKV has become endemic in most South- and Central American countries causing millions of disease cases in humans. In 2007, the virus was detected for the first time in Europe (Italy) to be involved in an active transmission cycle [5]. Additional small CHIKV outbreaks have been reported since then in France [6]. As of 2025, the virus remains a major public health problem in Asia and the Americas [7]. Typical disease symptoms in humans caused by CHIKV include sudden high-grade fever; (severe) arthralgia in

Claudia Filomatori and Eliana F. Castro (eds.), *Chikungunya Virus*, Methods in Molecular Biology, vol. 3067, https://doi.org/10.1007/978-1-0716-5496-5_21,

fingers, wrists, ankles, and elbows; myalgia; rash; fatigue; and headache [1]. CHIKV-caused arthralgia can be highly debilitating and persisting for several months. In 2023, two vaccines against CHIKV, Ixchiq (Valneva), and VIMKUNYA (Bavarian Nordic), have become available to the general public. Ixchiq is a live-attenuated vaccine, which in 2025 has raised safety concerns in the U.S., while VIMKUNYA is a virus-like particle (VLP) vaccine [8, 9].

Like all other alphaviruses, CHIKV is mosquito-borne. In sylvatic cycles, the virus has a non-human primate reservoir, while in the epidemiologically relevant urban disease cycles, the virus circulates between *Aedes aegypti* and/or *Ae. albopictus* mosquitoes and human hosts [1, 10–13]. *Ae. aegypti* is a highly anthropophilic day-biting mosquito species, whereas *Ae. albopictus* is an omnivorous biter using a wide range of animal species as bloodmeal sources, including humans [13, 14]. Similar to other arboviruses, CHIKV must persistently and systemically infect its arthropod vector to maintain the viral transmission cycle [15]. Following the ingestion of a viremic bloodmeal from a human host, the virus, along with the bloodmeal, enters the mosquito's midgut lumen [16]. Virions then infect the mosquito's midgut epithelium, where viral replication is initiated. De novo synthesized virions disseminate from the midgut to secondary tissue, including fat body, ovaries, neuro tissue, and the salivary glands. Once the latter are infected, the female mosquito is releasing CHIKV along with saliva during probing/feeding on a vertebrate host. Vertical CHIKV transmission from female to offspring has been frequently reported; however, the observed incidences in the field and experimental work in the laboratory show a great level of variability so that efficiency and importance of CHIKV vertical transmission in the virus' life cycle remain unclear [17, 18].

Studying the vector interactions of CHIKV is highly relevant for several reasons: (1) the virus is spreading at an epidemic scale within/across several continents, and vector (mosquito) populations are highly divergent [19]; (2) Mosquito vector competence for CHIKV can be variable: not every strain/population of *Ae. aegypti* or *Ae. albopictus* transmits each CHIKV strain with a similar efficiency [20, 21]; (3) an adaptation between a CHIKV strain and a vector population is possible, which then can result in a more rapid and efficient transmission of a particular virus strain [22, 23]; (4) The role of vertical mosquito transmission in regard to CHIKV epidemiology is still not understood [17, 18]. Here, we describe several basic experimental designs for the purpose of studying CHIKV–mosquito interactions in the laboratory. Our experimental description starts with the propagation of *Ae. aegypti* mosquitoes and the cultivation of CHIKV in cell culture. We then show how the virus can be administered to the mosquitoes followed by a description of cell culture-based methods for the detection of CHIKV in infected mosquitoes. Tissue/organ dissections from individual mosquitoes are not shown here as they have been demonstrated elsewhere [24, 25].

2 Materials

2.1 Preparing Aedes aegypti *Mosquitoes for CHIKV Infection in the Laboratory*

All developmental stages of mosquitoes are maintained in an insectary or insect rearing (environmental growth) chamber.

1. Egg-liner of *Aedes aegypti*, strain Higgs White Eye (HWE), stored at 28°C, 80% humidity for 4–5 months.
2. Distilled (deionized) water.
3. Desiccator, ID 273 mm.
4. Short Pasteur pipet for distribution of larvae.
5. Plastic shoe boxes, 205 mm × 140 mm × 120 mm (larva hatch pans) with lids.
6. Tropical fish-food—Tetra Cichlid Flakes, stored at room temperature.
7. 64 oz. (1893 mL) hot food cups with Vented Paper Lids, used as mosquito cartons.
8. 200 mm rubber bands for cartons.
9. 4 oz. (118 mL) or 2 oz. (59.1 mL), 0.5 oz. (14.8 mL) OP cups.
10. 1.0 mL polyethylene transfer pipette, for pupae picking.
11. Mesh fabric, white, polyester, for mosquito carton lids.
12. Water cups for mosquito cartons using 0.5 oz. (14.8 mL) OP cups covered by a 4 × 4 nonwoven gauze. The gauze is held in place with a small (20 mm) rubber band.
13. Raisins as a sugar source for mosquitoes (five to ten raisins per 100 females).

2.2 Cultivation of CHIKV in Vero Cells

1. Class II biosafety cabinet.
2. CO_2 incubator at 37°C.
3. Vero cells CCL-81, ATCC, online order at atcc.org.
4. CHIKV strain 37997 (GenBank accession #: AY726732.1) or LR2006_OPY1 (GenBank accession #: DQ443544.2).
5. Inverted microscope.
6. T-25, T-75 culture flasks.
7. 24-well culture plates with lids.
8. CHIKV growth medium: GlutaMax Dulbecco's Modified Eagle's Medium (DMEM), supplemented with 3%–7% fetal bovine serum (FBS), 100 U/mL penicillin/streptomycin. Keep at 4°C.
9. FBS-free DMEM. Keep at 4°C.
10. Pipettors and tips: 20, 200, 1000 mL.

2.3 Preparing and Administering a CHIKV Containing Artificial Bloodmeal to Mosquitoes

1. Class II biosafety cabinet.
2. 2 mL glass feeders.
3. Parafilm for glass feeders.
4. Immersion circulator SC100 and 5 L bucket.
5. Tubing, ID: 7 mm, OD: 10 mm.
6. Defibrinated sheep blood. Keep at 4°C.
7. 15 or 50 mL polypropylene centrifuge tube for bloodmeal mixture.
8. 1000 μL tips and pipet.
9. Laboratory labeling tape.
10. Forceps, Dumont 5 and 5SF.
11. 140 mm glass Petri dish with lid.
12. Laboratory ice bucket.
13. Glovebox for work with winged arthropods.
14. 70% (v/v) ethanol.
15. Stereomicroscope.
16. 60 cm × 38 cm x 34 cm large transparent bin, large rubber band, white mesh fabric.
17. Raisins (five to ten raisins per 100 females).
18. 0.5 oz. (14.8 mL) water cups, covered by a 4 × 4 non-woven gauze.

2.4 Intrathoracic Injection of CHIKV into Mosquitoes

1. Programmable needle puller P-97 (Sutter Instrument Inc., Novato, CA, USA) or similar.
2. 89-mm-length glass capillary.
3. Nanoject II Micro-Injector (Drummond Scientific, Broomall, PA, USA) or similar.
4. Kim-wipe tissues.
5. Hamilton needle.
6. Cargille Type B immersion oil.
7. ~10^7 PFU/mL frozen CHIKV stock.
8. Glovebox for work with winged arthropods.
9. 64 oz. (1893 mL) mosquito cartons.
10. 10-mm-diameter rubber stopper.
11. Self-made carton holding device.
12. Stereomicroscope.
13. Laboratory ice bucket.
14. 140 mm glass Petri dish with lid.

2.5 Maintaining CHIKV-Infected Mosquitoes

1. Insect Rearing Chamber IN084 (Darwin Chambers Co., St. Louis, MO, USA) or similar.
2. 60 cm × 38 cm x 34 cm large transparent bin, large rubber band, white mesh fabric.
3. Raisins (five to ten raisins per 100 females).
4. 0.5 oz. (14.8 mL) water cups, covered by a 4×4 non-woven gauze.

2.6 Sample Preparation for Plaque Assay–Based Virus Detection

1. Class II biosafety cabinet.
2. Adhesive laboratory labels.
3. Autoclaved 1 mL Eppendorf centrifuge tubes.
4. Eppendorf tube racks.
5. 140 mm glass Petri dish.
6. Laboratory ice bucket.
7. Dumont 5 forceps.
8. Sample processing medium: DMEM supplemented with 7% FBS, 1% (10 mM) HEPES.
9. Handheld motorized pellet grinder.
10. Disposable plastic pellet pestles. Pestles can be autoclaved and reused several times.
11. Insulin syringes, BD 1 mL with U-100 slip tip.
12. Sterile Acrodisc 2 mm HT Supor PES membrane syringe filter, 13 mm diameter.

2.7 Conduction of a Plaque Assay for CHIKV Detection

1. Class II biosafety cabinet.
2. CO_2 incubator at 37°C.
3. Vero cells CCL-81, ATCC, online order at atcc.org.
4. T-75 flasks.
5. 500 mL 1× phosphate-buffered saline (PBS).
6. 100 mL 1× trypsin/EDTA, 0.05% trypsin/0.53 mM EDTA in HBSS.
7. 24-well culture plates with lids.
8. Sample processing medium: DMEM supplemented with 7% FBS, 10 mM HEPES.
9. 96-well culture plates with lids.
10. Eight-channel pipette, 20 µL.
11. Pipet tips, 20 µL.
12. Pipet tips, 200 µL.
13. Serological pipette, 1 mL.
14. Vacuum pump and aspirator.

15. E1-ClipTip electronic adjustable tip spacing multichannel equalizer pipette.
16. Microwave oven.
17. Digital water bath.
18. 2% (w/v) agarose in autoclaved distilled water.
19. 2× nutrient solution: 20% Medium 199, 14% (v/v) FBS, 1% (v/v) MEM non-essential amino acids, 1% (v/v) MEM vitamin solution, 0.0075% (w/v) $NaHCO_3$.
20. Serological pipette, 25 mL.
21. Electronic pipette controller.
22. Cell staining solution (MTT): 0.5% (w/v) 3-(4,5-dimethylthiazol-2-yl)-2,5-diphenyltetrazolium bromide.
23. Light table.

2.8 Testing for the Presence of CHIKV in Mosquito Saliva

1. Class II biosafety cabinet.
2. CO_2 Incubator at 37°C.
3. Vero cells CCL-81, ATCC, online order at atcc.org.
4. 24-well culture plates with lids.
5. Glass capillaries, ID: 0.5 mm.
6. Cargille Type B immersion oil.
7. 140 mm glass Petri dish with lid.
8. Dumont 5SF forceps.
9. 1 mL Eppendorf centrifuge tubes.
10. Eppendorf tube racks.
11. Refrigerated tabletop centrifuge.
12. Saliva sample medium: DMEM supplemented with 7% FBS.
13. Insulin syringes, BD 1 mL with U-100 slip tip.
14. Sterile Acrodisc 2 mm HT Supor PES membrane syringe filter, 13 mm diameter.
15. Inverted microscope.

3 Methods

3.1 Preparation of *Aedes aegypti* Mosquitoes for CHIKV Infection in the Laboratory

For many of our virus challenge experiments, we use the laboratory-adapted "Higgs White Eye" strain of *Ae. aegypti* [26] and the "Lake Charles" strain of *Ae. albopictus* [27].

1. For *Ae. aegypti*, choose a piece of an egg-liner at <5 months old post-oviposition (Fig. 1a), place it into a 4 oz. OP cup filled

with distilled water and insert it into a sealed desiccator that will be connected to a vacuum source (*see* **Note 1**). Turn on the vacuum and let the vacuum hatch of L1 larvae proceed for 1–4 h. (For *Ae. albopictus*, typically choose an egg-liner that is <3 months old post-oviposition.)

2. Following vacuum hatching, once sufficient numbers of L1 larvae are visible in the OP cup when viewed under a stereomicroscope, transfer them in numbers of ~100 each to a plastic shoe box that is filled with ~800 mL distilled water. For larva transfer, use a short Pasteur pipet and try to adjust the larva uptake in such a way that each drop from the pipette yields around ten L1 larvae (ten drops would then amount to ~100 larvae/shoe box). Provide larvae in each shoe box with ~0.15 g of tropical fish food (*see* **Note 2**). Observe larvae daily until pupation to make sure that they have sufficient food available. There are four larvae stages (Fig. 1b) before the pupa stage is reached (Fig. 1c) at which no more fish food is taken up.

3. Estimate the approximate number of larvae for a typical CHIKV challenge experiment: assume that 20 virus-challenged female mosquitoes per timepoint and treatment need to be analyzed to ensure relatively robust statistics. Consider ~50% of the hatched larvae will be developing into males, which do not take bloodmeals; thus, 40 larvae per time point/treatment need to be considered. Double the number to 80 larvae per time point/treatment when estimating that perhaps just 50%

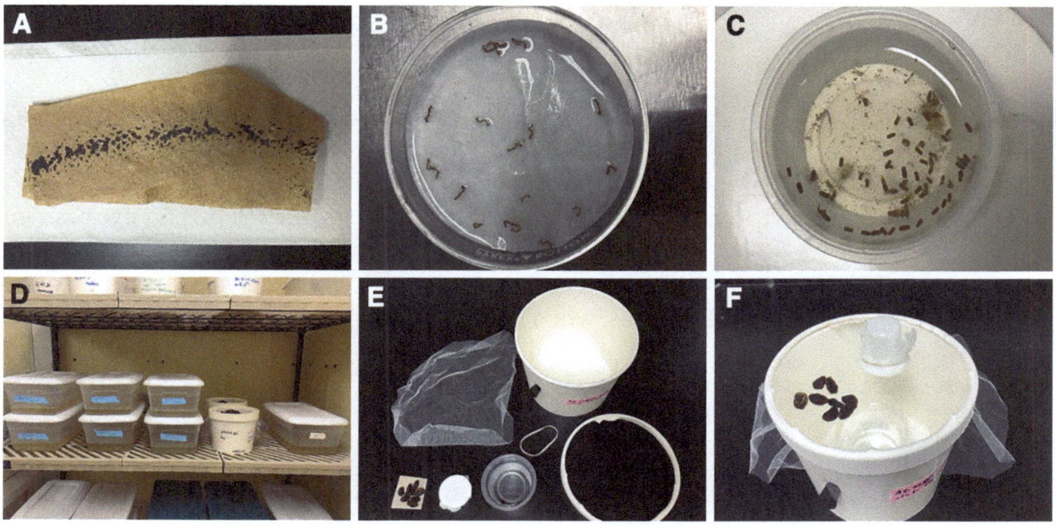

Fig.1 Developmental stages and containers for the rearing of *Aedes aegypti*. (**a**) A typical egg-liner (*Ae. aegypti*) to hatch out larvae. (**b**) L3/L4 larvae of *Ae. aegypti* XE "Aedes aegypti infection." (**c**) *Ae. aegypti* pupae in a pupa cup. (**d**) Larva hatch pans and mosquito cartons in the insectary. (**e**) Assembly of a mosquito carton: 64 oz. carton, fine mesh, lid rim, rubber band, 4 oz. (or 2 oz.) pupa (OP) cup, 0.5 oz. water cup, raisins. (**f**) The assembled carton

of the resulting females might fully ingest a virus-containing bloodmeal. This means that four hatch pans (Fig. 1d) will generate a number of females sufficient for five treatments/time points (20 replicates per timepoint/treatment) in a standard experiment.

4. Once larvae are metamorphosing into pupae, collect ~100 female pupae and ten male pupae per 4 oz. (118.3 mL) OP cup and place each single OP cup into a 64 oz. (1893 mL) carton (*see* **Note 3**). Use a modified lid: cut out the inner part of the lid, leaving the rim intact, and replace the opening with a dense (<0.5 mm) polyester mesh (Fig. 1e). Place the mesh-rim lid on top of the 64 oz. carton containing the OP cup and secure it with a strong rubber band. Add ~ten raisins on top of the mesh as sugar source for the emerging adults (Fig. 1f). Maintain mosquitoes in these cartons until females are at least 6 days old post-emergence before bloodfeeding/virus challenge.
5. At least 2 days before the bloodfeeding experiment or when all pupae in the OP cup have been metamorphosed into adults, remove the OP cup (water-containing pupae cup) from each carton. To safely remove the OP cup, immobilize mosquitoes by placing the carton in a 4°C refrigerator for 5–10 min (*see* **Note 4**). Quickly open the carton, remove the OP cup, then close the carton and secure the lid with a rubber band (*see* **Note 5**). Following 4°C chilling (or CO_2 knockout, *see* **Note 4**), adults typically remain immobilized for 2–3 min. After the OP cup has been removed from inside the carton, place a 0.5-oz. (14.8 mL) water cup covered with 4×4 non-woven gauze on top of the mesh to serve as water source for adults.

3.2 Propagation of CHIKV for a Mosquito Challenge Experiment

We typically use the 37997 strain (West-African genotype) or the La Reunion LR2006_OPY1 strain (East-Central-South-African genotype) of CHIKV for mosquito challenge experiments, making sure that viruses have a rather low passage history (less than five passages in cell culture) (*see* **Note 6**). We first amplify CHIKV in cell culture instead of directly using frozen stocks of the virus because freshly amplified virus, when administered in artificial bloodmeals, typically yields a higher mosquito infection rate than non-amplified virus from frozen stocks. All cell culture work must be conducted in a biosafety cabinet.

1. Two days before the CHIKV challenge experiment, seed Vero cells (ATCC # CCL-81) in four T-25 flasks at a density of ~28,000–35,000 cells/cm^2 using 5 mL CHIKV growth medium per flask. Incubate the cells at 37°C with 5% CO_2 supplement until they have reached at least 90% confluency on the day of virus infection. Use three of the four T-25 flasks for

CHIKV infection and designate one flask as the noninfected control.

2. At 30–36 h prior to the challenge experiment, infect the 90% confluent Vero cells with CHIKV at a multiplicity of infection (MOI) of 0.01 (*see* **Note 7**). Remove and discard the original CHIKV growth medium and add 1 mL of pure DMEM containing CHIKV (prepared from a virus stock aliquot) to each of the three infection flasks to achieve the target MOI. For the noninfected control flask, add 1 mL of pure DMEM, without virus. Incubate the flasks for 1 h at 37°C with 5% CO_2 supplement. Thereafter add 5 mL of CHIKV growth medium to each flask and return the flasks to the incubator.
3. Monitor virus infection and the development of cytopathic effects (CPE) daily until at least 50% (no more than ~80%, compared to the noninfected control) of the cells show strong CPE. This is typically observed to occur between 30 and 36 h post-infection (*see* **Note 8**).

3.3 Bloodfeeding Preparation and Mosquito Infection

We use custom-made, handblown 2 mL glass feeders and defibrinated sheep blood for mosquito bloodfeeding.

1. Remove raisins from the mosquito cartons 6–12 h prior to bloodfeeding, and the small water cups 1–2 h prior to bloodfeeding. Ensure that mosquitoes remain mildly starved but not debilitated to increase bloodfeeding efficiency.
2. Use six to eight cartons (64 oz./1893 mL), each containing 100 females (~1 week old) for simultaneous bloodfeeding. Assemble a "chain" of six to eight glass feeders (one feeder per carton). Connect the glass feeders using elastic tubing to form a "water jacket" system. Attach the tubing system to a heatable water circulator set to 37°C to maintain an appropriate blood temperature (Fig. 2a).
3. Cover the open end of each glass feeder (mosquito-facing side) with thinly stretched Parafilm® or a piece of hog gut (*see* **Note 9**), which is affixed to each feeder using a small rubber band. If possible, perform mosquito bloodfeeding on a bench top in a small, secluded anteroom within the containment facility (*see* **Note 10**). Temporarily, affix each glass feeder to the rim of the carton's lid using strips of laboratory tape. Prepare the artificial bloodmeal by mixing defibrinated sheep blood 1:1 with cell culture supernatant harvested from the CHIKV-infected Vero cells (*see* **Note 11**). Carefully, pipet up to 2 mL of the cell culture/blood mixture into each feeder. Allow females to feed; under optimal conditions, they will actively ingest blood within the first 10 min. Continue feeding for up to 1 h, ensuring that most females become fully engorged (Fig. 2a) (*see* **Note 12**).

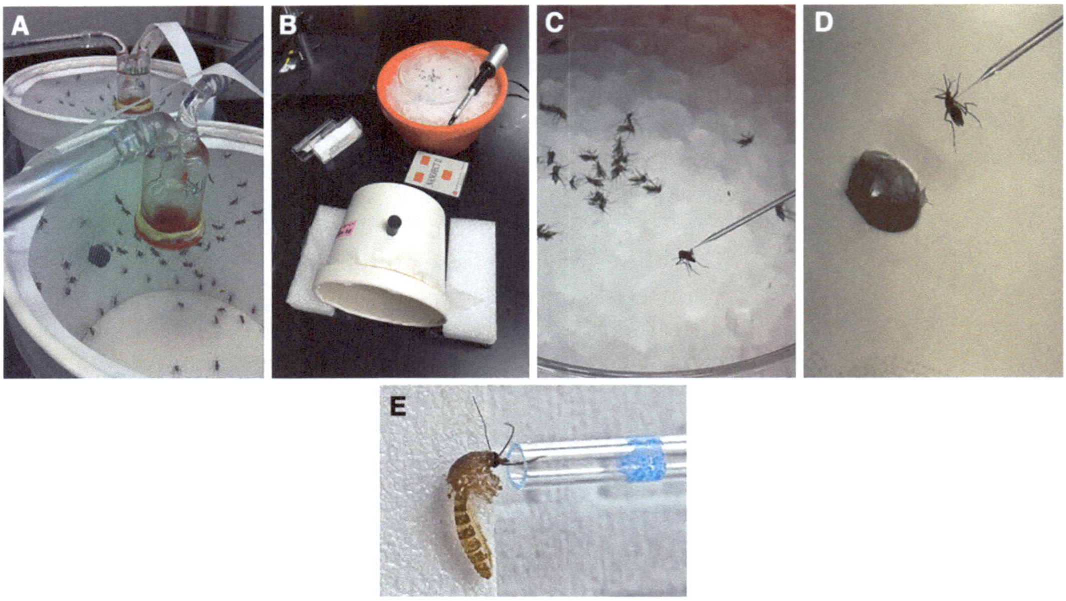

Fig. 2 Assessing the vector competence of *Aedes aegypti* XE "*Aedes aegypti* infection" for CHIKV: virus acquisition via bloodfeeding XE "Infection: of mosquitoes by bloodfeeding" or intrathoracic injection XE "Infection: of mosquitoes by intrathoracic injection", and testing saliva for the presence of the virus. (**a**) One-week-old females (strain: HWE) are receiving a bloodmeal containing 10^7 PFU XE "Plaque: forming-units (PFU)"/mL CHIKV (strain: 37997). Each carton is provided with a 2-mL glass feeder, which is connected to a water jack providing 37°C ambient bloodmeal temperature. The two glass feeders shown are fitted with a hog gut membrane through which the mosquitoes are feeding. Alternatively, a thin-stretched Parafilm membrane can be used. Using laboratory tape, the feeders are temporally affixed to the cartons. (**b**) Setup for mosquito injections: 64 oz. carton with "drop" hole plugged by a small rubber stopper. The carton sits on its side in foam material (formerly packaging material) to prevent the carton from moving. The Nanoject II control pad and a box of fabricated glass capillaries are shown, as well as an ice bucket containing a large glass Petri dish with mosquitoes. The Nanoject needle holder, in which a capillary has been inserted (and which is connected to the control pad), rests on top of the ice bucket. (**c**) Picking up a female by gently pressing the capillary into the thorax of the mosquito below the site of wing attachment (note: in the demonstration, the capillary does not contain any material). (**d**) Dropping the female, which is still attached to the capillary, through the "drop" hole of the carton (rubber stopper is removed). (**e**) A female whose wings and legs have been clipped off salivating into the immersion oil meniscus within a glass capillary. Note that the mosquito's proboscis is inserted into the capillary. (For better visibility, there is no immersion oil within the capillary.)

4. Once bloodfeeding is completed, collect a 100- to 200-μL sample of the bloodmeal from a feeder and store it at −80°C for later determination of the viral titer by plaque assay. Chill the cartons containing the fed mosquitoes at 4°C for up to 10 min until all females are immobilized at the bottom. Inside a glovebox designed for insect work, quickly open the carton and transfer all mosquitoes to a large glass Petri dish kept on ice. Immediately, place the glass lid on the Petri dish to prevent escape. *Ae. aegypti* mosquitoes can remain alive lying on ice for ~30 min, allowing time for the next manipulation step.

5. Remove the ice bucket containing the Petri dish from the glove box and place it under a stereomicroscope. Using fine forceps, select and count bloodfed females, and discard non-fed females and all males in 70% ethanol. Return the bloodfed mosquitoes to the original cartons. Close cartons with their mesh/lids, and provide raisins and small water cups. Place up to eight cartons into large bins covered with mesh held in place by a large rubber band. Transfer the large bins inside the insect rearing chamber to allow systemic virus infection of mosquitoes until further processing.

3.4 Alternative Infection of Ae. aegypti Females by Intrathoracic Injection of CHIKV

Intrathoracic injection of CHIKV into females bypasses the mosquito's midgut tissue, which is the initial organ in which the virus replicates following its acquisition from a viremic bloodmeal. In general, the mosquito midgut plays a determining role in the mosquito's vector competence for a virus. When applying intrathoracic injection, each mosquito can be infected with an identical, defined amount (concentration) of virus. The intrathoracic injection procedure is carried out in a secluded anteroom within containment.

1. Using a programmable needle puller, pull the capillaries in such a way that they have extra-fine tips, while the shanks of the capillaries toward their tips should have relatively shallow angles (Fig. 2c, 2d).
2. For intrathoracic injections, frozen CHIKV stocks with a concentration of up to 10^7PFU/mL are used. Slowly thaw the vial containing CHIKV at room temperature and then place it on ice. Conduce the intrathoracic injections of females using a stereomicroscope.
3. The 64 oz. (1893 mL) cartons each containing 100 ~5- to 7-day-old females need to be modified by having a single, small drop-hole on their side, which is plugged by a removable small rubber stopper. Punch the hole into the carton using fine, sharp scissors. The function of this opening is to drop individual injected mosquitoes into the carton while leaving the mesh-covered top of the carton untouched. Then, place the carton on its side into a self-made "carton holding device" (Fig. 2b).
4. As described under Sect. 3.3.4, carefully load females from a carton onto an ice-chilled large Petri dish. Mosquitoes should not lay on top of each other. Close the Petri dish with its lid.
5. The pulled capillaries are used in combination with a micro-injector (Fig. 2b). Slightly blunt/open the fine-pulled tip of a fabricated capillary by gently pushing the needle toward a sheet of Kim-Wipe tissue. Using a Hamilton needle, backfill the capillary with Cargill Type B immersion oil. Insert the capillary

into the injector's needle holder according to the manufacturer's description. Using the injector's control pad, push the mineral oil out of the capillary until it forms a meniscus near the tip.

6. Meanwhile, place the thawed cryo-tube containing the CHIKV stock into the same ice bucket containing the Petri dish with the mosquitoes. Under the stereomicroscope, carefully lower the capillary within the injector's needle holder into the cryo-tube and take up the virus suspension by operating the injector's control pad.
7. Once the capillary has been filled with virus stock, remove the lid from the Petri dish and, under the stereomicroscope, slowly/gently insert the capillary (no more than 1 mm deep) into the thorax of a female below the site of wing-attachment. Gently pick up the mosquito by the capillary and, operating the injector's control pad, inject between ~70 nL (a single key push) and ~210 nL (three key pushes) of virus stock into the insect's thorax (Fig. 2c). Carefully move the capillary holding the mosquito to the carton, remove its rubber-stopper and gently brush the mosquito off from the capillary so that it drops into the carton (Fig. 2d). Immediately, place the rubber-stopper back into the hole (*see* **Note 13**). Using this method, each mosquito can be injected with the same concentration of CHIKV, which (typically) will systemically and productively infect the mosquito's secondary tissues. For additional information, see also [28].

3.5 Maintenance of CHIKV-Infected Mosquitoes

Maintain the CHIKV-infected mosquitoes until their further analysis in an insect rearing chamber. Therefore, place six to eight large (64 oz.) cartons into a large bin, which is covered with a mesh held in place with a large rubber band. Our insect-rearing chamber can hold up to three of these large bins. The chamber settings are: 28°C, 80% humidity, and a 12-h light:12-h dark cycle. Supply each mosquito carton with a 0.5 oz. (14.8 mL) water cup and raisins, which should be carefully placed on top of the mesh cover. Refill water cups after 2–3 days, replenish/change raisins after 5–6 days. Mosquitoes can be collected at any time point to analyze virus infection. Typical time points for CHIKV testing in *Ae. aegypti* are: 24, 48, 96, and 168 h post-infection.

3.6 Sample Preparation for Plaque Assay–Based Virus Detection

This step involves preparation of sample tubes, mosquito/tissue homogenization, and sample sterile filtration. Conduct all cell culture work in a biosafety cabinet. Treat each sample individually as follows.

1. Prepare labels for 1 mL Eppendorf centrifuge tubes providing the following information on each label: mosquito strain, virus strain, tissue type, time/day post-infection, date of tissue harvest

(*see* **Note 14**). Use two batches of 1 mL tubes: (1) RNAse-free tubes that have not been autoclaved and (2) RNase-free tubes that underwent an additional autoclavation step (*see* **Note 15**). Place individual labels onto the non-autoclaved tubes only. Place the tubes into 1 mL tube racks. According to the number of mosquito samples to be processed, thoroughly wash the required number of reusable plastic pestles with dish washer detergent, spray with 70% ethanol, and then place them into a 200-mL glass beaker. Rinse the pestles inside the beaker with sterile (RNAse-free) water and thereafter, cover the pestle containing beaker with aluminum foil so that it can be autoclaved.

2. As described under Sect. 3.3.4, load the CHIKV-infected females (and/or non-infected controls) from each carton (each carton representing a different treatment) onto a separate large pre-chilled glass Petri dish sitting on ice. If no further dissections of individual mosquito tissues are required, carefully pick up each individual female from the chilled Petri dish using forceps and place the mosquito into the correctly labeled tube. The mosquito-containing tubes can be stored at −80°C until further processing.
3. Otherwise, fill each mosquito-containing tube with 300-μL sample processing medium and place it on ice. Using a handheld motorized pellet grinder with an attached reusable plastic pestle, homogenize each mosquito body (~1 min each) until no more intact body parts are visible. Following homogenization, add 700 μL sample processing medium to each 1 mL tube. Use each plastic pestle to process only a single sample to avoid cross-contamination of samples.
4. For plaque assays, sterile-filter the homogenized mosquito samples to avoid any bacterial/fungal contamination of the plaque assay's cell monolayer. Take a tube's entire sample homogenate up using a 1-mL insulin syringe. Then, manually attach a 0.2-mm Acrodisc HT Supor syringe filter to the filled syringe. Push the sample homogenate through the filter into an autoclaved 1-mL tube. Thereafter, switch the label of the source tube to the autoclaved (new) sample tube now containing ready-to-use sample material, which can be stored at −80°C until further use.

3.7 Plaque Assay for CHIKV Detection

For CHIKV plaque assays, we use 24-well culture plates. Perform six 10-fold serial dilutions of each mosquito sample generated under Sect. 3.6.4. A 24-well plate can accommodate four mosquito samples. Thus, for example, 80 mosquito samples would require 20 24-well plates. Conduct all cell culture work in a biosafety cabinet.

1. Cultivate Vero cells (CCL-81) in T-75 source flasks to 90%–100% confluency (~8×10^6 Vero CCL-81 cells/flask) in 10 mL

CHIKV growth medium per flask. Vero cells need to be trypsin treated to transfer them from the T-75 flasks to 24-well plates. Therefore, remove the growth medium from the flask and replace it with 10–15 mL (1×) PBS to gently rinse off remaining growth medium from the cells.

2. Remove the PBS from the flask and replace it with 3–4 mL Trypsin/EDTA. Then, incubate the T-75 flasks at 37°C for 5 min or longer until all cells have detached from the inner surface of the flask.
3. Add 100–105 mL DMEM (supplemented with 7% FBS) to each "trypsinized" T-75 flask and mix the medium by pipetting up and down a few times. Mix again the cell suspension in the T-75 flask, whenever starting a new 24-well culture plate to be filled with cells. An evenly mixed cell suspension of each T-75 can supply four 24-well plates containing 1 mL medium/well (1×10^5 Vero CCL-81 cells/well). After 2–3 days, when cells in the 24-well plates have reached 90% confluency, the cells are ready for the plaque assay experiment.
4. Conduct serial sample dilution in 96-well culture plates. Each plate can accommodate 16 10-fold serially diluted samples. Fill the plates with 180 μL/well of DMEM supplemented with 3% FBS except the wells in columns A1–H1 and A7–H7 into which the individual samples (200 μL vol. each) are added (A1–H1 = samples 1–8; A7–H7 = samples 9–16). Using an eight-channel pipette (20 μL vol), serially dilute samples by 10-fold from column A1–H1 to column A6–H6 and from column A7–H7 to column A12–H12 (*see* **Note 16**). The 96-well plates containing serially diluted samples can be stored for a few hours at 4°C.
5. Meanwhile, using a 1-mL serological pipet connected to a vacuum pump, aspirate the CHIKV growth medium from the 24-well plates. Using an electronic adjustable tip spacing 8-channel pipet, transfer 150 μL of the diluted samples from the 96-well culture plate in the same order to four 24-well plates containing the confluent Vero cells (*see* **Note 17**). Incubate the inoculated 24-well plates for 1 h at 37°C (with 5% CO_2 supplement) and gently shake them manually every 10–15 min (*see* **Note 18**).
6. Using a microwave oven, melt 2% agarose (in distilled water) in a bottle. The total volume is dictated by the number of 24-well plates (12 mL/plate). Cool down the agarose to 38°C–42°C by placing the bottle in a preheated water bath. Incubate the plaque assay 2x nutrient solution (12 mL/plate) at 37°C. Just before adding to the cells in the 24-well plates (following 1 h incubation), combine nutrient solution and agarose solution at a 1:1 ratio and maintain the mix in a water bath at 38°C–42°C.

Using a 25-mL serological pipet, add 1 mL of the agarose-nutrient solution to each well of each 24-well plate (*see* **Note 19**). Once added to the wells, let the agarose solidify for 30 min at room temperature. Incubate the plates upside-down for 2 days at 37°C (with 5% CO_2 supplement).

7. Stain the cells by adding 150–500 µL/well MTT and incubate overnight (24 h) at 37°C (with a 5% CO_2 supplement). The following day, CHIKV plaques will be visible, which can be counted when placing the 24-well plate on a light table. Count the plaques visible in the highest dilution and multiply by the dilution factor to calculate plaque-forming units per mL (PFU/mL) for each sample/mosquito (*see* **Note 20**).

3.8 CHIKV Detection in Mosquito Saliva

Detecting CHIKV in mosquito saliva is an indication that the female is potentially able to transmit the virus to a vertebrate host, which would conclude the horizontal transmission cycle in the vector. Typically, saliva is collected from CHIKV-infected females at 1, 2, 4, and/or 7 days post-infection via an infectious bloodmeal. This assay, based on cell-mediated virus amplification, is highly sensitive and provides a qualitative result whether or not a mosquito has released CHIKV in its saliva. Conduct all cell culture work in a biosafety cabinet.

1. Seed Vero cells into 24-well plates as described under Sects. 3.7.1–3.7.3. Consider one well of cells for each saliva sample to be tested. In addition, include several [6–8] "mock" infected wells to act as negative controls because by the end of the assay, CPE will be compared between saliva-inoculated and mock-inoculated cells.
2. For each time point and treatment, cut 10–15 readymade glass capillaries (0.5 mm ID) in half-length and fill one of their ends with 1–3 µL Cargille Type B immersion oil by dipping the bundle of capillaries into a 1-mL tube containing the oil.
3. Anesthetize CHIKV-infected females on a chilled glass Petri dish and under a stereomicroscope, carefully remove their wings and legs using fine forceps. This way, the mosquitoes stay immobilized and alive at room temperature.
4. Lay ten of the mineral oil–filled capillaries out onto a large (non-chilled) glass Petri dish. Carefully pick up each mosquito by its abdominal tip using fine forceps and insert the mosquito's proboscis into the oil meniscus of each capillary (Fig. 2e). Observe mosquitoes frequently under the stereomicroscope to ensure that their proboscises are still inserted into the mineral oil meniscuses (*see* **Note 21**).
5. After ~45 min of probing, collect the saliva from those capillaries in which droplets of saliva exuding from the mosquito's proboscises are visible (*see* **Note 22**). Carefully remove

the mosquito carcasses from the capillaries. Place the carcasses individually into labeled 1 mL tubes for later testing (i.e., via plaque assay) to assess their virus titers.

6. Recover saliva samples from each capillary by placing the saliva containing capillary fragment into a 1-mL Eppendorf tube containing 300 μL saliva sample medium and centrifugate the tube at 3000 × *g* for 10 min at 4°C to force the mineral oil (along with saliva) out of the capillary.
7. Following centrifugation, add another 200 μL saliva sample medium to each tube. Using forceps, carefully remove the capillaries from each tube and discard them into a sharps container. Sterile-filter the medium of each tube as described under Sect. 3.6.4.
8. Remove the CHIKV growth medium from the confluent Vero cells prior to their inoculation with 180 μL of sterile-filtrated sample (one sample/per well). Following 1 h incubation at 37°C (with 5% CO_2 supplement) and periodic shaking (as described under Sect. 3.7.5) of the sample inoculated 24-well plates, add 1 mL saliva sample medium to each well.
9. Maintain the 24-well plates at 37°C (with 5% CO_2 supplement) and observe them daily for a period of up to 1 week under an inverted stereomicroscope for the development of CPE indicative of CHIKV infection. This assay, based on cell-mediated virus amplification, is highly sensitive and provides a qualitative result whether or not a mosquito has released CHIKV in its saliva (*see* **Note 23**).

4 Notes

1. An egg-liner is a strip of paper towel covered with several 100 to several 1000 desiccated, fertilized eggs. Maintain all developmental stages of mosquitoes in an insectary or insect rearing (environmental growth) chamber.
2. Pulverize the tropical fish-food flakes into extra fine (1–2 mm) flakes using a large mortar and pestle to ensure that it dissolves properly in the hatch pan. Adding too much fish food (so that the water has a reddish hue) leads to excessive bacteria buildup and can become harmful for the mosquito larvae. Providing insufficient amounts of food (the water looks completely clear the next day) will lead to larva starvation. The number of larvae per shoe box should not exceed 120; otherwise, overcrowding effects would become apparent resulting in tiny pupae, which do not allow a clear distinction between the sexes based on pupa size.

3. The pupa stage lasts for ~36 h before winged adults start to emerge. If pupae are not picked from the larva hatch pans, they will eventually be populated by emerged (winged) adults, which will fly out as soon as the lid of the larva hatch pan is removed. We add ten males into each carton containing 100 females so that post-emergence, the latter will have been mated at the time point of bloodfeeding, which strongly increases female bloodfeeding efficiency. Female pupae can be relatively easily distinguished from male pupae as the former have a considerably larger size on average than the latter. Also, if all larvae were hatching at the same time, male pupae would typically emerge ~1 day earlier than the female pupae.
4. Alternatively, adults can be immobilized/anesthetized by providing CO_2 from a tank via a Drosophila CO_2 Fly Pad, which is held over the carton for 1–3 min until all adults have dropped to the bottom of the carton.
5. This operation can also be performed in a glove box that is especially designed for work with winged insects to restrict the movement of any accidental escapee. Escaped mosquitoes can be caught/eliminated using a spray bottle containing 70% ethanol or using an electric mosquito swatter. Any adults that landed into the water containing OP cup during the immobilization step need to be swiftly collected (before they "wake up") and placed into the carton using forceps.
6. In the U.S., CHIKV is classified by the CDC/NIH as a risk group 3/biosafety level 3 agent, which means that any mosquito experiment in which infectious virus is involved must be conducted in an Arthropod Containment Laboratory 3 (ACL-3) environment following the Personnel Protection Equipment (PPE) and containment requirements set up by the responsible Institutional Biosafety Committee.
7. The volume of required viral inoculum can be calculated as follows: (the number of target cells in the flask x desired MOI)/virus titer (PFU/mL) in stock vial. The answer is in mL.
8. There is a time window of ~6–9 h between 50% CPE and 80% CPE during which the artificial, virus-containing bloodfeeding experiment should be performed.
9. Alternatively, a piece of hog gut as typically used for sausage skin can be used. Sausage skin can be obtained at the butcher shop and needs to be kept at 4°C in 1× PBS. Little squares of hog gut that match the glass feeder can be cut out using fine scissors. The hog gut needs to be fixed to the glass feeder using small (20 mm) rubber bands. The hog gut should be slightly moist but not wet when used for feeding. Otherwise,

mosquitoes tend to "drink" the residual PBS solution from the hog gut instead of imbibing the bloodmeal.

10. The bloodfeeding assemblage on the bench top can be covered by a hood made of fine mesh material. Alternatively, bloodfeeding can be set up inside a biosafety cabinet with the sash lowered; however, in that case the blower of the cabinet should be switched off; otherwise, mosquitoes may get disoriented/distracted by the air stream and won't feed efficiently.
11. The defibrinated sheep blood is fragile and needs to be handled very carefully. Vortexing, vigorous shaking, up-and-down pipetting all should be avoided to prevent the red blood cells from bursting, which can negatively affect mosquito feeding.
12. Depending on the mosquito strain and laboratory environment, feeding of mosquitoes can be poor, i.e., mosquitoes are still not attracted to the bloodmeal in the feeders after ~20 min feeding time. A possibility to increase feeding enthusiasm is to add 10 mM ATP to the bloodmeal (before filling it into the feeders) and/or to place tiny pieces of dry ice (mimicking breath-like release of CO_2) near the cartons on the bench top. These efforts may not always help but they can improve feeding.
13. Inside the carton, the mosquitoes should wake up and move/fly around within 10–20 min post-injection. However, if by then, they are still lying motionless inside the carton, it is an indication that the injection procedure was too harsh, and the mosquitoes may not recover anymore. Possible reasons: the capillary may have been too wide at its tip, causing severe injury to the mosquito; the capillary was inserted too deep into the mosquito thorax; too much volume of material was injected into the mosquito; the mosquitoes stayed on ice for a prolonged period (>1 h).
14. Each Tough-Tag sheet contains 75 labels, which can be printed in color using an inkjet or laser printer.
15. For this, the tubes need to be closed with their lids.
16. Following each pipetting/dilution step, the pipet tips need to be changed; otherwise, the dilution series will be inaccurate due to residual sample carryover.
17. For example: samples/dilutions A1–A6, B1-B6–, C1–C6, D1–D6 from 96-well plate to 24-well plate 1; 96-well samples/dilutions E1–E6, F1–F6, G1–G6, H1–H6 to 24-well plate 2, etc., with A1–A6 representing the 10-fold dilution series of sample 1. The sample containing 24-well plates need to be individually numbered and labeled so that each sample

position and dilution (sample identity) can be clearly tracked later on during plaque counting.

18. The periodic agitation of the inoculum is important to avoid drying out of cells at the fringes of each well. Dried-out cells will often detach when staining/fixing the cells.
19. The pipetting needs to be performed swiftly and gently to avoid (1) solidifying of the agarose in the bottle or serological pipet; (2) disturbance of the cell monolayer. Solidified nutrient solution cannot be heated up again as it would cause the denaturation of the nutrients. The water bath temperature should not exceed 42°C. To avoid pipetting damage to the monolayer, the serological pipet should be held almost horizontally over each well, with its tip pointing to the periphery of the well to be filled.
20. Calculation: PFU/mL = (# plaques × dilution factor)/volume plated (in mL).
21. The glass Petri dish with the immobilized mosquitoes needs to be maintained at room temperature to ensure salivation activity.
22. It may be a good idea to collect and test material from every capillary even if no saliva "bubble" is visible. It should be recorded for each sample whether or not bubbles were visible.
23. Plaque assays, as a quantitative alternative, may sometimes not be sensitive enough to reliably detect miniscule quantities of virus in mosquito saliva.

References

1. Silva LA, Dermody TS (2017) Chikungunya virus: epidemiology, replication, disease mechanisms, and prospective intervention strategies. J Clin Invest 127(3):737–749
2. de Lima Cavalcanti TYV, Pereira MR, de Paula SO et al (2022) A review on chikungunya virus epidemiology, pathogenesis and current vaccine development. Viruses 14(5):969
3. Thiberville SD, Moyen N, Dupuis-Maguiraga L et al (2013) Chikungunya fever: epidemiology, clinical syndrome, pathogenesis and therapy. Antiviral Res 99(3):345–370
4. Leparc-Goffart I, Nougairede A, Cassadou S et al (2014) Chikungunya in the Americas. Lancet 383(9916):514
5. Liumbruno GM, Calteri D, Petropulacos K et al (2008) The Chikungunya epidemic in Italy and its repercussion on the blood system. Blood Transfus 6(4):199–210
6. Fournier L, Durand GA, Cochet A et al (2025) Multiple early local transmissions of chikungunya virus, Mainland France, from May 2025. Euro Surveill 30(32):2500545
7. WHO (2025) Chikungunya virus disease – global situation. (Accessed 08 Feb 2026). https://www.who.int/emergencies/disease-outbreak-news/item/2025-DON581
8. Weber WC, Streblow DN, Coffey LL (2024) Chikungunya virus vaccines: a review of IXCHIQ and PXVX0317 from pre-clinical evaluation to licensure. BioDrugs 38(6):727–742
9. Richardson JS, Anderson DM, Mendy J et al (2025) Chikungunya virus virus-like particle vaccine safety and immunogenicity in adolescents and adults in the USA: a phase 3, randomised, double-blind, placebo-controlled trial. Lancet 405(10487):1343–1352
10. Diallo M, Thonnon J, Traore-Lamizana M et al (1999) Vectors of chikungunya virus in Senegal: current data and transmission cycles. Am J Trop Med Hyg 60(2):281–286

11. Diallo D, Sall AA, Buenemann M et al (2012) Landscape ecology of sylvatic chikungunya virus and mosquito vectors in southeastern Senegal. PLoS Negl Trop Dis 6(6):e1649
12. Broeckel R, Haese N, Messaoudi I et al (2015) Nonhuman primate models of chikungunya virus infection and disease (CHIKV NHP model). Pathogens 4(3):662–681
13. Little EAH, Harriott OT, Akaratovic KI et al (2021) Host interactions of *Aedes albopictus*, an invasive vector of arboviruses, in Virginia, USA. PLoS Negl Trop Dis 15(2):e0009173
14. Facchinelli L, Badolo A, McCall PJ (2023) Biology and behaviour of *Aedes aegypti* in the human environment: opportunities for vector control of arbovirus transmission. Viruses 15(3):636
15. Franz AW, Kantor AM, Passarelli AL et al (2015) Tissue barriers to arbovirus infection in mosquitoes. Viruses 7(7):3741–3767
16. Dong S, Kantor AM, Lin J et al (2016) Infection pattern and transmission potential of chikungunya virus in two New World laboratory-adapted *Aedes aegypti* strains. Sci Rep 6:24729
17. Agarwal A, Dash PK, Singh AK et al (2014) Evidence of experimental vertical transmission of emerging novel ECSA genotype of Chikungunya Virus in *Aedes aegypti*. PLoS Negl Trop Dis 8(7):e2990
18. Honório NA, Wiggins K, Eastmond B et al (2019) Experimental vertical transmission of chikungunya virus by Brazilian and Florida *Aedes albopictus* populations. Viruses 11(4):353
19. Bartholomeeusen K, Daniel M, LaBeaud DA et al (2023) Chikungunya fever. Nat Rev Dis Primers 9(1):17
20. Vega-Rúa A, Zouache K, Girod R et al (2014) High level of vector competence of *Aedes aegypti* and *Aedes albopictus* from ten American countries as a crucial factor in the spread of Chikungunya virus. J Virol 88(11):6294–6306
21. Vega-Rúa A, Marconcini M, Madec Y et al (2020) Vector competence of Aedes albopictus populations for chikungunya virus is shaped by their demographic history. Commun Biol 3(1):326
22. Powers AM, Logue CH (2007) Changing patterns of chikungunya virus: re-emergence of a zoonotic arbovirus. J Gen Virol 88(Pt 9):2363–2377
23. de Lamballerie X, Leroy E, Charrel RN et al (2008) Chikungunya virus adapts to tiger mosquito via evolutionary convergence: a sign of things to come?. Virol J 5:33
24. Coleman J, Juhn J, James AA (2007) Dissection of midgut and salivary glands from *Ae. aegypti* mosquitoes. J Vis Exp 5:228
25. Tutagata J, Pocquet N, Trouche B et al (2024) Dissection of mosquito ovaries, midgut, and salivary glands for microbiome analyses at the organ level. J Vis Exp (212)
26. Wendell MD, Wilson TG, Higgs S et al (2000) Chemical and gamma-ray mutagenesis of the white gene in *Aedes aegypti*. Insect Mol Biol 9(2):119–125
27. Darby CS, Featherston KM, Lin J et al (2023) Detection of La Crosse virus in situ and in individual progeny to assess the vertical transmission potential in *Aedes albopictus* and *Aedes aegypti*. Insects 14(7):601
28. Sanchez-Vargas I, Williams AE, Franz AWE et al (2021) Intrathoracic inoculation of Zika Virus in *Aedes aegypti*. Bio Protoc 11(18):e4165

Chapter 22

Global Genomic Surveillance

Francesco Branda, Massimo Ciccozzi and Fabio Scarpa

Abstract

Global genomic surveillance has emerged as a foundational pillar of public health in the twenty-first century, enabling real-time tracking of pathogen evolution and informing outbreak response. This chapter examines the strategic architecture of global genomic surveillance, focusing on its application to arboviruses such as chikungunya virus (CHIKV). It explores the integration of genomic data with epidemiological, clinical, and environmental information within a One Health framework, while addressing critical challenges in governance, equity, and interoperability. The discussion covers the entire genomic surveillance workflow, from sample collection and sequencing to bioinformatic analysis and phylogenetic inference, and highlights the transformative role of artificial intelligence (AI) in predictive surveillance. By analyzing global initiatives, operational barriers, and emerging technologies, this chapter underscores the necessity of sustainable, equitable, and interoperable genomic systems to proactively address current and future infectious disease threats.

Keywords Genomic surveillance, Public health, Arboviruses, Chikungunya virus, Molecular epidemiology, Next-generation sequencing, Bioinformatics, Phylogenetics, Artificial intelligence, One Health

1 Introduction

Throughout the twenty-first century, global genomic surveillance, i.e., the systematic analysis of the genomes of emerging and circulating pathogens worldwide, has emerged as a transformative and fundamental pillar of public health, thanks to the advancement of sequencing technologies and their integration into infectious disease emergency responses. In an era where human mobility, environmental changes, and selective pressures, such as vaccines, drugs, and population immunity, continuously drive microbial evolution, the integrated use of genomics is one of the most powerful tools for anticipating, monitoring, and responding promptly to epidemics and pandemics. This approach is particularly critical not only for respiratory viruses but also for vector-borne pathogens, including arboviruses such as chikungunya, dengue, and Zika viruses, whose

Claudia Filomatori and Eliana F. Castro (eds.), *Chikungunya Virus*, Methods in Molecular Biology, vol. 3067,
https://doi.org/10.1007/978-1-0716-5496-5_22,

frequent emergence in new geographic areas, complex transmission dynamics, and sensitivity to ecological and climatic changes make them an increasing challenge for global public health and highlight the need for proactive genomic monitoring [1, 2], capable not only of rapidly identifying the pathogen but also of tracking genomic evolution in real time, estimating transmission patterns, evaluating the effectiveness of vaccines and control measures, and guiding informed public health decisions.

The COVID-19 pandemic has represented an unprecedented acceleration in this area: in the first 33–34 months of SARS-CoV-2 virus circulation, over 14 million viral sequences were collected and shared in public databases, highlighting the ability of genomics to track the emergence of variants in real time, modulate the health response, and effectively guide public policy [3]. However, the transition from emergency response to sustainable, routine genomic surveillance poses several operational and infrastructural challenges. Sample collection, sequencing, bioinformatics analysis, metadata integration, and data sharing require complex and coordinated workflows, and each step can be a critical point. Recent studies highlight how unrepresentative sampling methods, global disparities in bioinformatics, delays in sequence publication, and gaps in the use of data in public health have compromised the overall effectiveness of surveillance [4]. In particular, there are significant inequalities between high-income countries and countries with limited resources. A study conducted in South and Southeast Asia showed that, although all countries had Next-Generation Sequencing (NGS) capacity, only a few had an integrated strategic plan for genomics, the median time between sampling and reporting was estimated at 18 days, and laboratories used only about half of their potential capacity [5]. Beyond Asia, similar disparities are even more pronounced in Africa, where major gaps persist in the genomic surveillance of vector-borne and zoonotic pathogens. Recent studies on emerging viruses have highlighted how limited sequencing infrastructure, fragmented surveillance networks, and uneven data sharing practices hinder the early detection and genomic characterization of outbreaks across the continent [6]. These data highlight the need to develop modular and flexible systems that can be quickly reconfigured for different types of pathogens, moving from the surveillance of respiratory viruses to that of vector-borne pathogens and zoonoses, with a One Health approach. Genomic surveillance of arboviruses, in particular, requires specific approaches due to challenges such as low viral load in samples, the need for enrichment, vector and environmental surveillance, and infrastructure distributed even in rural or resource-limited settings. At the same time, it has become clear that integrating genomic data with clinical, epidemiological, and environmental data is essential for translating information into effective public health actions, as demonstrated by the epidemic genomics dashboards developed during the pandemic [7].

This chapter explores the main applications of genomic surveillance, from the development of modular networks for emerging pathogens to the integration of multisource data and the use of emerging technologies, such as artificial intelligence, for predicting viral properties, while addressing key infrastructure and governance challenges. Furthermore, the work emphasizes the importance of extending surveillance to vector-borne pathogens and the environment, highlighting the need for a comprehensive, equitable, and interoperable approach that allows for monitoring spillover dynamics and timely intervention to prevent health emergencies. In particular, the chapter focuses on arboviruses, analyzing how the application of genomics to these pathogens can improve understanding of their evolutionary and transmission dynamics, facilitate the early detection of emerging variants, and strengthen response capabilities in different geographical contexts. In summary, the chapter describes how genomic surveillance represents a cross-cutting axis of modern public health, capable of integrating technology, data, governance, and health strategy to support preparedness and response to emerging and future infectious threats.

2 The Strategic Architecture of Global Genomic Health: Governance, Viral Evolution, and Predictive Surveillance

In recent years, genomics has emerged as one of the most powerful tools for global public health, connecting molecular biology to computational epidemiology and providing real-time insight into the evolution of pathogens. However, its application on a global scale raises crucial questions about governance, equity, and interoperability. The COVID-19 pandemic has been a turning point: on the one hand, it has highlighted the extraordinary ability of the scientific community to generate and share millions of sequences in unprecedented time; on the other, it has exposed the fragility of a system that is still highly asymmetrical, in which the ability to produce, analyze, and interpret genomic data is not distributed equitably.

2.1 Establishing a Global Vision: The WHO Genomic Surveillance Strategy

Aware of these challenges, the World Health Organization (WHO) promoted the Global Genomic Surveillance Strategy for Pathogens with Pandemic and Epidemic Potential (2022–2032) in 2022, with the aim of "making genomics an integral part of the health response in every country" [8]. The strategy identifies four priority areas: (i) Structural integration between traditional surveillance systems and genomics; (ii) Equitable access to sequencing technologies and analysis infrastructure; (iii) Transparent data sharing based on principles of solidarity and reciprocity; and (iv) Training and capacity building at the local and regional levels. These principles are part of a complex geopolitical context, where genomic data is not only scientific information but also a strategic resource with economic and diplomatic implications [9]. The experience of the Global Influenza Surveillance and Response System (GISRS) network, originally dedicated to influenza and subsequently expanded to other respiratory viruses, shows that institutional trust and transparency in data sharing remain fundamental preconditions for the functioning of global networks.

At the operational level, genomic surveillance is based on the integration of three dimensions: the wet lab (data production), the dry lab (bioinformatic analysis), and data governance (data management, sharing, and ownership). Each of these components presents specific challenges. In high-capacity laboratories, data generation is now facilitated by automated platforms and NGS systems with high throughput. However, as reports from the Africa CDC Pathogen Genomics Initiative [10] and PAHO BioHub [11] show, many low-income countries continue to depend on external reference centers, slowing down the flow of information and reducing the timeliness of health responses. The difficulties do not only concern material resources but also the ability to process and interpret data. Recent studies [12, 13] emphasize that data latency, i.e., the time between the generation and publication of genomic data, is a critical factor in the detection of new variants. The gap is therefore not only technological but also epistemological: genomics only becomes useful when integrated into a digital ecosystem capable of connecting genomes with epidemiological, clinical, and geographic metadata, thus enabling truly predictive surveillance. The structure of genomic surveillance systems across international, regional, national, and local levels is summarized in Table 1.

One of the emerging challenges concerns the ethical and political management of genomic data. The debate on data sovereignty, i.e., the right of countries to control their own biological and genomic data, has intensified since the pandemic. Several countries, especially in Africa and Southeast Asia, have called for data shared with global platforms to be associated with mechanisms for the equitable return of benefits, such as co-authorship in scientific works, resource sharing, or technical training [14]. The concept of "genomic solidarity," as outlined in the 2021 UNESCO/International Bioethics Committee report [15], which calls for the promotion of international cooperation to ensure that genome technologies serve the common good and are taken up, for example, by Zhang et al. [16] as an alternative to the extractive approach of "genomic sovereignty," represents an attempt to establish a cooperative model based on transparency, accountability, and reciprocity.

A significant contribution to the democratization of data has come from open-source platforms such as Nextstrain [17], Pathogenwatch [18], and CZ ID [19], which integrate phylogenetic analysis, real-time visualization, and clinical metadata into interactive

Table 1
Operational levels of genomic surveillance and associated challenges

Level	Main objective	Examples of infrastructures or programs	Key challenges
International	Coordination, governance, and standardization	WHO Pandemic Hub (Berlin), GISRS, GISAID, GenBank, ENA	Asymmetries in data access and ownership criteria; lack of semantic interoperability
Regional	Development of technical capacity, infrastructure, and training	Africa CDC Pathogen Genomics Initiative, PAHO BioHub, WHO South-East Asia Region (WHO SEARO)	Disparities in computational resources; limited financial sustainability
National	Sequencing, surveillance, and rapid reporting	National public health institutes, reference laboratories	Shortage of bioinformatics personnel and cloud infrastructure
Local	Sample collection and shipment	Hospitals, peripheral laboratories, diagnostic centers	Fragile transport chains; absence of digital data tracking

environments. These tools encourage the active participation of scientific communities in pathogen surveillance, the creation of new analyses, and the transparent communication of results. However, the sustainability of such platforms remains a challenge, especially in low- and middle-income countries, where limited analytical capacity, infrastructure, and financial resources hinder the full implementation of genomic surveillance [20]. In addition, issues related to the protection of sensitive data, ethical disparities, and the heterogeneity of available data persist, which can reduce the effectiveness of global surveillance and support for health decisions. Despite these challenges, global genomic surveillance has already changed the logic behind how health emergencies are addressed: from an event-centered response to a vision oriented toward evolutionary risk. For example, the WHO Pandemic Hub in Berlin plays a pivotal role in advancing pandemic and epidemic intelligence by promoting the integration of genomic, epidemiological, and environmental data into interoperable systems. Through initiatives such as the Epidemic Intelligence from Open Sources (EIOS) platform and collaborations with partners including the Global Initiative on Sharing All Influenza Data (GISAID) [21], the Hub leverages artificial intelligence and data analytics to strengthen early warning capacities and predictive modeling for health threats [22, 23, 24].

Among these, the chikungunya virus (CHIKV) represents a paradigmatic case. Its rapid geographical expansion, vector adaptation, and recent epidemics demonstrate how genomic data, when linked to epidemiological and environmental information, can shed light on viral evolution in real time. Recent genomic studies have revealed that the Indian subcontinent and Southeast Asia serve as major hubs for the recent global spread of CHIKV, highlighting the role of international genomic surveillance in tracking viral lineage movement and mutation hotspots [25]. Similarly, large-scale genomic investigations in Brazil have demonstrated the co-circulation of CHIKV and dengue viruses, highlighting the importance of integrated arbovirus surveillance for monitoring concurrent epidemics and genomic recombination events [26]. Furthermore, the increasing availability of complete genomic sequences from recent epidemics in regions such as Pakistan has confirmed the circulation of the East/Central/South African (ECSA) genotype, highlighting the global interconnectedness of arboviral evolution and the adaptive potential of CHIKV to new ecological and vector contexts [27].

2.2 Molecular Epidemiology and Genomic Insights into the Chikungunya Virus

Genomic surveillance has emerged as an indispensable tool for understanding the complex transmission dynamics, evolutionary trajectories, and molecular epidemiology of CHIKV [28]. Advances in sequencing technologies now allow near real-time monitoring of viral lineages, shedding light on how subtle genetic changes can influence transmission efficiency, virulence, and vector competence [29, 30]. When integrated with epidemiological and ecological data, these genomic insights provide actionable guidance for outbreak preparedness, vector control, and vaccine development, highlighting the transformative potential of molecular epidemiology in public health [17, 21].

2.2.1 Genome Organization and Genetic Diversity

CHIKV carries a single-stranded, positive-sense RNA genome of approximately 11.8 kilobases, organized into two polyproteins: the nonstructural (nsP1–4) and structural (C, E1, E2, E3, and 6K) proteins [31, 32]. The nonstructural proteins are primarily involved in viral replication, transcription, and assembly, whereas the structural proteins form the virion and mediate host cell entry. Among these, the E1 and E2 envelope glycoproteins are particularly critical, facilitating receptor binding and membrane fusion and playing a pivotal role in vector specificity and transmission potential [32, 33].

The genetic diversity of CHIKV is structured into several well-characterized lineages, including the West African, East/Central/South African (ECSA), and Asian lineages. More recently, the Indian Ocean Lineage (IOL), derived from ECSA, has been implicated in large-scale outbreaks across the Indian Ocean islands and parts of Asia [31, 32, 34]. Each lineage exhibits characteristic single-nucleotide polymorphisms (SNPs) and amino acid substitutions, some of which confer enhanced vector adaptability or altered virulence. For example, the E1-A226V mutation has been associated with increased fitness in *Aedes albopictus*, demonstrating the interplay between viral evolution and ecological adaptation [32].

Understanding these lineage-specific variations is crucial not only for epidemiological tracking but also for informing vaccine design and anticipating potential shifts in transmission dynamics. The accumulation of mutations within the CHIKV genome, typical of RNA viruses, underscores the importance of continuous genomic surveillance to detect emerging variants and adapt public health strategies accordingly [28].

2.2.2 Sample Collection and Nucleic Acid Extraction

High-quality sample collection forms the foundation of effective genomic surveillance [29, 30]. Clinical specimens generally include serum, plasma, or whole blood obtained from infected individuals, whereas entomological surveillance focuses on mosquito pools containing potential vectors such as *Aedes aegypti* and *Aedes albopictus* [28]. Given the inherent instability of RNA viruses, preserving RNA integrity is paramount. Samples should be collected following standardized protocols, stored at −80°C for long-term preservation, and transported under controlled conditions to prevent degradation.

Viral RNA extraction typically relies on commercially available kits based on silica columns or magnetic beads, which provide consistent yield and purity [30]. In situations where viral titers are low, enrichment or concentration steps may be employed to increase the proportion of viral RNA relative to host material. Prior to sequencing, rigorous quality control through spectrophotometric or fluorometric quantification, as well as assessment of RNA integrity, ensures reliable downstream analyses [30].

2.2.3 Sequencing Approaches

Once RNA extraction is complete, the choice of sequencing strategy depends on study objectives, sample quality, and available resources [29, 30]. While Sanger sequencing remains useful for targeted genomic regions, NGS platforms have become the mainstay of CHIKV genomic surveillance [29, 30].

Illumina sequencing provides high accuracy and depth, making it ideal for detecting low-frequency variants and conducting population-level studies [29]. Long-read platforms such as Oxford Nanopore and PacBio are particularly valuable for assembling complete genomes and identifying structural variations [29, 30]. Library preparation may involve direct RNA sequencing, reverse transcription followed by PCR amplification, or hybrid capture-based enrichment to increase the proportion of viral reads [30].

Coverage depth and uniformity are critical considerations. Insufficient coverage can obscure minor variants, while uneven read distribution may lead to assembly gaps. Optimizing primer design, implementing multiplexed sequencing strategies, and performing targeted enrichment are practical approaches to ensure comprehensive and reliable genome reconstruction [30, 35].

2.2.4 Bioinformatic Analysis

After sequencing, computational pipelines transform raw data into interpretable genomic information [36, 37]. Initial steps typically include quality assessment and trimming of reads using tools such as FastQC and Trimmomatic [37]. Clean reads are then aligned to a reference genome using software such as BWA or Bowtie2, with variant calling performed through pipelines such as GATK or FreeBayes [36]. For studies requiring de novo assembly, tools such as SPAdes and Canu are widely employed to reconstruct complete viral genomes [37].

Once assembled, viral genomes undergo annotation and comparative analyses to identify lineage-defining mutations, recombination events, and adaptive changes [28]. Public repositories such as GenBank and GISAID enable the sharing of newly generated sequences, contributing to global surveillance efforts [21]. Analytical frameworks such as Nextstrain facilitate visualization of CHIKV phylogenetic relationships, providing dynamic insights into viral evolution and geographic spread [17].

2.2.5 Phylogenetics and Molecular Epidemiology

Phylogenetic reconstruction offers a powerful approach for tracing the evolutionary history and transmission patterns of CHIKV [38, 39]. Multiple sequence alignment using MAFFT or MUSCLE sets the stage for tree inference through maximum likelihood methods (RAxML, IQ-TREE) or Bayesian approaches (BEAST) [38, 39, 40].

Integrating genetic, temporal, and geographic information allows researchers to elucidate patterns of viral introduction, expansion, and local circulation [29]. Minor genetic changes, such as the E1-A226V substitution, can have profound effects on vector competence and outbreak dynamics [32]. Phylogenetic and population-based analyses distinguish between local persistence and new introductions, informing timely public health interventions [29]. Linking these genomic insights with epidemiological data enables targeted vector control, efficient resource allocation, and tailored outbreak preparedness strategies [29]. A summarized overview of the CHIKV genomic surveillance workflow is summarized in Table 2.

2.3 Integrating New Technologies for Arbovirus Genomic Surveillance: The Role of Artificial Intelligence

The enormous volume and inherent complexity of genomic and epidemiological data generated within global surveillance systems require the application of advanced computational methods. In this context, artificial intelligence (AI) has become a key tool for exploiting such data and transforming surveillance from a reactive and descriptive science into a proactive and predictive discipline, particularly with regard to arboviruses. The use of AI in viral genomics is redefining the paradigm of molecular surveillance, enabling a shift from retrospective analysis of variants to the ability to predict the evolutionary pathways of pathogens. Recent advances in protein language models (pLMs) [41] and deep learning algorithms make it possible to encode the information contained in protein sequences as

Table 2
Overview of the methodological pipeline for CHIKV genomic surveillance

Phase	Main Objective	Key techniques/Tools
Collection/Extraction	RNA integrity; yield and purity from clinical and entomological samples	Storage at −80°C; magnetic bead-based kits; fluorometric QC; Sample types: serum/plasma, *Aedes* spp. pools
Sequencing (NGS)	Real-time monitoring; high accuracy for variants; complete genome assembly	Illumina (accuracy/depth); Oxford Nanopore/PacBio (long-read, structural variations); RT-PCR or hybrid capture (library preparation)
Bioinformatic Analysis	Transform raw data into interpretable genomic information	FastQC/Trimmomatic (QC/trimming); BWA/Bowtie2 (alignment); GATK/FreeBayes (variant calling); SPAdes/Canu (de novo assembly)
Phylogenetics	Track evolutionary and transmission history; distinguish local persistence vs. introductions	MAFFT/MUSCLE (alignment); RAxML/IQ-TREE (maximum likelihood); BEAST (Bayesian inference); Nextstrain (visualization/dissemination)

a true "biological language," capable of revealing signs of adaptation and selective pressures before they translate into epidemiological phenomena. A pioneering example of this approach is the EVEscape model, which, by integrating evolutionary and structural data, accurately predicted the most relevant immune escape mutations that emerged during the COVID-19 pandemic, anticipating variants of interest such as Omicron and its subvariants [42]. Such predictive frameworks are now being extended to other rapidly evolving viruses, including arboviruses such as Zika virus and Chikungunya virus, to estimate the potential fitness of mutations, predict host changes, and classify new emerging lineages according to evolutionary risk. This approach marks a conceptual shift: while traditional genomic surveillance, as in the case of the E1-A226V mutation of CHIKV, detected viral adaptations only after they had spread, AI models can now serve as early warning systems, providing predictive information on the likelihood that a given mutational combination will result in increased transmissibility or virulence.

In the context of arboviruses, AI has immediate applications in two main areas. Firstly, in supporting rapid differential diagnosis and classification. Many arbovirus diseases share similar clinical manifestations and circulate simultaneously in the same geographical areas, making early diagnosis difficult. Numerous studies show how machine learning (ML) models can support the distinction between these diseases based on clinical symptoms and laboratory results. For example, a Colombian study developed two models (Decision Tree and Random Forest) to classify dengue, Zika, and chikungunya, achieving accuracies of up to 99.7% for chikungunya, 99.1% for dengue, and 98.8% for Zika [43]. However, a systematic review on the use of ML in tropical diseases points out that there is still a lack of studies that consider multiple arboviruses simultaneously [44]. These tools provide clinical decision support in resource-limited settings, where specialized molecular tests are not always available. Secondly, AI and ML contribute significantly to epidemic prediction and modeling of transmission dynamics. By analyzing complex datasets—including climatic, environmental, demographic, mobility, and case history variables—it is possible to estimate the imminent risk of increased cases or epidemic emergence. A 2022 review in the context of arboviruses included 139 studies published between 2015 and 2020, highlighting challenges and gaps in spatiotemporal models for these viruses [45]. For example, a study on dengue prediction in Colombia compared Random Forest and artificial neural networks (ANN), finding that Random Forest outperformed ANN, with environmental and climatic variables being more important for short time horizons and sociodemographic variables for longer horizons [46]. More recently, a 2025 study comparing various models (statistical and ML) for weekly dengue prediction in Rio de Janeiro—including climatic variables—found that LSTM (Long Short-Term Memory) networks with climatic covariates were the most accurate, albeit computationally more expensive [47]. These methodologies make it possible to anticipate epidemic peaks, optimize resource allocation, and plan vector control interventions. To accurately predict the behavior of complex vector-borne pathogens, AI architectures must be able to integrate the diverse data streams required by modern surveillance strategies. Temporal and spatiotemporal modeling plays a crucial role, since the transmission dynamics of arboviruses are intrinsically linked to environmental and climatic variables [48]. Therefore, effective predictive frameworks require sophisticated architectures such as recurrent neural networks (RNNs) and LSTM networks, which excel at handling sequential data over time. These tools can predict zoonotic diseases such as Rift Valley fever and West Nile fever by integrating climate data with behavioral inputs from vectors. Although explicit modeling of wildlife movement and environmental factors is critical to understanding spillover, it presents significant computational challenges due to the large volume of data involved. In addition, ML methods for viral genomic fitness and host-shift prediction are emerging as powerful instruments for understanding viral adaptation. The BERT-infect model, for instance, leverages large language models trained on extensive nucleotide sequences to predict the potential for zoonotic spillover across numerous viral families [49]. For evolutionary analysis, tree-based

ML methods can assess Viral Codon Fitness (VCF) by analyzing the relative synonymous codon usage (RSCU) of the viral genome, generating a Human Virus Codon Fitness (HVCF) score that serves as a molecular readout of viral adaptation to the human host. This approach provides insights into evolutionary trajectories, even those driven by silent synonymous mutations [50].

These diverse AI architectures contribute synergistically to predictive arbovirus surveillance by enabling early variant detection, outbreak forecasting, and assessment of viral adaptation potential. The main categories, data inputs, and functional roles of these models are summarized in Table 3, which illustrates how different AI approaches support proactive public health interventions across multiple levels of surveillance. Although AI models can achieve high predictive accuracy, their adoption in public health practice depends on the ability to address the "black box" problem. Public health decision-makers require clear justifications for critical decisions, such as the allocation of resources for vector control, beyond a simple accuracy score. For this reason, it is essential to integrate explainable AI (XAI) frameworks, such as SHapley Additive exPlanations (SHAP) and Local Interpretable Model-agnostic Explanations (LIME). These techniques allow the results of models, even those of highly complex architectures such as deep neural networks or ensemble methods, to be translated into understandable biological, ecological, or clinical explanations, identifying which variables have the greatest weight, how they influence the prediction, which "what-if" scenarios can

Table 3
Applications and architectures of AI models in arbovirus surveillance

AI model type	Primary function/application	Key data inputs integrated	Role in proactive surveillance
pLMs/EVEscape	Prediction of viral properties, immune escape mutations, and lineage fitness	Genomic sequences, evolutionary selection pressure, structural data	Anticipates variant emergence (e.g., novel E1-A226V-like mutations)
RNN/LSTM	Forecasting temporal disease outbreaks (time-series)	Climate/environmental data, epidemiological time-series, vector behavior	Predicts outbreak timing and intensity based on ecological risk factors
Decision Trees/ Random Forests	Differential clinical diagnosis and classification	Clinical symptoms, laboratory results (non-genomic)	Supports rapid and accurate public health triage in co-circulating epidemics
VCF models	Predicting virus host codon fitness and zoonotic spillover risk	Viral codon usage biases (RSCU), nucleotide sequences	Prioritizes surveillance for arboviruses with high human infectivity potential

be simulated, and what the limitations of the prediction may be. For example, XAI allows key characteristics to be highlighted (e.g., "2°C increase in average temperature combined with increased urban mobility" → increased risk of transmission) and explains to the decision-maker why a particular prediction was generated. It also helps to verify the biological and epidemiological consistency of the model beyond its numerical accuracy, thus promoting acceptance among non-technical professionals. Furthermore, it promotes accountability: when deciding on an intervention based on the model's results, XAI allows the underlying "chain of reasoning" to be traced, which is essential for public health and vector control. Importantly, it can also help identify and mitigate biases (e.g., socioeconomic variables influencing predictions) and spurious dependencies that would otherwise remain hidden within a "black-box" model. However, it should be noted that XAI tools have certain limitations: they can be sensitive to the underlying model, the presence of collinearity between variables, or missing data, and sometimes the explanation provided may oversimplify the actual complexity of the model. Therefore, XAI should be combined with good modeling practices, including model validation, documentation, and the development of intuitive interfaces that facilitate interpretation by nontechnical stakeholders.

In the context of arbovirus surveillance, where decision-makers may include national or regional health authorities and interventions range from vector control campaigns to diagnostic testing allocation and early warning system implementation, the lack of model explainability may hinder its adoption. In contrast, an AI system that provides both a prediction and a rationale (e.g., "expected increase in transmission due to favorable weather conditions, reduced vaccination coverage, and increased human mobility") will be much more credible, actionable, and implementable in real-world settings. In the medical and healthcare field, systematic reviews have shown that XAI methods such as SHAP and LIME are critical for building end-user trust in AI systems [51]. In the context of infectious disease surveillance, recent studies have shown that XAI can clarify the relationships between model predictions and spatial or environmental characteristics, for example, in dengue risk modeling [52]. At the same time, methodological analyses highlight that XAI methods themselves are subject to inherent limitations (e.g., dependence on the underlying model, sensitivity to feature collinearity), thus requiring careful design, validation, and clear communication to nontechnical stakeholders [53].

3 Conclusions and Future Trajectory

Global health continues to face the evolving threat of pathogens, particularly arboviruses such as chikungunya, and requires a unified strategy that integrates policy, molecular epidemiology, and advanced predictive technologies. Effective genomic surveillance of these pathogens is critical to proactively monitor, predict, and mitigate the impact of emerging epidemics.

First and foremost, the success of genomic surveillance depends largely on strong global governance that ensures equitable access to sequencing technologies and promotes trust among nations. The WHO Global Genomic Surveillance Strategy (2022–2032) emphasizes the need to ensure timely and universal access to sequencing and calls for greater interoperability between national health systems. However, the sustainability of this global approach is fragile and depends on protecting transparency and providing safeguards against political or economic retaliation, as demonstrated by the delayed disclosure of the Omicron variant. Without such safeguards, the incentive structure will continue to favor opacity, which can leave critical gaps in global surveillance efforts. It is therefore imperative to create systems that promote transparency, trust, and cooperation in genomic data sharing, especially considering the complex geopolitical context. Second, molecular epidemiology, as demonstrated by the CHIKV case study, provides high-resolution information for immediate risk assessment. Genomic surveillance allows for the monitoring of phylogenetic changes and key mutations that may signal increased epidemic potential, such as the spread of the ECSA lineage in the Americas and the vector-adaptive E1-A226V mutation. To maximize the effectiveness of intervention measures, this molecular understanding of viral behavior must be combined with real-time entomological and environmental surveillance, following the One Health approach. This integrated approach, which recognizes the interconnection between human health, animal health, and environmental factors, is essential for responding effectively to arboviral threats. Thirdly, AI has become a key enabling technology for transforming genomic surveillance from a descriptive to a predictive tool. The application of AI models, such as pLMs and VCF models, has revolutionized the ability to predict viral fitness and the emergence of new variants. These technologies enable proactive public health decisions by predicting and mitigating potential outbreaks. However, for these predictive tools to be widely adopted and considered reliable, it is essential to integrate XAI frameworks, such as SHAP and LIME, which help make complex AI outputs understandable and actionable to decision-makers. The incorporation of XAI ensures that predictive outputs are not only accurate but also transparent and accountable, thereby promoting trust among health authorities and ensuring the implementation of successful genomic surveillance systems. One of the main challenges for the future is overcoming the limitations of current AI models, particularly with regard to understanding complex host–vector interactions and viral infectivity. Future research should focus on integrating host immunity data, such as population seroprevalence, to refine selective pressure models. This will improve our ability to predict and control outbreaks in real time. Furthermore, to improve the predictive power of AI models, it is essential to continue collecting comprehensive datasets and ensure their representativeness. Particularly in LMICs, increasing access to low-cost, portable sequencing technologies will be critical to ensuring the continuous flow of genomic data needed to support AI surveillance systems.

Finally, integrating genomic surveillance across sectors—human health, animal health, and environmental monitoring—is a key challenge. A One Health strategy, supported by collaborative international initiatives, is essential for monitoring and responding to arboviral threats. Networks such as CREATE-NEO [54] and initiatives such as EpiArbo [55] and ArboItaly [56] are working to harmonize data from different surveillance sources and create integrated platforms for analyzing genomic data alongside entomological and environmental information. These efforts will help build a more robust and comprehensive global surveillance system, providing a solid foundation for addressing future arboviral threats. In conclusion, the future trajectory of global genomic surveillance depends on strengthening collaboration between data generation, analysis, and governance, ensuring equitable access to sequencing technologies, and integrating advanced predictive tools such as AI. This approach will be essential for predicting, monitoring, and responding to the dynamic and evolving challenges posed by arboviruses and other emerging infectious diseases. The success of this global surveillance strategy will ultimately depend on continued investment in technology, infrastructure, and training, particularly in resource-limited settings, to ensure that genomic surveillance remains a powerful tool for protecting global public health.

References

1. Patiño L, Benítez AD, Carrazco-Montalvo A et al (2024) Genomics for arbovirus surveillance: considerations for routine use in public health laboratories. Viruses 16(8):1242
2. Vieira CJ, Onn MB, Shivas MA et al (2024) Long-term co-circulation of multiple arboviruses in southeast Australia revealed by xeno-monitoring and viral whole-genome sequencing. Virus Evol 10:1
3. Tosta S, Moreno K, Schuab G et al (2023) Global SARS-CoV-2 genomic surveillance: what we have learned (so far). Infect Genet Evol 108:105405
4. Ling-Hu T, Rios-Guzman E, Lorenzo-Redondo R et al (2022) Challenges and opportunities for global genomic surveillance strategies in the COVID-19 era. Viruses 14(11):2532
5. Getchell M, Wulandari S, de Alwis R et al (2024) Pathogen genomic surveillance status among lower resource settings in Asia. Nat Microbiol 9(10):2738–2747
6. Moir M, Sitharam N, Hofstra LM et al (2025) Identifying genomic surveillance gaps in Africa for the global public health response to West Nile virus: a systematic review. Lancet Microbe 6(10):101176
7. Sitharam N, Tegally H, Silva DD et al (2024) SARS-CoV-2 genomic epidemiology dashboards: a review of functionality and technological frameworks for the public health response. Genes 15(7):876
8. Carter LL, Yu MA, Sacks JA et al (2022) Global genomic surveillance strategy for pathogens with pandemic and epidemic potential 2022–2032. Bull World Health Organ 100(4):239
9. McKibbin K, Shabani M (2023) Genomic data as a national strategic resource: implications for the genomic commons and international data

sharing for biomedical research and innovation. J Law Med Ethics 51(2):301–313
10. Africa Centres for Disease Control and Prevention (Africa CDC) The Africa Pathogen Genomics Initiative (Africa PGI). Available from:https://africacdc.org/download/the-africa-pathogen-genomics-initative-africa-pgi-integrated-genomic-surveillance-for-outbreak-detection-detect/. Accessed on 4 Nov 2025
11. Pan American Health Organization (PAHO). Human genomics for health: enhancing the impact of effective research. Report of the first regional meeting for the Americas. Available from: https://iris.paho.org/handle/10665.2/62584. Accessed on 4 Nov 2025
12. Brito AF, Semenova E, Dudas G et al (2022) Global disparities in SARS-CoV-2 genomic surveillance. Nat Commun 13(1):7003
13. Chen Z, Azman AS, Chen X et al (2022) Global landscape of SARS-CoV-2 genomic surveillance and data sharing. Nat Genet 54(4):499–507
14. Mboowa G, Kakooza F, Egesa M et al (2024) The rise of pathogen genomics in Africa. F1000Research 13:468
15. UNESCO. Report of the International Bioethics Committee (IBC) on the principle of protecting future generations. Available from: https://unesdoc.unesco.org/ark:/48223/pf0000378723. Accessed on 4 Nov 2025
16. Zhang JY (2023) Commoning genomic solidarity to improve global health equality. Cell Genomics 3:10
17. Hadfield J, Megill C, Bell SM et al (2018) Nextstrain: real-time tracking of pathogen evolution. Bioinformatics 34(23):4121–4123
18. Argimón S, David S, Underwood A et al (2021) Rapid genomic characterization and global surveillance of Klebsiella using Pathogenwatch. Clin Infect Dis 73(4):S325–S335
19. Lu D, Kalantar KL, Glascock AL et al (2025) Simultaneous detection of pathogens and antimicrobial resistance genes with the open source, cloud-based, CZ ID platform. Genome Med 17(1):46
20. Alakija A (2023) Leveraging lessons from the COVID-19 pandemic to strengthen low-income and middle-income country preparedness for future global health threats. Lancet Infect Dis 23(8):e310–317
21. Shu Y, McCauley J (2017) GISAID: Global initiative on sharing all influenza data–from vision to reality. Eurosurveillance 22(13):30494
22. World Health Organization. WHO Hub for pandemic and epidemic intelligence. Available online: https://pandemichub.who.int/. Accessed on 4 Nov 2025
23. World Health Organization. The WHO Hub in Berlin: driving innovation to make the world safer from health threats. News release, 17 June 2025. Available online: https://www.who.int/news/item/17-06-2025-the-who-hub-in-berlin-driving-innovation-to-make-the-world-safer-from-health-threats. Accessed on 4 Nov 2025.
24. Contributors to the Pandemic T, Morgan O, Redies I et al (2022) Innovations in public health surveillance: updates from a forum convened by the WHO Hub for pandemic and epidemic intelligence. Eurosurveillance 27(15):2200302
25. Qiu S, Guo J, Li P et al (2021) Source-tracking of the Chinese Chikungunya viruses suggests that Indian subcontinent and Southeast Asia act as major hubs for the recent global spread of Chikungunya virus. Virol J 18(1):203
26. de Jesus AC, Fonseca PL, Alves HJ et al (2024) Retrospective epidemiologic and genomic surveillance of arboviruses in 2023 in Brazil reveals high co-circulation of chikungunya and dengue viruses. BMC Med 22(1):546
27. Umair M, Yasir M, Jamal Z et al (2025) Whole-Genome sequencing of Chikungunya Virus (CHIKV) from Pakistan: detection of the East/Central/South African (ECSA) genotype during the 2024 outbreak in Mansehra. PLoS ONE 20(8):e0329856
28. Weaver SC, Forrester NL (2015) Chikungunya: evolutionary history and recent epidemic spread. Antivir Res 120:32–39
29. Quick J, Loman NJ, Duraffour S et al (2016) Real-time, portable genome sequencing for Ebola surveillance. Nature 530(7589):228–232
30. Quick J, Grubaugh ND, Pullan ST et al (2017) Multiplex PCR method for MinION and Illumina sequencing of Zika and other virus genomes directly from clinical samples. Nature Protoc 12(6):1261–1276
31. Schuffenecker I, Iteman I, Michault A et al (2006) Genome microevolution of chikungunya viruses causing the Indian Ocean outbreak. PLoS Med 3(7):e263
32. Tsetsarkin KA, Vanlandingham DL, McGee CE et al (2007) A single mutation in chikungunya virus affects vector specificity and epidemic potential. PLoS Pathog 3(12):e201
33. Tsetsarkin KA, Chen R, Leal G et al (2011) Chikungunya virus emergence is constrained in Asia by lineage-specific adaptive landscapes. Proc Natl Acad Sci U S A 108(19):7872–7877
34. Giovanetti M, Vazquez C, Lima M et al (2023) Rapid epidemic expansion of chikungunya virus East/Central/South African lineage, Paraguay. Emerg Infect Dis 29(9):1859
35. Grubaugh ND, Gangavarapu K, Quick J et al (2019) An amplicon-based sequencing framework for accurately measuring intrahost virus

diversity using PrimalSeq and iVar. Genome Biol 20(1):8
36. Li H, Durbin R (2009) Fast and accurate short read alignment with Burrows–Wheeler transform. Bioinformatics 25(14):1754–1760
37. Bankevich A, Nurk S, Antipov D et al (2012) SPAdes: a new genome assembly algorithm and its applications to single-cell sequencing. J Comput Biol 19(5):455–477
38. Stamatakis A (2014) RAxML version 8: a tool for phylogenetic analysis and post-analysis of large phylogenies. Bioinformatics 30(9):1312–1313
39. Suchard MA, Lemey P, Baele G et al (2018) Bayesian phylogenetic and phylodynamic data integration using BEAST 1.10. Virus Evol 4(1):vey016
40. Katoh K, Misawa K, Kuma KI et al (2002) MAFFT: a novel method for rapid multiple sequence alignment based on fast Fourier transform. Nucleic Acids Res 30(14):3059–3066
41. Lytras S, Lamb KD, Ito J et al (2025) Pathogen genomic surveillance and the AI revolution. J Virol 99(2):e0160124.
42. Thadani NN, Gurev S, Notin P et al (2023) Learning from prepandemic data to forecast viral escape. Nature 622(7984):818–825
43. Arrubla-Hoyos W, Gómez JG, De-La-Hoz-Franco E (2024) Differential classification of dengue, zika, and chikungunya using machine learning—random forest and decision tree techniques. InInformatics 11(3):69
44. Chitrapady S, Rajendran R, Rashid M et al (2025) Machine-learning-based artificial intelligence tools for the diagnosis of tropical fevers: a systematic review and meta-analysis protocol of diagnostic test accuracy. BMJ Open 15(8):e102158
45. Lima CL, da Silva AC, Moreno GM et al (2022) Temporal and spatiotemporal arboviruses forecasting by machine learning: a systematic review. Front Public Health 10:900077
46. Zhao N, Charland K, Carabali M et al (2020) Machine learning and dengue forecasting: Comparing random forests and artificial neural networks for predicting dengue burden at national and sub-national scales in Colombia. PLoS Negl Trop Dis 14(9):e0008056
47. Chen X, Moraga P (2025) Assessing dengue forecasting methods: a comparative study of statistical models and machine learning techniques in Rio de Janeiro, Brazil. Trop Med Health 53(1):52
48. Carbone G, Boiardi G, Infantino C et al (2025) Vectors on the move: how climate change fuels the spread of arboviruses in Europe. Microorganisms 13(9):2034
49. Kawasaki J, Suzuki T, Hamada M (2025) Hidden challenges in evaluating spillover risk of zoonotic viruses using machine learning models. Commun Med 5(1):187
50. Su S, Ni Z, Lan T et al (2025) Predicting viral host codon fitness and path shifting through tree-based learning on codon usage biases and genomic characteristics. Sci Rep 15(1):12251
51. Ali S, Akhlaq F, Imran AS et al (2023) The enlightening role of explainable artificial intelligence in medical & healthcare domains: a systematic literature review. Comput Biol Med 166:107555
52. Yang H, Nguyen TN, Chuang TW (2023) An integrative explainable artificial intelligence approach to analyze fine-scale land-cover and land-use factors associated with spatial distributions of place of residence of reported dengue cases. Trop Med Infect Dis 8(4):238
53. Sathyan A, Weinberg AI, Cohen K (2022) Interpretable AI for bio-medical applications. Complex Eng Syst 2(4):18
54. Vasilakis N, Hanley KA (2023) The coordinating research on emerging arboviral threats encompassing the neotropics (CREATE-NEO). Zoonoses (Burlington, Mass.) 3(1):16
55. Wallau GL, Abanda NN, Abbud A et al (2023) Arbovirus researchers unite: expanding genomic surveillance for an urgent global need. Lancet Glob Health 11(10):e1501–e1502
56. Branda F, Giovanetti M, Ceccarelli G et al (2024) ArboItaly: leveraging open data for enhanced arbovirus surveillance in Italy. Front Pharm 15:1459408

Check for updates

Chapter 23

Nanopore Sequencing for Chikungunya Virus: Principles and Application

Talita Émile Ribeiro Adelino, Joilson Xavier, Vagner Fonseca and Marta Giovanetti

Abstract

Nanopore sequencing is transforming viral genomics through real-time, portable, long-read analysis of RNA and DNA. Unlike traditional short-read platforms, it detects nucleotide sequences by measuring ionic current changes as nucleic acids pass through nanoscale pores, enabling direct single-molecule sequencing and base modification detection. Its simplicity, flexibility, and capacity for ultra-long reads make it ideal for resolving complex genomic regions, structural variants, and full viral genomes. These advantages have accelerated its use in pathogen surveillance and outbreak response, especially in resource-limited settings. For chikungunya virus (CHIKV), nanopore sequencing allows rapid, culture-independent recovery of complete genomes from clinical and vector samples, enabling real-time tracking of viral diversity, evolution, and spread. Experiences from Ebola, Zika, and COVID-19 have demonstrated the power of portable sequencing, now applied to CHIKV monitoring. Advances in tools such as Guppy, Dorado, Minimap2, and Medaka enhance read quality, consensus accuracy, and downstream analyses. Despite challenges in basecalling and error correction, robust quality control pipelines ensure reliable results. Ongoing improvements in chemistry, flow cell design, and machine learning will further enhance fidelity and throughput, establishing nanopore sequencing as a cornerstone of CHIKV genomic surveillance and epidemic preparedness.

Keywords Chikungunya virus, Nanopore sequencing, Genomic surveillance, Viral evolution, Arbovirus monitoring

As nanopore technology opened the door to real-time, portable sequencing, it transformed 'what if' into 'what now,' turning curiosity into capability and making genomics more immediate, accessible, and adaptive than ever before.

1 Introduction

Over the past two decades, genomics has undergone a transformative evolution, driven by advances that expanded the scale, speed, and resolution of nucleic acid analysis. From the foundational Sanger method [1] to the dominance of short-read platforms such as Illumina [2] and the emergence of long-read technologies like PacBio [3],

Claudia Filomatori and Eliana F. Castro (eds.), *Chikungunya Virus*, Methods in Molecular Biology, vol. 3067,
https://doi.org/10.1007/978-1-0716-5496-5_23,

the sequencing landscape has been shaped by the pursuit of comprehensive, efficient, and accessible methodologies. A major recent innovation is nanopore sequencing, which offers real-time, portable, long-read capabilities through a fundamentally novel biophysical approach [4]. Unlike optical and synthesis-based platforms, nanopore technology relies on the direct detection of changes in ionic current as individual nucleic acid molecules pass through a nanometer-scale pore embedded within a membrane [5]. These nanopores, typically protein channels, allow ions to flow across the membrane under an applied voltage. As a DNA or RNA strand translocates through the pore, each nucleotide—including chemically modified bases—alters the ionic current in a sequence-specific manner. Electrical disruptions are decoded in real time by computational algorithms to reconstruct nucleotide sequences [6], enabling sequencing without amplification or labeling and allowing direct detection of epigenetic modifications [7]. The versatility of nanopore sequencing confers significant advantages: ultra-long reads, direct RNA sequencing, and workflows deployable in decentralized or field settings [8–10]. These attributes have facilitated rapid pathogen surveillance and outbreak response, including during Ebola, Zika, and COVID-19 emergencies, and are increasingly applied to arboviruses, such as chikungunya virus (CHIKV). For CHIKV, the capacity to recover complete genomes directly from clinical and vector-derived samples supports real-time monitoring of viral evolution, genotype replacement, and international spread, enhancing preparedness in endemic and at-risk regions. The accessibility and scalability of devices such as the MinION and PromethION extend the technology from high-throughput laboratories to field investigations of CHIKV epidemics. Early limitations—especially basecalling accuracy and sequencing error rates—posed analytical challenges. Initial iterations were characterized by relatively high insertion, deletion, and substitution error rates, limiting some high-precision applications. Substantial improvements followed via refinements in nanopore chemistry, flow cell design, and advanced neural-network basecalling [11], coupled with robust error-correction and consensus strategies that have markedly improved reliability. A growing suite of bioinformatics tools tailored to nanopore data—covering signal-level basecalling, real-time read classification, high-confidence variant calling, and structural variant detection—has further matured the platform. As genomic science demands greater speed, flexibility, and adaptability, nanopore sequencing is shifting the paradigm from centralized, resource-intensive sequencing to real-time, distributed genomic analysis. Its capacity to operate across contexts—from remote arbovirus surveillance to personalized medicine—positions it as a pivotal tool for the next generation of genomic inquiry. Continued integration with improvements in accuracy, throughput, and real-time analytics will consolidate its role in shaping the future of CHIKV genomics and epidemic preparedness.

2 Overview of Nanopore Sequencing for Chikungunya Virus Genomics

Nanopore sequencing is a third-generation technology enabling direct, real-time analysis of native DNA or RNA molecules as they translocate through a nanoscale biological pore [12]. Unlike traditional methods that require amplification and complex chemical modifications, nanopore technology interrogates long nucleic acid strands with minimal sample preparation while preserving native epigenetic modifications [13]. The core principle is the detection of ionic current changes as single-stranded nucleic acids pass through a nanopore embedded in a synthetic lipid bilayer or solid-state membrane [14, 15]. Because different nucleotide sequences modulate the current in distinct ways, these electrical fluctuations can be decoded to infer sequence (Fig. 1) [16]. For CHIKV, this approach enables rapid, culture-independent sequencing of complete viral genomes, supporting phylodynamic reconstruction of outbreak dynamics, detection of lineage diversification, and real-time assessment of viral introduction routes in endemic and emerging regions.

The conceptual and experimental foundations of nanopore sequencing were laid in the early 1990s by David Deamer, George Church, and Daniel Branton. Early studies demonstrated electrophoretic translocation of single-stranded nucleic acids through nanopores [17]. Subsequent advances included enzymatic control of translocation at single-nucleotide resolution [18, 19] and nanopores with sensitivity sufficient to distinguish individual bases [5, 20, 21]. These breakthroughs culminated in commercial development by Oxford Nanopore Technologies (ONT), which released the portable MinION in the mid-2010s, followed by higher-throughput platforms such as PromethION [22]. Early implementations used α-hemolysin, a pore-forming protein from *Staphylococcus aureus*, for base discrimination [17]. This was superseded by MspA, an engineered nanopore from *Mycobacterium smegmatis* with a narrow constriction (~1.2 nm) improving base discrimination [23, 24]. Modern ONT devices utilize optimized versions of the curli transport lipoprotein CsgG, selected for enhanced structural stability and signal-to-noise [25, 26]. A transformative innovation was the integration of processive motor enzymes (helicases or DNA polymerases) to control translocation in a stepwise fashion, essential for accurate signal interpretation [24, 27]. Correspondingly, ONT developed basecallers such as Guppy and Dorado to convert raw current signals into nucleotide sequences with increasing accuracy [28].

2.1 Library Preparation

ONT supports three main workflows. Ligation-based protocols (end-repair, dA-tailing, and adapter ligation) suit high-throughput applications and ultra-long read generation (>4 Mb). Rapid kits use transposase-mediated fragmentation and adapter tagging for fast, low-input preparation, with trade-offs in read length and yield. Direct

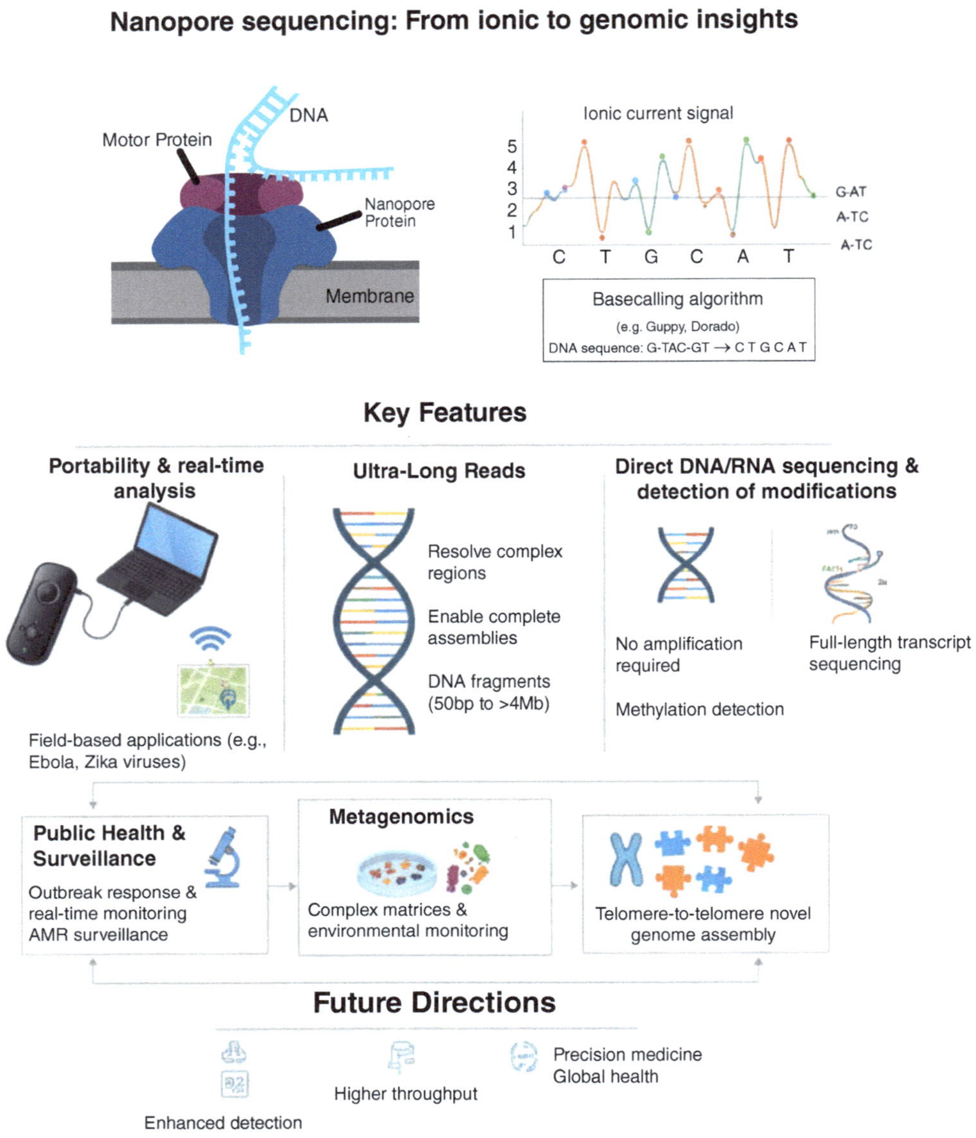

Fig. 1 Nanopore sequencing: from ionic signals to genomic applications. Schematic overview of the nanopore sequencing workflow and its applications. An applied voltage drives DNA or RNA molecules through a nanopore protein, generating characteristic ionic current signals that are translated into nucleotide sequences by basecalling algorithms (e.g., Guppy, Dorado). Key features include portability and real-time analysis, ultra-long reads (>4 Mb), and direct sequencing of DNA/RNA with detection of modifications. Applications range from outbreak response and antimicrobial resistance surveillance to metagenomics and complete genome assemblies. Future directions focus on enhanced detection, higher throughput, and integration into precision medicine and global health

RNA sequencing enables analysis of native polyadenylated RNA while preserving base modifications. All workflows attach motor proteins and sequencing adapters to ensure controlled translocation and platform compatibility [29, 30]. Multiplexing is supported via barcoded adapters or native barcoding for up to 96 samples per run [27]. Under a constant electric field, negatively charged DNA/RNA molecules are driven through the pore, producing characteristic current fluctuations reflecting nucleotide composition at the sensing region [17, 31]. Notably, PCR-free options minimize bias and preserve epigenetic information [30, 32–34]. For CHIKV, these workflows allow complete genome recovery directly from serum or vector-derived RNA, enabling rapid genomic surveillance without the need for viral culture.

2.2 Capabilities and Positioning

Nanopore sequencing enables real-time analysis, direct DNA/RNA sequencing, and modification detection (e.g., 5-methylcytosine) without chemical treatments [7, 35]. Ultra-long reads excel at resolving repetitive regions, phasing haplotypes, and identifying large structural variants [4, 36, 37]. Devices like MinION and Flongle provide portability for outbreak settings, whereas PromethION supports large-scale sequencing—spanning clinical diagnostics, environmental surveillance, and complex genome assemblies [4, 37–40]. Adaptive sampling allows real-time enrichment or depletion of specific regions during sequencing [41, 42]. For CHIKV, this means rapid identification of viral lineages, monitoring of co-circulating genotypes, and real-time phylodynamic reconstruction during epidemic expansion. Compared with Illumina (high-accuracy short reads without long-range or epigenetic resolution) and PacBio (high-fidelity long reads with higher input/cost), ONT offers an accessible, scalable, versatile alternative [27, 37, 43].

2.3 Limitations

Despite improvements in R10.4.1 chemistry and basecalling [44–47], raw error rates remain higher than other platforms [48, 49]. Flow cell performance varies with storage, handling, and pore degradation [50]. Basecalling and signal processing are computationally intensive and typically benefit from GPU acceleration. Large data volumes pose storage and data-management challenges [51, 52]. Ongoing advances continue to improve performance, positioning nanopore sequencing as an increasingly powerful tool for chikungunya virus genomics and outbreak response.

3 Applications of Nanopore Sequencing in Genomics

Nanopore sequencing portability and streamlined workflows enable applications from controlled laboratories to remote field settings. During the 2015 West Africa Ebola epidemic, a UK–Germany team implemented a MinION-based surveillance system in Guinea, transporting three MinIONs, laptops, a thermocycler, and reagents in a standard suitcase. Using an amplicon-based whole-genome protocol, they sequenced 142 ebolavirus samples with >25× coverage, resolving two circulating lineages (GN1 and SL3) and detecting cross-border transmission between Guinea and Sierra Leone [38]. Building on this, an international team investigated the origin and spread of the 2016 Zika outbreak in Brazil and the Americas [53]. Using a mobile genomics laboratory (a converted trailer) in Northeast Brazil, amplicon-based nanopore sequencing produced 54 complete or partial genomes, and phylogeographic analysis indicated months of cryptic transmission before wider spread. Subsequent studies extended to healthcare-associated SARS-CoV-2 infections in the UK [54], chikungunya virus transmission in Latin America [9, 55], methicillin-resistant *Staphylococcus aureus* outbreaks in Far North Queensland and Papua New Guinea [56], and rapid detection of uncommon fungal pathogens in South Korea [57]. These works collectively illustrate versatility across biological systems and the value of nanopore sequencing for rapid outbreak response.

3.1 Metagenomics and Diagnostics

Metagenomic approaches characterize all genetic material present in samples. In 2015, Greninger et al. [58] demonstrated unbiased, rapid detection of viruses (chikungunya, Ebola, hepatitis C) directly from clinical blood using nanopore metagenomic sequencing, delivering actionable results in under 6 h despite then-high per-read error rates (~24%). Improvements in chemistry and accuracy have driven continued development of nanopore-based metagenomic protocols for rapid, accurate diagnosis in clinical samples [59–61].

3.2 Environmental Monitoring and AMR

Nanopore sequencing detects pathogenic species and contextualizes antibiotic resistance (AR) genes in environmental samples. Yang et al. [62] reported high abundance of AR genes (e.g., *fexA*, *floR*, *optrA*, *poxtA*) in livestock fecal samples, with long reads resolving mobile genetic elements that disseminate resistance. Mills et al. [63] analyzed river biofilms, describing seasonal and spatial changes in microbial communities and the potential for resistome transmission.

3.3 Reference Genomes and Assemblies

Nanopore long reads (including ultra-long) facilitate assembly of large, repetitive genomes. Loman et al. [64] produced a complete *E. coli* K-12 genome in a single 4.6 Mb contig using only MinION reads. Ultra-long reads (up to 882 kb) enabled contiguous human genome assemblies (NG50 6.4 Mb) with 99.96% accuracy and closure of 12 >50 kb reference gaps [4]. The 2020 complete X chromosome assembly closed 29 previously unresolved gaps [65]. Nanopore data contributed to the telomere-to-telomere, gapless human reference (25 chromosomes) published in 2022 [66, 67] and to assemblies across taxa—*Begonia fimbristipula* [68], *Musa acuminata* [69], African swine fever virus [70], and the hourglass dolphin [71]—highlighting nanopore sequencing role in rapidly generating new references.

4 Downstream Analysis and Bioinformatics

4.1 Basecalling Nanopore Reads

ONT devices produce raw ionic-current signals (FAST5 or POD5) that require basecalling to yield sequences with quality scores (FASTQ). Modern basecallers use deep neural networks and benefit from GPU acceleration [11, 72]. ONT's Guppy and newer Dorado are state-of-the-art for MinION data. Guppy (since 2018) introduced GPU-accelerated basecalling with multiple accuracy tiers (fast, HAC, SUP) [73, 74]. Dorado (2023) features improved architectures and optimized GPU utilization, delivering higher throughput and per-read accuracy than Guppy under equivalent settings [75, 76]. Dorado performs modified-base detection entirely on GPU, eliminating slower CPU steps required by Guppy [77], and can significantly outpace Guppy in 5 hmC calling (~3.8× throughput in one benchmark) [74]. These gains—now exceeding ~99% single-read identity in the latest SUP models [78, 79]—have been pivotal for downstream assembly quality.

4.2 Demultiplexing Multiplexed Libraries

Multiplexing with barcode adapters enables parallel processing of multiple samples, increasing throughput and cost-efficiency. Demultiplexing assigns reads to samples based on barcodes [80]. Early tools like Porechop trimmed adapters and segregated reads but have been supplanted by integrated or specialized solutions. Guppy's *guppy_barcoder* dominated for years [81]. Stand-alone tools include qcat (basecalled FASTQ) and Deepbinner (raw signals), the latter trading speed for accuracy [80, 82]. Dorado, now ONT's primary basecaller, integrates demultiplexing in real time or as a post-step [83]. It supports multiple ONT barcode kits and features, such as dual-barcode enforcement and configurable parameters. Initial benchmarks suggest demultiplexing performance comparable to or better than Guppy, with early versions being more conservative in classifications [84, 85]. ONT now endorses Dorado for both basecalling and demultiplexing, marking a shift away from legacy tools [83].

4.3 Genome Assembly Strategies

Long ONT reads change assembly fundamentals. Two broad strategies are used: (1) Reference-guided assembly: Align reads to a related reference and infer a consensus. This is less compute-intensive and yields high quality when a suitable reference exists but can mask novel sequences and structural variation; and (2) De novo assembly: Build the genome without a reference, crucial for discovering novel sequences and avoiding reference bias. Long reads span repeats, yielding more contiguous assemblies [4, 86, 87].

4.3.1 De Novo Assembly with Long Reads

Most long-read assemblers follow either an Overlap-Layout-Consensus (OLC) paradigm or a long-read–adapted graph approach. Popular tools include Canu/HiCanu, Flye, Raven, Shasta, and Redbean (wtdbg2) [88–90]. Flye uses a repeat-graph strategy, scales from bacterial to human genomes, and is noted for speed and consensus accuracy, albeit with higher RAM [88, 89]. Canu/HiCanu provides sophisticated overlap/consensus algorithms and handles plasmids well, though runtime is longer [89]. Raven targets speed and low memory, producing highly contiguous assemblies [90]. Shasta employs a sparse, marker-based de Bruijn approach to achieve extreme speed on large genomes [89]. Redbean/wtdbg2 rapidly drafts assemblies without pre-correction, typically followed by polishing.

Polishing is essential (see below). It is often pragmatic to try multiple assemblers when results are suboptimal; each handles repeats/coverage biases differently. Ultra-fast scaffolding with Miniasm (overlap-only) plus polishing (e.g., Racon) can be useful when compute resources are limited [89].

4.3.2 Reference-Guided Assembly

Long-read aligners underpin mapping-first approaches. GraphMap pioneered high-sensitivity mapping for raw ONT reads [91]. Minimap2 now dominates due to speed/accuracy and supports genomic and spliced modes [92]. Winnowmap improves mapping in repetitive regions [93]. A common workflow maps reads with Minimap2 and generates consensus via pileups/variant callers or dedicated polishers (e.g., Medaka). For CHIKV, reference-guided strategies provide rapid genome recovery for epidemiological surveillance, although unmapped reads should still be screened to detect novel lineages or recombinant events. Reference-guided scaffolding (e.g., RaGOO) can order and orient contigs, but novel sequences should be recovered from unmapped reads to avoid reference bias.

4.4 Error Correction and Polishing

Raw long-read assemblies contain base-level errors requiring polishing to reach publication-quality consensus.

Long-read polishing. Racon uses partial-order alignment of reads to contigs for fast consensus correction and is often run iteratively [94]. Medaka (ONT) applies neural-network consensus on read pileups; it is typically more accurate than Racon alone and ~50× faster than early signal-level methods, making it a common final step after Racon [95]. Running Racon after Medaka can reintroduce errors; apply the highest-accuracy polisher last [95]. Nanopolish aligns raw signals to assemblies for signal-aware correction, excelling at homopolymers but at significantly higher computational cost [7, 96]. As basecalling improved, Nanopolish is now used selectively when signal-level detail is needed.

Other methods (e.g., MarginPolish/HELEN) combine signal and base information; however, Racon→Medaka balances speed and accuracy for many ONT assemblies [95].

4.5 Read Alignment and Variant Calling

Minimap2 and Winnowmap are pivotal for alignment, enabling variant callers to detect SNVs, indels, and structural variants. Neural-network approaches (Medaka, Clair/Clair3, PEPPER-DeepVariant) integrate basecall probabilities and haplotype context to achieve high accuracy in diploid genomes [92, 97]. Long-read SV callers (Sniffles2, SVIM) leverage long-range information to identify insertions, deletions, inversions, and translocations that challenge short-read methods [4, 86, 98]. Combining robust mapping (e.g., Winnowmap) with advanced callers (e.g., PEPPER-DeepVariant) improves performance in difficult regions [99, 100].

4.6 Data Visualization and Quality Control

Reliable analysis requires rigorous QC at both read and assembly levels. pycoQC and NanoPlot provide read-level dashboards (length distributions, per-channel yield, quality profiles) for early detection of issues [101]. At the assembly level, standard metrics (N50, L50, consensus QV) are commonly reported; tools such as QUAST are widely used to compute these metrics [102]. BUSCO assesses gene-content completeness across taxonomic lineages. Visualization tools like IGV and JBrowse enable manual inspection of alignments for variant validation and misassembly detection [103, 104]. Consistent QC/visualization practices are essential across clinical diagnostics, microbial surveillance, and ecological metagenomics [105].

5 Future Perspectives and Challenges in CHIKV Genomics

The continued evolution of nanopore sequencing is driven by needs for improved accuracy, higher throughput, and broader integration into real-world applications—especially in outbreak-prone, resource-limited settings where chikungunya virus (CHIKV) circulates. For arbovirus genomics, progress can be considered across five strategic dimensions: accuracy, throughput, real-time surveillance, biomolecular scope, and data analysis (Table 1).

One of the most critical frontiers is the enhancement of sequencing accuracy. Although nanopore sequencing has significantly improved—largely through advancements in basecalling algorithms, pore chemistry, and motor proteins—challenges remain when distinguishing closely related CHIKV genotypes or detecting minority variants during co-infection. Improving resolution of homopolymers, minimizing insertion/deletion errors, and enabling robust identification of single-nucleotide variants are essential for tracking CHIKV evolution and lineage diversification [11, 97, 106]. Machine learning–based correction strategies, coupled with optimized library preparation protocols, are expected to increase per-read accuracy and reduce artifacts. Parallel to accuracy gains is the need for higher sequencing throughput to support population-level CHIKV studies. Optimized barcoding kits, efficient multiplexing, and improved pore stability are critical for scaling surveillance across multiple sites and time points. Such developments will be particularly important in monitoring simultaneous circulation of different arboviruses, where

Table 1

Future perspectives and challenges in applying nanopore sequencing for chikungunya virus (CHIKV) genomic surveillance. The table summarizes key areas of development, outlining current technological directions and major challenges that influence the accuracy, scalability, and implementation of nanopore sequencing in CHIKV monitoring

Key area	Future directions	Challenges
Accuracy improvement	Enhanced basecalling algorithms (e.g., transformer-based models); improved pore chemistry and motor proteins; better signal processing	Homopolymer resolution; high per-read error rates; consistent detection of SNVs and lineage-defining variants in CHIKV genomes
Throughput and scalability	Higher pore-density flow cells; optimized barcoding/multiplexing; longer run times without signal degradation	Balancing throughput and data quality; variability across runs; cost per base for large-scale CHIKV surveillance
Real-time genomic surveillance	Integration with cloud/edge computing; deployment in mobile/public-health units; rapid sequencing pipelines tailored to arboviruses	Infrastructure in low-resource regions; interoperability of genomic data; ethical and regulatory challenges in outbreak contexts
Single-molecule and multiomics	Direct RNA sequencing of CHIKV genomes and transcriptomes; detection of epigenetic modifications; potential integration with proteomic readouts	Pore stability for RNA/protein analytes; signal discrimination in mixed infections (e.g., CHIKV + DENV/ZIKV); lack of standardized arbovirus protocols
Bioinformatics and data management	Automated outbreak workflows; integration with AI/ML for lineage classification and transmission inference; cloud-native analysis pipelines; FAIR genomic datasets	High computational demand in outbreak regions; long-term archiving of FAST5/POD5 data; reproducibility of genomic surveillance; cybersecurity and privacy concerns

both depth and speed of sequencing are essential for public health response [29, 107].

Another transformative frontier is the integration of nanopore sequencing into real-time genomic surveillance of CHIKV outbreaks. Portable devices have already proven invaluable in epidemic contexts, and future deployments will likely leverage automated, cloud-connected pipelines for on-site consensus generation, lineage classification, and phylodynamic analysis. For CHIKV-endemic regions, embedding nanopore sequencing into national surveillance systems could provide near real-time insight into viral spread, facilitating earlier interventions and guiding vector-control strategies [64, 108]. Looking beyond genomics, nanopore sequencing's capacity for direct RNA sequencing opens opportunities for analyzing CHIKV transcriptomes, capturing subgenomic RNA and epigenetic modifications that may influence viral replication and pathogenesis. Future developments in pore engineering and sensor miniaturization may extend to protein-level readouts, with the potential to characterize CHIKV-host protein interactions in real time. Despite these advances, several challenges persist: ensuring reproducibility across labs, maintaining pore stability in tropical field deployments, and overcoming data storage and transfer barriers in regions with limited computational infrastructure. At the policy level, interoperable data-sharing frameworks and ethical standards are needed to ensure transparency and equitable use of CHIKV genomic data. Overall, the convergence of biochemical innovation, AI-driven analytics, and decentralized sequencing positions nanopore technology as a cornerstone of future chikungunya virus surveillance. By addressing current limitations while extending into multiomic applications, nanopore sequencing is poised to become a foundational tool in arbovirus genomics—adaptable to the realities of both large-scale research and rapid-response outbreak settings [109–111].

References

1. Sanger F, Nicklen S, Coulson AR (1977) DNA sequencing with chain-terminating inhibitors. Proc Natl Acad Sci U S A 74:5463–5467
2. Bentley DR, Balasubramanian S, Swerdlow HP, Smith GP, Milton J, Brown CG et al (2008) Accurate whole-genome sequencing using reversible terminator chemistry. Nature 456:53–59
3. Korlach J, Bjornson KP, Chaudhuri BP, Cicero RL, Flusberg BA, Gray JJ et al (2010) Real-time DNA sequencing from single polymerase molecules. Methods Enzymol 472:431–455
4. Jain M, Koren S, Miga KH, Quick J, Rand AC, Sasani TA et al (2018) Nanopore sequencing and assembly of a human genome with ultra-long reads. Nat Biotechnol 36:338–345
5. Branton D, Deamer DW, Marziali A, Bayley H, Benner SA, Butler T et al (2008) The potential and challenges of nanopore sequencing. Nat Biotechnol 26:1146–1153
6. Deamer D, Akeson M, Branton D (2016) Three decades of nanopore sequencing. Nat Biotechnol 34:518–524
7. Simpson JT, Workman RE, Zuzarte PC, David M, Dursi LJ, Timp W (2017) Detecting DNA cytosine methylation using nanopore sequencing. Nat Methods 14:407–410
8. Giovanetti M, Slavov SN, Fonseca V, Xavier J, Pereira L, Mello C et al (2022) Genomic epidemiology of the SARS-CoV-2 epidemic in Brazil. Nat Microbiol 7:1490–1500
9. Xavier J, Alcantara LCJ, Fonseca V, Faria N, Caleiro R, Cerqueira D et al (2023) Increased interregional virus exchange and nucleotide diversity outline the expansion of chikungunya virus in Brazil. Nat Commun 14:4413
10. Giovanetti M, Pinotti F, Zanluca C, Fonseca V, Nakase T, Koishi AC et al (2023) Genomic epidemiology unveils the dynamics and spatial corridor behind the Yellow Fever virus outbreak in Southern Brazil. Sci Adv 9:eadg9204.
11. Wick RR, Judd LM, Holt KE (2019) Performance of neural network basecalling tools for Oxford Nanopore sequencing. Genome Biol 20:129
12. Garalde DR, Snell EA, Jachimowicz D, Sipos B, Lloyd JH, Bruce M et al (2018) Highly parallel direct RNA sequencing on an array of nanopores. Nat Methods 15:201–206
13. Wanunu M (2012) Nanopores: a journey towards DNA sequencing. Phys Life Rev 9:125–158
14. Butler TZ, Pavlenok M, Derrington IM, Niederweis M, Gundlach JH (2008) Single-molecule DNA detection with an engineered MspA protein nanopore. Proc Natl Acad Sci U S A 105:20647–20652
15. Bayley H (2015) Nanopore sequencing: from imagination to reality. Clin Chem 61:25–31
16. Laszlo AH, Derrington IM, Ross BC, Brinkerhoff H, Adey A, Nova IC et al (2014) Decoding long nanopore sequencing reads of natural DNA. Nat Biotechnol 32:829–833
17. Kasianowicz JJ, Brandin E, Branton D, Deamer DW (1996) Characterization of individual polynucleotide molecules using a

membrane channel. Proc Natl Acad Sci U S A 93:13770–13773
18. Cherf GM, Lieberman KR, Rashid H, Lam CE, Karplus K, Akeson M (2012) Automated forward and reverse ratcheting of DNA in a nanopore at 5-Å precision. Nat Biotechnol 30:344–348
19. Ayub M, Bayley H (2012) Individual RNA base recognition in immobilized oligonucleotides using a protein nanopore. Nano Lett 12:5637–5643
20. Oxford Nanopore Technologies (ONTa). Company History. Available at: https://nanoporetech.com/about/history. Accessed on 7 June 2025
21. Faller M, Niederweis M, Schulz GE (2004) The structure of a mycobacterial outer-membrane channel. Science 303:1189–1192
22. Manrao EA, Derrington IM, Laszlo AH, Langford KW, Hopper MK, Gillgren N et al (2012) Reading DNA at single-nucleotide resolution with a mutant MspA nanopore and phi29 DNA polymerase. Nat Biotechnol 30:349–353
23. Brown CG, Clarke J (2016) Nanopore development at Oxford Nanopore. Nat Biotechnol 34:810–811
24. Van der Verren SE, Van Gerven N, Jonckheere W, Hambley R, Singh P, Kilgour J, Jordan M, Wallace EJ, Jayasinghe L, Remaut H (2020) A dual-constriction biological nanopore resolves homonucleotide sequences with high fidelity. Nat Biotechnol 38:1415–1420
25. MacKenzie M, Argyropoulos C (2023) An introduction to nanopore sequencing: past, present, and future considerations. Micromachines (Basel) 14:459
26. Liu-Wei W, van der Toorn W, Bohn P, Hölzer M, Smyth RP, von Kleist M (2024) Sequencing accuracy and systematic errors of nanopore direct RNA sequencing. BMC Genomics 25:528
27. Jain M, Olsen HE, Paten B, Akeson M (2016) The Oxford Nanopore MinION: delivery of nanopore sequencing to the genomics community. Genome Biol 17:239
28. Wang Y, Zhao Y, Bollas A, Wang Y, Au KF (2021) Nanopore sequencing technology, bioinformatics and applications. Nat Biotechnol 39:1348–1365
29. Shi W, Friedman AK, Baker LA (2017) Nanopore sensing. Anal Chem 89:157–188
30. Oikonomopoulos S, Bayega A, Fahiminiya S, Djambazian H, Berube P, Ragoussis J (2020) Methodologies for transcript profiling using long-read technologies. Front Genet 11:606
31. Grünberger F, Ferreira-Cerca S, Grohmann D (2022) Nanopore sequencing of RNA and cDNA molecules in *Escherichia coli*. RNA 28:400–417
32. Oxford Nanopore Technologies (ONTb) (2025) Prepare. Available at https://nanoporetech.com/products/prepare. Accessed on 7 June 2025
33. Chera A, Stancu-Cretu M, Zabet NR, Bucur O (2024) Shedding light on DNA methylation and its clinical implications: the impact of long-read-based nanopore technology. Epigenetics Chromatin 17:39
34. Amarasinghe SL, Su S, Dong X, Zappia L, Ritchie ME, Gouil Q (2020) Opportunities and challenges in long-read sequencing data analysis. Genome Biol 21:30
35. Logsdon GA, Vollger MR, Eichler EE (2020) Long-read human genome sequencing and its applications. Nat Rev Genet 21:597–614
36. Adelino TÉR, Giovanetti M, Fonseca V, Xavier J, de Abreu ÁS, do Nascimento VA et al (2021) Field and classroom initiatives for portable sequence-based monitoring of dengue virus in Brazil. Nat Commun 12:2296
37. Yeow D, Rudaks LI, Davis R, Ng K, Ghaoui R, Cheong PL, Ravenscroft G, Kennerson M, Deveson I, Kumar KR (2025) Long-read sequencing for diagnosis of genetic myopathies. BMJ Neurol Open 7:e000990
38. Quick J, Loman N, Duraffour S, Quartier L, Sissoko D, O'Shea MK et al (2016) Real-time, portable genome sequencing for Ebola surveillance. Nature 530:228–232
39. Loose M, Malla S, Stout M (2016) Real-time selective sequencing using nanopore technology. Nat Methods 13:751–754
40. Payne A, Holmes N, Clarke T, Munro R, Debebe BJ, Loose M (2021) Readfish enables targeted nanopore sequencing of gigabase-sized genomes. Nat Biotechnol 39:442–450
41. Rhoads A, Au KF (2015) PacBio sequencing and its applications. Genomics Proteomics Bioinformatics 13:278–289
42. Zhang T, Li H, Ma S, Cao J, Liao H, Huang Q, Chen W (2023) The newest Oxford Nanopore R10.4.1 full-length 16S rRNA sequencing enables the accurate resolution of species-level microbial community profiling. Appl Environ Microbiol 89:e006052
43. El-Lagta N, Truong L, Ayora F, Mobegi F, Bruce S, Martinez P, D'Orsogna L, De Santis D (2024) Revolutionising high resolution HLA genotyping for transplant assessment: validation, implementation and challenges of Oxford Nanopore Technologies Q20+ sequencing. HLA 104:e15725
44. Ratcliff JD, Merritt B, Gooden H, Siegers JY, Srikanth A, Yann S, Kol S, Sin S, Tok S, Karlsson EA, Thielen PM (2024) Improved resolution of avian influenza virus using Oxford Nanopore R10 sequencing chemistry. Microbiol Spectr 12:e0188024

45. Oxford Nanopore Technologies (ONTc) (2025) Accuracy. Available at https://nanoporetech.com/platform/accuracy. Accessed on 8 June 2025
46. Cook R, Brown N, Rihtman B, Michniewski S, Redgwell T, Clokie M, Stekel DJ, Chen Y, Scanlan DJ, Hobman JL, Nelson A, Jones MA, Smith D, Millard A (2024) The long and short of it: benchmarking viromics using Illumina, Nanopore and PacBio sequencing technologies. Microb Genom 10:001198
47. Bejaoui S, Nielsen SH, Rasmussen A, Coia JE, Andersen DT, Pedersen TB, Møller MV, Kusk Nielsen MT, Frees D, Persson S (2025) Comparison of Illumina and Oxford Nanopore sequencing data quality for *Clostridioides difficile* genome analysis and their application for epidemiological surveillance. BMC Genomics 26:92
48. Oxford Nanopore Technologies (ONTd) (2025) Support. Available at https://nanoporetech.com/support/flow-cells/general-flow-cell-FAQs/how-do-i-store-my-flow-cells. Accessed on 8 June 2025
49. Delahaye C, Nicolas J (2021) Sequencing DNA with nanopores: troubles and biases. PLoS One 16:e0257521
50. Singh G, Alser M, Denolf K, Firtina C, Khodamoradi A, Cavlak MB, Corporaal H, Mutlu O (2024) RUBICON: a framework for designing efficient deep learning-based genomic basecallers. Genome Biol 25:49
51. Faria NR, Quick J, Claro IM, Thézé J, de Jesus JG, Giovanetti M, Kraemer MUG, Hill SC, Black A, da Costa AC et al (2017) Establishment and cryptic transmission of Zika virus in Brazil and the Americas. Nature 546:406–410
52. Meredith LW, Hamilton WL, Warne B, Houldcroft CJ, Hosmillo M, Jahun AS et al (2020) Rapid implementation of SARS-CoV-2 sequencing to investigate cases of health-care associated COVID-19: a prospective genomic surveillance study. Lancet Infect Dis 20:1263–1271
53. Fabbri C, Giovanetti M, Luppo V, Fonseca V, Garcia J, Barulli C et al (2024) Tracing the evolution of the chikungunya virus in Argentina, 2016-2023: independent introductions and prominence of Latin American lineages. Emerg Microbes Infect 13:2362941
54. Steinig E, Duchêne S, Aglua I, Greenhill A, Ford R, Yoannes M et al (2022) Phylodynamic Inference of Bacterial Outbreak Parameters Using Nanopore Sequencing. Mol Biol Evol 39:msac040.
55. Kim S, Kim N, Park WB, Kang CK, Park JH, Lee ST et al (2024) Real-time application of ITS and D1-D3 nanopore amplicon metagenomic sequencing in fungal infections: enhancing fungal infection diagnostics. Int J Med Microbiol 316:151630
56. Greninger AL, Naccache SN, Federman S, Yu G, Mbala P, Bres V et al (2015) Rapid metagenomic identification of viral pathogens in clinical samples by real-time nanopore sequencing analysis. Genome Med 7:119
57. Gu W, Deng X, Lee M, Sucu YD, Arevalo S, Stryke D et al (2021) Rapid pathogen detection by metagenomic next-generation sequencing of infected body fluids. Nat Med 27:115–124
58. Nielsen ME, Søgaard KK, Karst SM, Krarup AL, Albertsen M, Nielsen HL (2025) Application of rapid Nanopore metagenomic cell-free DNA sequencing to diagnose bloodstream infections: a prospective observational study. Microbiol Spectr 13:e03295–24.
59. Sung YH, Ju YK, Lee HJ, Park SM, Suh JW, Kim JY et al (2025) Clinical performance of real-time nanopore metagenomic sequencing for rapid identification of bacterial pathogens in cerebrospinal fluid: a pilot study. Sci Rep 15:3789
60. Yang X, Zhang T, Lei CW, Wang Q, Huang Z, Chen X, Wang HN (2022) Florfenicol and oxazolidone resistance status in livestock farms revealed by short- and long-read metagenomic sequencing. Front Microbiol 13:1018901
61. Mills M, Wittum T, Lee J (2024) Dynamic microbiome and mobile resistome are revealed in river biofilms from a multi-use watershed through long-read sequencing. Front Environ Sci 12:1440635
62. Loman NJ, Quick J, Simpson JT (2015) A complete bacterial genome assembled de novo using only nanopore sequencing data. Nat Methods 12:733–735
63. Miga KH, Koren S, Rhie A, Vollger MR, Gershman A, Bzikadze A et al (2020) Telomere-to-telomere assembly of a complete human X chromosome. Nature 585:79–84
64. Nurk S, Koren S, Rhie A, Rautiainen M, Bzikadze AV, Mikheenko A et al (2022) The complete sequence of a human genome. Science 376:44–53
65. Rhie A, Nurk S, Cechova M, Hoyt SJ, Taylor DJ, Altemose N et al (2023) The complete sequence of a human Y chromosome. Nature 621:344–354
66. Xiao TW, Wang ZF, Yan HF (2025) A chromosomal-level genome assembly of *Begonia fimbristipula* (Begoniaceae). Sci Data 12:35
67. Belser C, Baurens FC, Noel B, Martin G, Cruaud C, Istace B et al (2021) Telomere-to-telomere gapless chromosomes of banana using nanopore sequencing. Commun Biol 4:1018

68. Phadphon P, Sonthirod C, Thaweerattanasinp T, Shearman JR, U-Thoomporn S, Saenboonrueng J et al (2025) Hairpin loop to hairpin loop: a full-length assembly of the ASFV genome using Oxford Nanopore long-read sequencing. Front Microbiol 16:1615977
69. Mcgrath N, Le Roux J, Whibley A, Alexander A, Stewart RO, Johnstone M et al (2025) A high-quality Oxford Nanopore assembly of the hourglass dolphin (*Lagenorhynchus cruciger*) genome. G3 (Bethesda) 15:jkaf044
70. Neumann D, Reddy ASN, Ben-Hur A (2022) RODAN: a fully convolutional architecture for basecalling nanopore RNA sequencing data. BMC Bioinformatics 23:142
71. Murigneux V, Roberts LW, Forde BM, Phan MD, Nhu NTK, Irwin AD et al (2021) MicroPIPE: validating an end-to-end workflow for high-quality complete bacterial genome construction. BMC Genomics 22:474
72. Ferguson S, McLay T, Andrew RL, Bruhl JJ, Schwessinger B, Borevitz J et al (2022) Species-specific basecallers improve actual accuracy of nanopore sequencing in plants. Plant Methods 18:137
73. Diensthuber G, Milenkovic I, Llovera L, Milovanovic A, Pelizzari F, Novoa EM (2025) A systematic benchmarking of basecalling models for RNA modification detection with highly-multiplexed nanopore sequencing. bioRxiv. Available at https://doi.org/10.1101/2025.07.11.663424. Accessed on 10 September 2025
74. Prior K, Becker K, Brandt C, Cabal Rosel A, Dabernig-Heinz J, Kohler C et al (2025) Accurate and reproducible whole-genome genotyping for bacterial genomic surveillance with Nanopore sequencing data. J Clin Microbiol 63:e0036925
75. Diensthuber G, Pryszcz LP, Llovera L, Lucas MC, Delgado-Tejedor A, Cruciani S et al (2024) Enhanced detection of RNA modifications and read mapping with high-accuracy nanopore RNA basecalling models. Genome Res 34:1865–1877
76. Vereecke N, Yoon TB, Luo TL, Corey BW, Lebreton F, Mc Gann PT et al (2025) An open-source nanopore-only sequencing workflow for analysis of clonal outbreaks delivers short-read level accuracy. J Clin Microbiol 63:e0066425
77. Baeza JA, Minish JJ, Michael TP (2025) Assembly of mitochondrial genomes using nanopore long-read technology in three sea chubs (Teleostei: Kyphosidae). Mol Ecol Resour 25:e14034
78. Wick RR, Judd LM, Holt KE (2018) Deepbinner: demultiplexing barcoded Oxford Nanopore reads with deep convolutional neural networks. PLoS Comput Biol 14:e1006583
79. Hasan A, Hossain MM, Zamil MF, Trina AT, Hossain MS, Kumkum A et al (2025) Concurrent transmission of Zika virus during the 2023 dengue outbreak in Dhaka, Bangladesh. PLoS Negl Trop Dis 19:e0012866
80. Serna C, Calderón Bernal JM, Torre-Fuentes L, García Muñoz Á, Díez Guerrier A, Hernández M et al (2025) Integrative and conjugative elements associated with antimicrobial resistance in multidrug resistant Pasteurella multocida isolates from bovine respiratory disease (BRD)-affected animals in Spanish feedlots. Vet Q 45:1–15
81. Hamraoui A, Onfroy A, Senamaud-Beaufort C, Coulpier F, Lemoine S, Jourdren L et al (2025). A systematic benchmark of bioinformatics methods for single-cell and spatial RNA-seq Nanopore long-read data. bioRxiv. Available at https://doi.org/10.1101/2025.07.21.665920v1. Accessed on 10 September 2025
82. Brown DV, Anttila CJA, Ling L, Grave P, Baldwin TM, Munnings R et al (2024) A risk-reward examination of sample multiplexing reagents for single cell RNA-Seq. Genomics 116:110793
83. Yu T, Ren Z, Gao X, Li G, Han R (2024) Generating barcodes for nanopore sequencing data with PRO. Fundam Res 4:785–794
84. Di Genova A, Buena-Atienza E, Ossowski S, Sagot MF (2021) Efficient hybrid de novo assembly of human genomes with WENGAN. Nat Biotechnol 39:422–430
85. Senol Cali D, Kim JS, Ghose S, Alkan C, Mutlu O (2019) Nanopore sequencing technology and tools for genome assembly: computational analysis of the current state, bottlenecks and future directions. Brief Bioinform 20:1542–159
86. Griffiths S (2021). Assembly tools and Flye. Available at https://epi2me.nanoporetech.com/assembly-flye/. Accessed on 10 September 2025
87. Espinosa E, Bautista R, Larrosa R, Plata O (2024) Advancements in long-read genome sequencing technologies and algorithms. Genomics 116:110842
88. Vaser R, Šikić M (2021) Time- and memory-efficient genome assembly with Raven. Nat Comput Sci 1:332–336
89. Sović I, Šikić M, Wilm A, Fenlon SN, Chen S, Nagarajan N (2016) Fast and sensitive mapping of nanopore sequencing reads with GraphMap. Nat Commun 7:11307
90. Li H (2018) Minimap2: pairwise alignment for nucleotide sequences. Bioinformatics 34:3094–3100

91. Jain C, Rhie A, Hansen NF, Koren S, Phillippy AM (2022) Long-read mapping to repetitive reference sequences using Winnowmap2. Nat Methods 19:705–710
92. Vaser R, Sović I, Nagarajan N, Šikić M (2017) Fast and accurate de novo genome assembly from long uncorrected reads. Genome Res 27:737–746
93. Luan T, Commichaux S, Hoffmann M, Jayeola V, Jang JH, Pop M et al (2024) Benchmarking short and long read polishing tools for nanopore assemblies: achieving near-perfect genomes for outbreak isolates. BMC Genomics 25:679
94. Nanopolish. Available at https://nanopolish.readthedocs.io/en/latest/quickstart_consensus.html. Accessed on 10 September 2025
95. Ahsan MU, Liu Q, Fang L, Wang K (2021) NanoCaller for accurate detection of SNPs and indels in difficult-to-map regions from long-read sequencing by haplotype-aware deep neural networks. Genome Biol 22:261
96. Chen Y, Wang Y, Chen W, Tan Z, Song Y, Human Genome Structural Variation Consortium et al (2023) Gene fusion detection and characterization in long-read cancer transcriptome sequencing data with FusionSeeker. Cancer Res 83:28–33
97. Olson ND, Wagner J, McDaniel J, Stephens SH, Westreich ST, Prasanna AG et al (2022) PrecisionFDA truth challenge V2: calling variants from short and long reads in difficult-to-map regions. Cell Genom 2:100129
98. Hadadian Nejad Yousefi M, Goudarzi M, Motahari AS (2019) IMOS: improved Meta-aligner and Minimap2 On Spark. BMC Bioinformatics 20:51
99. Gargis AS, Kalman L, Lubin IM (2016) Assuring the quality of next-generation sequencing in clinical microbiology and public health laboratories. J Clin Microbiol. 2016 54(12):2857–65
100. Ounit R, Wanamaker S, Close TJ, Lonardi S (2015) CLARK: fast and accurate classification of metagenomic and genomic sequences using discriminative k-mers. BMC Genomics 16:236
101. Zook JM, Chapman B, Wang J, Mittelman D, Hofmann O, Hide W et al (2014) Integrating human sequence data sets provides a resource of benchmark SNP and indel genotype calls. Nat Biotechnol 32:246–251
102. Roach MJ, Beecroft SJ, Mihindukulasuriya KA, Wang L, Paredes A, Cárdenas LAC et al (2024) Hecatomb: an integrated software platform for viral metagenomics. Gigascience 13:giae020
103. O'Rawe J, Jiang T, Sun G, Wu Y, Wang W, Hu J et al (2013) Low concordance of multiple variant-calling pipelines: practical implications for exome and genome sequencing. Genome Med 5:28
104. Rang FJ, Kloosterman WP, de Ridder J (2018) From squiggle to basepair: computational approaches for improving nanopore sequencing read accuracy. Genome Biol 19:90
105. Davis S (2024) Transforming basecalling in genomic sequencing. Available at https://nanoporetech.com/blog/transforming-basecalling-in-genomic-sequencing. Accessed on 10 September 2025
106. Sahlin K, Tomaszkiewicz M, Makova KD, Medvedev P (2018) Deciphering highly similar multigene family transcripts from Iso-Seq data with IsoCon. Nat Commun 9:4601
107. Payne A, Holmes N, Clarke T, Munro R, Debebe BJ, Loose M (2021) Readfish enables targeted nanopore sequencing of gigabase-sized genomes. Nat Biotechnol 39:442–450
108. Brinkerhoff H, Kang ASW, Liu J, Aksimentiev A, Dekker C (2021) Multiple rereads of single proteins at single-amino acid resolution using nanopores. Science 374:1509–1513
109. Zlitni S, Bishara A, Moss EL, Tkachenko E, Kang JB, Culver RN et al (2020) Strain-resolved microbiome sequencing reveals mobile elements that drive bacterial competition on a clinical timescale. Genome Med 12:50
110. Ke S, Weiss ST, Liu YY (2022) Dissecting the role of the human microbiome in COVID-19 via metagenome-assembled genomes. Nat Commun 13:5235
111. Chen Z, Azman AS, Chen X, Wang J, Shi W, Lai S et al (2022) Global landscape of SARS-CoV-2 genomic surveillance and data sharing. Nat Genet 54:499–507

Check for updates

Chapter 24

Workflow for Long-Read Amplicon Sequencing of Chikungunya Virus Using Oxford Nanopore Technology

Francesco Branda , Massimo Ciccozzi and Fabio Scarpa

Abstract

This protocol provides a comprehensive, step-by-step workflow for whole-genome sequencing of Chikungunya virus (CHIKV) using an amplicon-based strategy optimized for Oxford Nanopore Technologies (ONT) platforms. The procedure includes detailed instructions for sample handling, viral RNA extraction, quality control, cDNA synthesis, multiplex PCR amplification, library preparation, sequencing, and primary bioinformatic processing. The protocol is designed to maximize reproducibility across laboratories and is suitable for genomic surveillance applications, including outbreak investigation and molecular epidemiology, even when working with low-to-moderate viral loads.

Keywords Chikungunya virus, Whole-genome sequencing, Oxford Nanopore Technologies, Amplicon-based sequencing, Genomic surveillance, Molecular epidemiology, Viral RNA, Next-generation sequencing, Protocol, Multiplex PCR

1 Introduction

Oxford Nanopore long-read sequencing enables rapid generation of viral genomes with flexible throughput and minimal infrastructure requirements, making it particularly suitable for real-time genomic surveillance of RNA viruses [1–3]. For RNA viruses, such as CHIKV, amplicon-based sequencing remains one of the most robust approaches, particularly for samples with limited RNA quantity or partial degradation [4, 5]. By selectively amplifying overlapping genomic regions, this strategy ensures uniform coverage and reduces host background, facilitating accurate consensus genome reconstruction.

This protocol describes a multiplex PCR amplicon workflow (Fig. 1) tailored for CHIKV whole-genome sequencing using ONT platforms, building on established approaches widely adopted for arbovirus and emerging virus surveillance [3–5].

Claudia Filomatori and Eliana F. Castro (eds.), *Chikungunya Virus*, Methods in Molecular Biology, vol. 3067,
https://doi.org/10.1007/978-1-0716-5496-5_24,

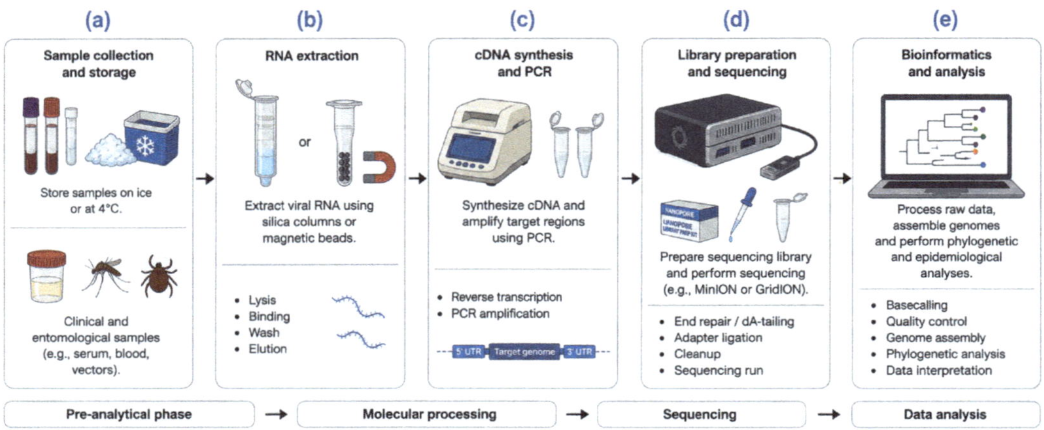

Fig. 1 Schematic overview of the amplicon-based Oxford Nanopore sequencing workflow for Chikungunya virus (CHIKV) genomic surveillance. The protocol comprises: (3.1) collection of clinical or entomological samples and storage under strict cold-chain conditions to preserve viral RNA integrity; (3.2) viral RNA extraction using silica column- or magnetic bead-based methods, followed by RNA quantification and quality assessment; (3.4, 3.5, and 3.6) reverse transcription and multiplex PCR amplification using overlapping primer pools covering the entire CHIKV genome; (3.7) library preparation through end-repair, adapter ligation, and optional native barcoding, optimized for Oxford Nanopore sequencing platforms; and (3.8 and 3.9) long-read sequencing, primary data processing, genome assembly, and consensus sequence generation for downstream phylogenetic and molecular epidemiology analyses

2 Materials

2.1 Sample Collection and Storage

1. Clinical specimens: serum or plasma collected in sterile tubes.
2. Entomological specimens: mosquito pools homogenized in viral transport medium (VTM).
3. Storage: −80°C freezer.
4. Transport: dry ice and insulated containers.

2.2 Viral RNA Extraction and Quality Control

1. Viral RNA extraction kit (silica column- or magnetic bead-based).
2. Carrier RNA (optional, for low-titer samples).
3. Nuclease-free water.
4. Fluorometric RNA quantification system (e.g., Qubit RNA HS Assay).
5. Automated electrophoresis system or spectrophotometer (optional).

Table 1
Primer sequences used for multiplex PCR amplification of the Chikungunya virus genome

Primer name	Sequence (5′ → 3′)	Pool	Amplicon position
CHIKV_F1	ATGCTGAACTGCGTAA	1	1–500
CHIKV_R1	TCCAGTCTGGTGTTGA	1	1–500
CHIKV_F2	GGAAGTCTCGTACGTT	2	400–900
CHIKV_R2	CTGAGGTTGACGAGTC	2	400–900

Primers are organized in pools for multiplexed amplification, generating overlapping amplicons of ~400–500 bp

2.3 cDNA Synthesis and Amplicon Generation

1. Reverse transcription kit.
2. Random hexamers or CHIKV-specific primers.
3. Multiplex PCR primer scheme covering the complete CHIKV genome (overlapping amplicons of ~400–500 bp) (see Table 1 for primer sequences).
4. High-fidelity DNA polymerase.
5. Thermal cycler.

2.4 Library Preparation and Sequencing

1. ONT ligation sequencing kit.
2. Native barcoding kit (optional).
3. AMPure XP beads or equivalent.
4. Magnetic rack.
5. ONT sequencing device (MinION, GridION, or equivalent).
6. R9.4.1 or compatible flow cells.

2.5 Bioinformatics

1. Basecalling software (e.g., Guppy) [6].
2. Read filtering and trimming tools (e.g., NanoFilt, Porechop, and Trimmomatic) [7–9].
3. Reference CHIKV genome [10].
4. Read mapping and consensus-generation software (e.g., Minimap2 and SAMtools/BCFtools) [11–13].

3 Methods

All procedures involving infectious material must be conducted according to institutional biosafety guidelines appropriate for arboviruses.

3.1 Sample Preparation

1. Thaw samples on ice and briefly centrifuge to collect liquid at the bottom of the tube.
2. Aliquot samples to avoid repeated freeze–thaw cycles.
3. Proceed immediately to RNA extraction.

3.2 Viral RNA Extraction

1. Use 140–200 μL of clinical sample or clarified mosquito homogenate.
2. Perform RNA extraction according to the manufacturer's instructions (*see* **Note 1**).
3. Elute RNA in 30–50 μL of nuclease-free water.
4. Keep RNA on ice and proceed immediately to quantification or store at −80°C.

3.3 RNA Quantification and Quality Assessment

1. Quantify RNA using a fluorometric assay; record RNA concentration (ng/μL).
2. RNA integrity was assessed using an Agilent Bioanalyzer, and samples with RIN ≥5 were used for downstream amplicon-based sequencing (*see* **Note 2**).
3. Samples with RNA concentrations below the detection limit may still be processed using increased PCR cycle numbers.

3.4 cDNA Synthesis

1. Prepare reverse transcription reactions using 5–10 μL of RNA in a final volume of 20 μL.
2. Use random hexamers or a combination of random and gene-specific primers.
3. Incubate according to kit instructions (typically 10 min at 25°C, 30–60 min at 42°C–50°C, followed by enzyme inactivation).
4. Store cDNA at −20°C if not immediately used.

3.5 Multiplex PCR Amplification

1. Divide primers into two or more nonoverlapping pools (*see* **Note 3**).
2. Prepare PCR reactions using high-fidelity polymerase and 2–5 μL of cDNA per reaction.
3. Typical cycling conditions: initial denaturation at 98°C for 30 s; 30–35 cycles of 98°C for 15 s and 65°C for 5 min; final hold at 4°C (*see* **Note 4**).
4. Pool PCR products from all primer pools for each sample.

3.6 Amplicon Clean-Up and Quantification

1. Purify pooled amplicons using magnetic beads at a 1× bead-to-sample ratio.
2. Wash beads twice with 70% ethanol without disturbing the pellet.
3. Elute DNA in 30 μL of nuclease-free water.
4. Quantify purified DNA and confirm expected amplicon size distribution if possible.

3.7 Library Preparation for Oxford Nanopore Sequencing

1. Use 50–100 ng of purified amplicons as input.
2. Perform end-repair and dA-tailing following the ONT ligation kit protocol.
3. Sequencing adapters and, when applicable, native barcodes for multiplexing were ligated according to the standard Oxford Nanopore Ligation Sequencing Kit protocol (SQK-LSK110/120). No additional ligation steps were performed.
4. Purify the final library using magnetic beads and elute in ONT elution buffer.

3.8 Sequencing

1. Flow cell quality was checked prior to sequencing to ensure a sufficient number of active pores (*see* **Note 5**).
2. Prime the flow cell according to the manufacturer's instructions.
3. Load the prepared library onto the flow cell.
4. Start the sequencing run and monitor performance in real time.
5. Stop the run once a minimum average genome coverage of 20× is achieved.

3.9 Primary Data Processing

1. Perform basecalling to generate FASTQ files.
2. Filter reads by minimum quality score (e.g., $Q \geq 7$) and length.
3. Reads were mapped to the reference CHIKV genome using Minimap2 [11], a long-read aligner suitable for Oxford Nanopore sequencing data.
4. Generate a consensus sequence, masking positions with coverage below 20× (*see* **Note 6**).

4 Notes

1. Contamination control: Include negative controls at RNA extraction and PCR stages.
2. RNA stability: Maintain a strict cold chain throughout the workflow to prevent RNA degradation.
3. Primer maintenance: Regularly update primer schemes to account for viral evolution and avoid amplicon drop-out.
4. PCR cycles: Excessive cycling may introduce bias; use the lowest number of cycles compatible with sufficient yield.
5. Flow cell quality: A minimum of 800–1000 active pores is generally recommended for optimal sequencing yield, according to Oxford Nanopore guidelines.
6. Coverage thresholds: Use conservative depth thresholds to ensure reliable consensus generation.

References

1. Jain M, Olsen HE, Paten B, Akeson M (2016) The Oxford Nanopore MinION: delivery of nanopore sequencing to the genomics community. Genome Biol 17:239
2. Quick J, Loman NJ, Duraffour S, Simpson JT, Severi E, Cowley L, Bore JA, Koundouno R, Dudas G, Mikhail A et al (2016) Real-time, portable genome sequencing for Ebola surveillance. Nature 530:228–232
3. Faria NR, Sabino EC, Nunes MRT, Alcantara LCJ, Loman NJ, Pybus OG (2016) Mobile real-time surveillance of Zika virus in Brazil. Genome Med 8:97
4. Grubaugh ND, Gangavarapu K, Quick J, Matteson NL, De Jesus JG, Main BJ, Tan AL, Paul LM, Brackney DE, Grewal S et al (2019) An amplicon-based sequencing framework for accurately measuring intrahost virus diversity. Genome Biol 20:8
5. Tyson JR, James P, Stoddart D, Sparks N, Wickenhagen A, Hall G, Choi JH, Lapointe H, Kamelian K, Smith AD et al (2020) Improvements to the ARTIC multiplex PCR method for SARS-CoV-2 genome sequencing using nanopore. BioRxiv. https://doi.org/10.1101/2020.09.04.283077
6. Wick RR, Judd LM, Holt KE (2019) Performance of neural network basecalling tools for Oxford Nanopore sequencing. Genome Biol 20:129
7. De Coster W, D'Hert S, Schultz DT, Cruts M, Van Broeckhoven C (2018) NanoPack: visualizing and processing long-read sequencing data. Bioinformatics 34:2666–2669
8. Wick RR, Judd LM, Gorrie CL, Holt KE (2017) Completing bacterial genome assemblies with multiplex MinION sequencing. Microb Genom 3:e000132
9. Bolger AM, Lohse M, Usadel B (2014) Trimmomatic: a flexible trimmer for Illumina sequence data. Bioinformatics 30:2114–2120
10. Khan AH, Morita K, Parquet MDC, Hasebe F, Mathenge EGM, Igarashi A (2002) Complete nucleotide sequence of Chikungunya virus and evidence for an internal polyadenylation site. J Gen Virol 83:3075–3084
11. Li H (2018) Minimap2: pairwise alignment for nucleotide sequences. Bioinformatics 34:3094–3100
12. Li H, Handsaker B, Wysoker A, Fennell T, Ruan J, Homer N, Marth G, Abecasis G, Durbin R (2009) The sequence alignment/map format and SAMtools. Bioinformatics 25:2078–2079
13. Danecek P, Bonfield JK, Liddle J, Marshall J, Ohan V, Pollard MO, Whitwham A, Keane T, McCarthy SA, Davies RM et al (2021) Twelve years of SAMtools and BCFtools. Gigascience 10:giab008

Chapter 25

Optimized Amplicon Strategy for Long-Read Sequencing of the Chikungunya Virus Genome

Chenyang Wu, Sara Ann Thannickal, Tommaso Cabani and Maria Gabriela Noval

Abstract

Chikungunya virus (CHIKV) is a positive-sense RNA alphavirus transmitted to humans primarily by *Aedes aegypti* and *Aedes albopictus* mosquitoes. Its global circulation and significant public health impact underscore the need to better understand the molecular mechanisms driving CHIKV pathogenesis and transmission. Although robust molecular biology methods exist for CHIKV genome sequencing, a major limitation for surveillance and research is the inability to determine whether two nucleotide variations co-occur within the same viral genome when they are separated beyond the span of typical short-read designs. Here, we describe an optimized approach for processing CHIKV RNA samples that generates large amplicons suitable for long-read nanopore sequencing. This protocol enables amplification of the complete CHIKV genome in only two or three amplicons and facilitates detection of co-occurring nucleotide variations across 4–7.5 kb within the same molecule, thereby simplifying sequencing workflows and improving resolution in studies of viral evolution.

Keywords Chikungunya virus, Alphavirus, Long-read sequencing, Nanopore sequencing, Amplicon sequencing, Viral evolution

1 Introduction

1.1 Chikungunya Virus Epidemiology and Symptoms

Chikungunya virus (CHIKV) is a positive-sense single-stranded RNA virus belonging to the alphavirus genus within the *Togaviridae* family and is primarily transmitted by *Aedes aegypti* and *Ae. albopictus* mosquitoes [1]. First identified in Tanzania in 1952, CHIKV had a reemergence in 2004 and has since spread to over 100 countries with an estimated ~35 million infections occurring annually [1–3]. Phylogenetically, CHIKV is classified into three major lineages: West African, East/Central/South African (ESCA), and Asian [4]. A prominent sublineage, the Indian Ocean Lineage (IOL), evolved from the ESCA lineage and was responsible for the large 2005–2006 outbreak

Claudia Filomatori and Eliana F. Castro (eds.), *Chikungunya Virus*, Methods in Molecular Biology, vol. 3067,
https://doi.org/10.1007/978-1-0716-5496-5_25,

in La Réunion, a French overseas territory in the Indian Ocean[5, 6]. CHIKV is endemic to South America, Africa, and Asia, and causes explosive outbreaks worldwide, including the 2024 resurgence in La Réunion and the 2025 circulation in France, Italy, and China [7].

Common symptoms of CHIKV infection include fever, myalgia, rash, and arthralgia, which can progress to chronic arthralgia lasting from months to years [1]. In 1%–13% of the cases, severe cardiovascular, hepatic, or neurological complications may occur, leading to hospitalization and even death [1, 8, 9]. As of 2025, there are no approved antiviral therapies for CHIKV. There are currently two CHIKV vaccines with regulatory approvals; however, their global availability remains limited [10].

1.2 Chikungunya Virus Genome Structure

The CHIKV genome is a 11- to 12-kb positive-sense single-stranded RNA with a 5′ cap (N7mGppp) and a 3′ poly(A) tail (Fig. 1a). It includes 5′ and 3′ untranslated regions (UTRs) flanking two open reading frames (ORFs) [11]. The first ORF, near the 5′ end, encodes four nonstructural proteins (nsP1–4). The second ORF, translated only from subgenomic RNA (sgRNA), encodes the structural proteins: capsid, E3, E2, 6 K/Transframe, and E1. A short UTR between the ORFs contains the sgRNA promoter. During infection, CHIKV produces multiple forms of RNA: genomic RNA (gRNA), a negative-sense RNA intermediate (a template for gRNA and sgRNA), and sgRNA itself[11] (Fig. 1b). Additionally, replication can lead to the accumulation of defective viral genomes (DVGs), truncated segments of the viral genome, which influence viral replication [12, 13].

Viral evolution, defined as the accumulation and selection of genetic changes within viral populations, is a key driver of emergence in RNA viruses [14]. Recent CHIKV outbreaks have been retrospectively linked to adaptive mutations in the CHIKV genome associated with increased transmission and adaptation to new vectors [6, 4]. Since genomic changes can occur throughout the entire genome, short-read sequencing is not suitable for determining whether two nucleotide variations separated by more than a few hundred base pairs (bp) occur within the same molecule. This limitation restricts our ability to define bona fide viral variants, as it cannot be determined whether distant nucleotide variations reside within the same or distinct viral genomes. Therefore, establishing efficient long-read sequencing strategies is essential to advance our understanding of viral evolution and identify drivers of transmission and pathogenesis.

1.3 Methods of Sequencing Chikungunya Virus

Several methods are used to sequence CHIKV RNA, including amplicon-based Illumina sequencing, Sanger sequencing, and amplicon-based or direct RNA nanopore sequencing. The choice of method depends on the application: nanopore sequencing is often employed during outbreak surveillance [15], Illumina for detecting minority variants in viral evolution studies [16–18], and targeted amplicon Sanger sequencing for consensus sequences [19, 20].

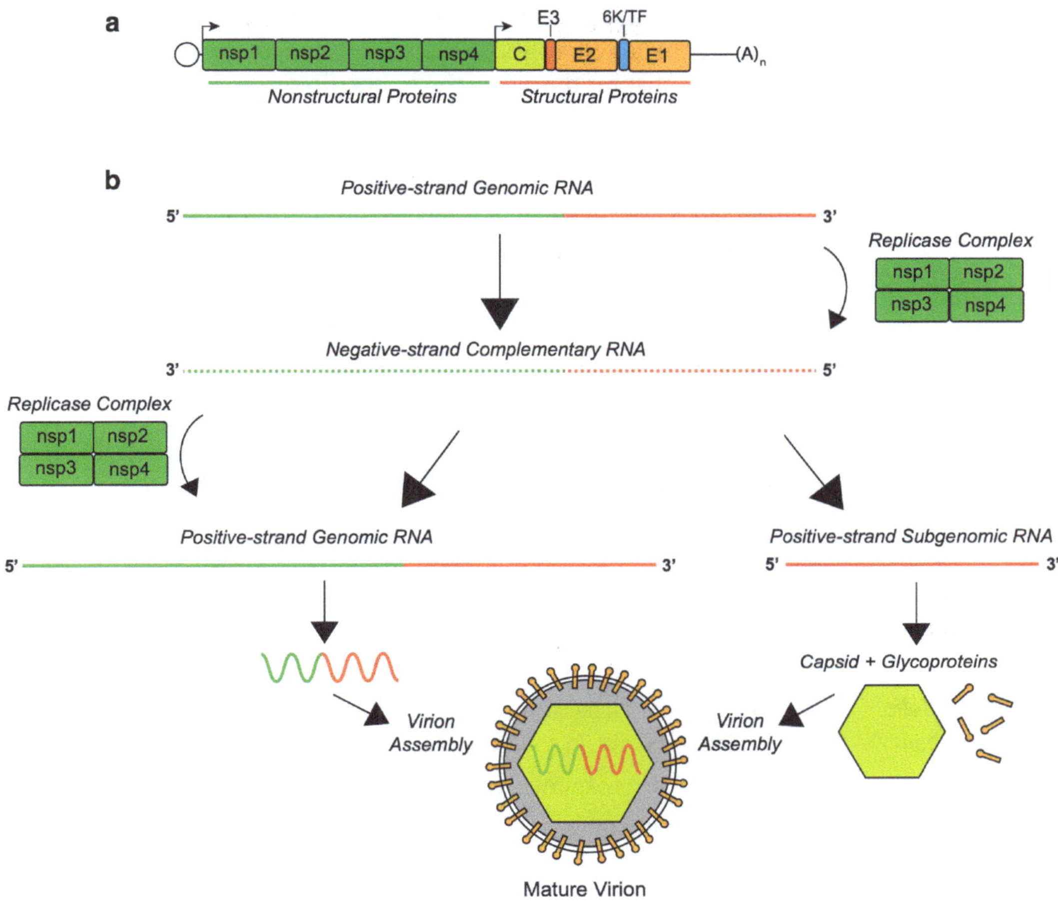

Fig. 1 Overview of CHIKV genome and replication. (**a**) Structure of the CHIKV genome. Beginning at the 5′ cap (open circle) and 5′ UTR and ending in the 3′ UTR and poly(A) tail, the genome comprises two open reading frames (ORFs): the first encodes the four nonstructural proteins (nsP1–4) and the second encodes the structural proteins capsid, E3, E2, 6 K/TF, and E1. A subgenomic promoter between ORFs drives transcription of the subgenomic RNA, from which the structural proteins are translated. (**b**). Replication cycleof CHIKV. Translation of nsPs from the positive-strand genomic RNA generates the replicase complex, which synthesizes the complementary negative-strand RNA. This negative-strand RNA serves as a template for production of both subgenomic RNA—translated into capsid and glycoproteins—and new positive-strand genomic RNA, which is subsequently encapsidated, assembled into virions, and released by budding at the plasma membrane

Most of these approaches rely on short-read sequencing, which produces reads under 800 bp and therefore does not require large amplicons. A major limitation of this strategy is the inability to determine whether nucleotide variations located in different amplicons occur in-phase within the same genome. Recent improvements in nanopore sequencing accuracy, particularly through duplex sequencing of complementary strands [21], now enable high-quality long-read sequencing of large amplicons [22]. This provides a unique advantage: the ability to identify phased nucleotide variants within individual viral genomes and the presence of DVGs at a cost and depth comparable to short-read deep sequencing methods.

This chapter describes the generation of 4.3–7.6 kb amplicons spanning the CHIKV genome in either two or three fragments and their preparation for downstream nanopore long-read sequencing. The methods covered here include: (1) RNA isolation and preparation from viral samples, (2) reverse transcription into cDNA [using either oligo(dT) or random primers], (3) polymerase chain reaction (PCR)-based amplicon generation using primer sets targeting regions conserved across different CHIKV lineages, and (4) amplicon fragment cleanup prior to sequencing. This approach enables efficient full-genome coverage of CHIKV using a minimal number of amplicons and is broadly applicable across CHIKV lineages, making it suitable for studies of viral evolution, molecular epidemiology, and outbreak surveillance. Practical tips and troubleshooting guidance are also provided.

2 Materials

2.1 RNA Isolation and Preparation

1. CHIKV virus samples (supernatant fluids, cells, tissues, blood samples, etc.) (*see* **Note 1**).
2. TRIzol™ reagent (*see* **Note 2**).
3. Chloroform (*see* **Note 2**).
4. GlycoBlue™ coprecipitant.
5. Molecular biology grade water.
6. 100% isopropanol.
7. 75% ethanol.
8. Consumables: Microcentrifuge tubes, micropipettes, pipette tips, ice.
9. Equipment: Centrifuge with refrigeration, microcentrifuge, chemical fume hood, vortex mixer, nanodrop spectrophotometer, or Qubit fluorometers.

2.2 cDNA Synthesis: Random Primer Strategy

1. Maxima™ H Minus First Strand cDNA Synthesis Kit (Table 1)
   a. 100 μM Random Hexamer Primers.
   b. 10 mM dNTP mix.
   c. 5× RT (reverse transcriptase) buffer.
   d. 200 U/μL Maxima™ H Minus Enzyme (*see* **Note 3**).
2. RNA from TRIzol™ extraction.
3. Molecular biology grade water.

Table 1
Conditions for cDNA synthesis using random primer strategy

	Reagents	Volume	Thermocycler
Mix 1 (1×) *15μL total*	100 μM random hexamer primers	1 μL (5 μM final)	(1) 65°C: 5 min (2) 4°C: hold
	10 mM dNTP mix	1 μL (0.5 mM final)	
	RNA	Up to 13 μL	
	Water	Remaining volume	
Mix 2 (1×) *5μL total*	5× RT buffer	4 μL (1× final)	(1) 25°C: 10 min (2) 50°C: 30 min (3) 85°C: 5 min
	Maxima™ H Minus Enzyme (200 U/μL)	1 μL (200 U)	
Final concentrations are in 20 μL			

4. Consumables: Microcentrifuge tubes, micropipettes, pipette tips, PCR tubes, ice.
5. Equipment: Thermal cycler, microcentrifuge.

2.3 cDNA Synthesis: Oligo (dT) Strategy

1. Induro® Reverse Transcriptase Kit (Table 2)
   a. Induro® Reverse Transcriptase (200 U/μL).
   b. 5× Induro® RT Reaction Buffer.
2. RNA from TRIzol™ extraction.
3. 10 mM dNTP mix.
4. 100 μM dT(23) GA CHIKV primers (*see* **Note 4**; Table 2).
5. Consumables: Microcentrifuge tubes, micropipettes, pipette tips, PCR tubes, ice.
6. Equipment: Thermal cycler, microcentrifuge.

2.4 PCR Amplicon Generation

1. Phusion™ High-Fidelity Kit.
   a. 5× Phusion™ HF Buffer.
   b. Phusion™ DNA Polymerase (2 U/μL) (*see* **Note 5**).
2. cDNA from prior synthesis.
3. dNTP mix.
4. Primer sets (Table 3).
5. Molecular biology grade water.
6. Consumables: Microcentrifuge tubes, micropipettes, pipette tips, PCR tubes, ice.
7. Equipment: Thermal cycler, microcentrifuge.

Table 2
Conditions for cDNA synthesis using oligo (dT) strategy

	Reagents	Volume	Thermocycler
Mix 1 (1×) *15μL total*	dT(23)_GA Primer (100 μM) Seq: TTTTTTTTTTTTTTTTTTTTTTT<u>GA</u>	1 μL (5 μM)	(1) 65°C: 5 min (2) 4°C: hold
	dNTP mix (10 mM)	1 μL (0.5 mM)	
	RNA	Up to 13 μL	
	Water	Remaining volume	
Mix 2 (1×) *5μL total*	5× Induro RT buffer	4 μL (1×)	(6) 55°C: 30 min (7) 95°C: 1 min
	Induro® Reverse Transcriptase (200 U/μL)	1 μL (200 U)	

Final concentrations are in 20 μL

3 Methods

3.1 RNA Isolation and Preparation

RNA from viruses such as CHIKV needs to be isolated before they can be sequenced. This RNA isolation protocol can be performed either through TRIzol™ extraction or by using commercially available RNA purification kits. After RNA isolation and purification, cDNA synthesis via reverse transcriptase is required to turn the genomic RNA into complementary DNA (cDNA) that can serve as a template for amplicon generation (Figs. 2 and 3).

1. Combine the CHIKV sample with equal volume TRIzol™ (*see* **Notes 1 and 6**).
2. Add one fifth of the volume of TRIzol™ in chloroform and gently vortex briefly.
3. Incubate at room temperature (20°C–25°C) for 5 min.
4. Centrifuge samples at 12,000 × g for 15 min at 4°C.
5. Prepare new tubes for each sample containing 500 μL of 100% isopropanol and 1 μL of Glycoblue™ each (*see* **Note 7**).
6. Following the spin, transfer the upper aqueous phase of each sample to the new tube containing the isopropanol mix. Gently vortex briefly (*see* **Note 8**).
7. Precipitate samples at −80°C for 30 min. After precipitation, all further steps must be performed on ice (*see* **Note 9**).
8. Vortex tubes briefly and spin down at 12,000 × *g* for 15 min at 4°C. A small blue pellet dyed by Glycoblue™ will confirm successful precipitation. Carefully discard supernatant (*see* **Note 10**).
9. Wash the RNA pellet with 700 μL of 75% ethanol. Gently vortex briefly (*see* **Note 11**).

Table 3

Primers for cDNA synthesis and PCR amplicon generation

Two-amplicon generation	Amplicon 1 ~7598 bp	Primer 5′ UTR-Fw	CACGTAGCCTACCAGTTTCTTA
		Primer F1a-Rv	AGGCTGGTACCTCCTATTGTA
	Amplicon 2 ~6643 bp	Primer F2a-Fw	ATAGAAGGAGTGCAAAAAGT
		Primer 3′ UTR-Rv	CCCTTTGAACTACTTCTGT
Three-amplicon generation	Amplicon 1 ~4204 bp	Primer 5′ UTR-Fw	CACGTAGCCTACCAGTTTCTTA
		Primer F1b-Rv	CACTGTTCTTAAAGGACTC
	Amplicon 2 ~4215 bp	Primer F2b-Fw	GCATACTCACCTGAAGTAGCC
		Primer F2b-Rv	GGCCATGGACATTACCACAACTGAT
	Amplicon 3 ~4353 bp	Primer F3b-Fw	TTAAACTGGGCAAACCGC
		Primer 3′ UTR-Rv	CCCTTTGAACTACTTCTGT

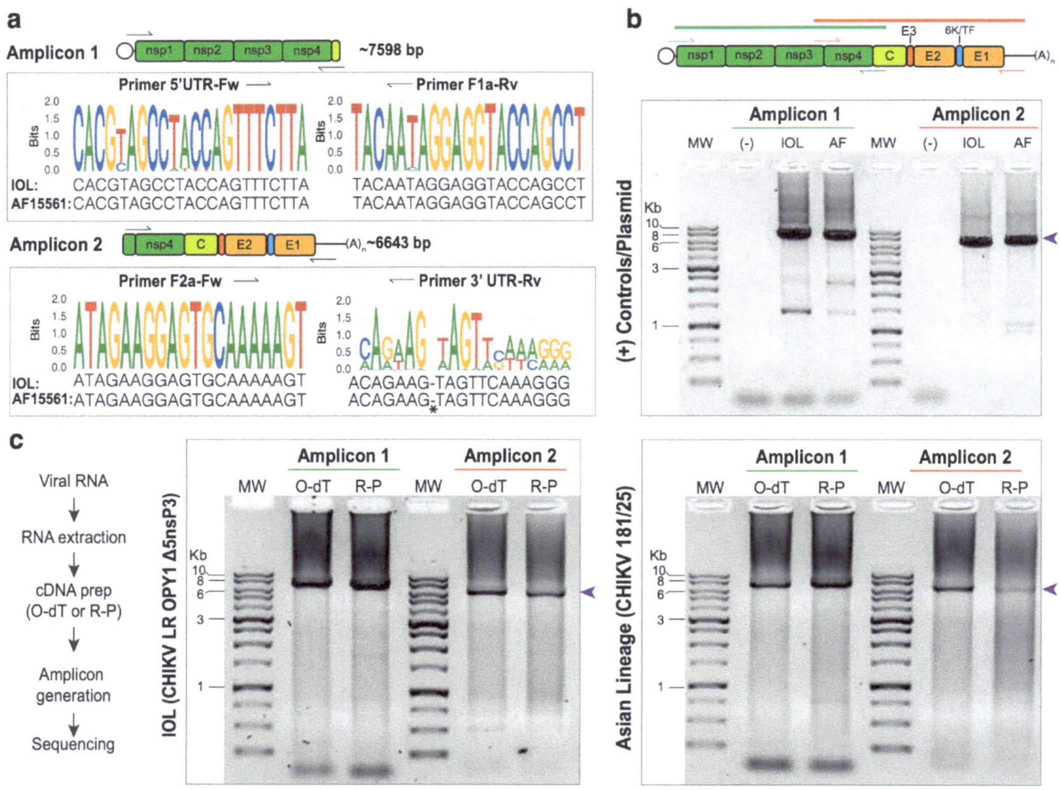

Fig. 2 Amplification of CHIKV genome using two-amplicon strategy. (**a**) Diagram showing the expected size and genome coverage of amplicons generated. Sequences of CHIKV IOL and Asian-Lineage AF15561 are shown at primer binding sites. Sequence logos illustrate conservation across 77 CHIKV genomes at these sites. (**b**) Representative PCR products generated from plasmid controls (CHIKV IOL and AF15561) using the two-amplicon approach. Amplicon size corresponds to predicted fragment lengths (purple arrowheads). (**c**) Workflow for amplicon generation from viral RNA (left panel). Examples of amplicons generated from CHIKV IOL Δ5nsP3 and Asian-lineage CHIKV 181/25 purified RNA using either oligo(dT) (O-dT) or random primers (R-P) during cDNA synthesis. Predicted band sizes are indicated (purple arrowheads). MW molecular weight ladder, (–) non-template control

10. Spin down the RNA at 12,000 × *g* for 15 min at 4°C. Carefully discard supernatant and repeat wash step.
11. Allow the pellet to air-dry for 10 min. Ensure all ethanol is evaporated before proceeding.
12. Resuspend RNA in 30–50 μL of molecular biology grade water. Pipette or vortex lightly until the Glycoblue™ pellet has disappeared, indicating proper homogenization (*see* **Note 12**).
13. Quantify RNA and store samples at −80°C (*see* **Note 13**).

3.2 cDNA Synthesis

Here, we propose two strategies for cDNA synthesis from viral RNA: one using random primers that are not template specific, and one

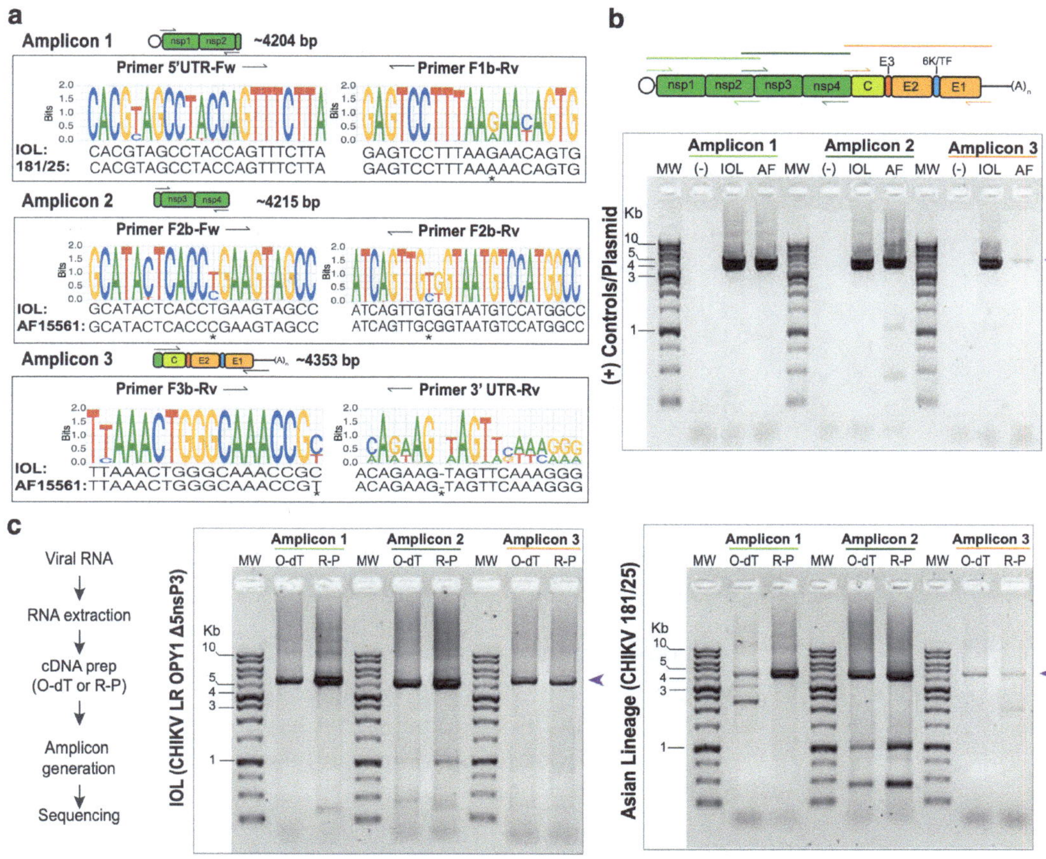

Fig. 3 Amplification of CHIKV genome using three-amplicon strategy. (**a**) Diagram showing the expected size and genome coverage of amplicons generated. Sequences of CHIKV IOL and Asian-Lineage AF15561 are shown at primer-binding sites. Sequence logos illustrate conservation across 77 CHIKV genomes at these sites. (**b**) Representative PCR products generated from plasmid controls (CHIKV IOL and AF15561) using the three-amplicon approach. Amplicon size corresponds to predicted fragment lengths (purple arrowheads). (**c**) Workflow for amplicon generation from viral RNA (left panel). Examples of amplicons generated from CHIKV IOL Δ5nsP3 and Asian-lineage CHIKV 181/25 purified RNA using either oligo(dT) (O-dT) or random priming (R-P) during cDNA synthesis. Predicted band sizes are indicated (purple arrowheads). MW molecular weight ladder, (–) non-template control

using an oligo (dT) primer targeting the poly(A) tail and two conserved bases at the 3′ end of CHIKV RNA (Tables 1 and 2). Factors such as expected viral RNA integrity, concentration, and experimental goals should guide the choice of priming strategy.

3.3 cDNA Synthesis: Random Primer Strategy

Random primers, typically a mixture of short oligonucleotides (e.g., hexamers), anneal non-specifically across the RNA template, producing a heterogeneous pool of cDNA fragments during reverse transcription. This strategy provides broad coverage of the viral genome and is particularly advantageous when RNA degradation is expected in the sample (*see* **Note 3**).

All steps of this reaction should be performed on ice. Additionally, reagents should be mixed and briefly spun down after thawing.

1. Make a master mix (Mix 1) of random primers, dNTPs, and water. Vortex and aliquot the required volume per reaction into PCR tubes (*see* **Notes 14 and 15**; Table 1).
   a. 100 μM random hexamer primers: 1 μL (Final concentration [Fc]: 5 μM).
   b. 10 mM dNTP mix: 1 μL ([Fc]: 0.5 mM).
2. Add up to 13 μL of RNA (with any remaining volume to be added as water) for a total of 15 μL per Mix 1 reaction (*see* **Note 16**; Table 1).
3. Place samples in the thermocycler and incubate for:
   a. Denaturation: 65°C for 5 min.
   b. 4°C hold.
4. Spin down samples and chill on ice for 2 min.
5. Make a master mix (Mix 2) of buffer and Maxima™ H Minus enzyme. Briefly, flick or pipette to mix, spin down, and aliquot 5 μL per Mix 1 reaction. These numbers are based on 20 μL final volume reactions (Table 1):
   a. 5× RT buffer: 4 μL.
   b. Maxima™ H Minus Enzyme: 1 μL (200 U).
6. Place samples in the thermocycler and incubate for (Table 1):
   a. First strand cDNA synthesis:
      i. 25°C for 10 min.
      ii. 50°C for 30 min.
      iii. Inactivation: 85°C for 5 min.
7. Store samples at −20°C until ready for amplicon generation.

3.4 cDNA Synthesis: Oligo (dT) Strategy

Oligo (dT) primers are designed to anneal to the poly(A) tails of RNA molecules. In this protocol, we use an anchored oligo (dT) primer d(T)23_GA (*see* **Note 4**), which targets both the poly(A) tail and the two conserved nucleotides immediately upstream of the poly(A) tract in CHIKV, conserved among alphaviruses [23] (Table 2). This design ensures specific priming from the 3′ end of the positive-sense viral RNA genome, favoring the synthesis of full-length cDNA compared to reactions primed with random hexamers. However, since reverse transcription efficiency decreases with increasing distance from the priming site, the resulting cDNA could be biased toward the 3′ end. Thus, when viral RNA levels are expected to be low, random primers are recommended to achieve more uniform coverage using the protocol described above. Here, reverse transcription was performed using Induro® Reverse Transcriptase, which supports cDNA synthesis of up to 12 kb.

All steps of this reaction should be performed on ice. Additionally, reagents should be mixed and briefly spun down after thawing.

1. Make a master mix (Mix 1) of dT(23) GA primer, dNTPs, and water. Vortex and aliquot the required volume per reaction into PCR tubes (*see* **Notes 14 and 15**; Table 2).
   a. 100 μM dT(23) GA primer: 1 μL ([Fc]: 5 μM).
   b. 10 mM dNTP mix: 1 μL ([Fc]: 0.5 mM).
2. Add up to 13 μL of RNA (with any remaining volume to be added as water) for a total of 15 μL per Mix 1 reaction (*see* **Notes 16 and 17**) (Table 2).
3. Place samples in the thermocycler and incubate for:
   a. Denaturation: 65°C for 5 min.
   b. 4°C hold.
4. Spin down samples and chill on ice until following step.
5. Make a master (Mix 2) mix of buffer and Induro® enzyme. Briefly, flick or pipette up and down to mix, spin down, and aliquot 5 μL per Mix 1 reaction. These numbers are based on 20 μL final volume reactions (Table 2).
   a. 5× Induro RT reaction buffer: 4 μL.
   b. Induro® Reverse Transcriptase: 1 μL.
6. Place samples in the thermocycler and incubate for:
   a. First strand cDNA synthesis: 55°C for 30 min.
   b. Inactivation: 95°C for 1 min.
7. Store samples at −20°C until ready for amplicon generation.

3.5 PCR Amplification Strategies

We propose two PCR amplicon strategies for amplification of the CHIKV genome from a cDNA template: one amplifying the genome in two overlapping amplicons (~7.5 kb fragments; Fig. 2) and another in three overlapping amplicons (~4.3 kb fragments; Fig. 3). Whereas previous studies have used 5–11 fragments to obtain full-length CHIKV genome information [18, 19, 24], this protocol provides an optimized approach for CHIKV amplicon generation. These strategies were validated using two attenuated CHIKV strains: LR2006_OPY1Δ5nsP3 [25] and 181/25 [26]. These attenuated strains were derived from the parental strains LR2006_OPY1 (GenBank accession no. DQ443544.2) and AF15561 (GenBank accession no. EF452493.1), which originated from the IOL and Asian CHIKV lineages, respectively.

Two primer sets are provided for the two-amplicon PCR strategy (Table 3). Primers for the two-amplicon design were optimized for both the IOL and Asian lineages of CHIKV, while those for the three-amplicon design were optimized for the IOL lineage. Primers were designed using MAFFT multiple sequence alignment to identify conserved regions of the genome, visualized as sequence logos (Figs. 2a and 3a).

If the sequence of the target strain is known, it is recommended to verify that each primer pair has no more than two mismatches with the template before proceeding with the PCR (*see* **Note 18**). If new CHIKV strains are being screened and the provided primers are ineffective, primers can be redesigned to target more conserved regions of the CHIKV genome. If both the two- and three-amplicon strategies yield insufficient product—such as in cases of low viral RNA input—smaller amplicons may be designed to improve efficiency and sensitivity.

3.5.1 Two-Amplicon Strategy

The two-amplicon strategy generates PCR products of approximately 7 kb each. However, due to reduced amplification efficiency associated with long targets, the final PCR yield is highly dependent on the amount of virus-specific cDNA template used as input. When cDNA input is expected to be high, such as when the original sample contains a high viral load, the two-amplicon strategy is highly effective. The coverage and expected amplicon sizes, along with sequence logos illustrating conservation at primer-binding sites across 77 CHIKV genomes, are shown in Fig. 2a. Representative examples of amplicon generation using control plasmids (Fig. 2b) and viral samples from which cDNA was synthesized using either oligo(dT) or random primers (R-P) are also presented (Fig. 2c).

3.5.2 Three-Amplicon Strategy

The three-amplicon strategy generates medium-sized amplicons of approximately 4 kb, offering improved amplification efficiency compared with the two-amplicon strategy due to the smaller fragment sizes (Fig. 3). When the amount of virus-specific cDNA template is expected to be low, the three-amplicon strategy is often more effective. The coverage and expected amplicon sizes, along with sequence logos illustrating conservation at primer-binding sites across 77 CHIKV genomes, are shown in Fig. 3a. Representative examples of amplicon generation using control plasmids (Fig. 3b) and viral samples from which cDNA was synthesized using either oligo(dT) or random primers (R-P) are also presented (Fig. 3c).

3.5.3 Amplicon Generation

A step-by-step PCR protocol optimized for CHIKV amplicon generation is shown below.

1. Make a master mix (45 μL per 1× reaction) of water, buffer, dNTPs, forward-primer, reverse-primer, and DNA polymerase,

flick or pipette to mix, and aliquot 45 μL per reaction into PCR tubes (Tables 3 and 4; *see* **Note 19**):

a. Water: 29.1 μL.

b. Phusion™ 5× HF Buffer: 10 μL (*see* **Note 5**).

c. dNTPs: 0.2 mM final concentration.

d. Forward primer: 0.5 μM final concentration.

e. Reverse primer: 0.5 μM final concentration.

f. Phusion™ DNA Polymerase: 0.5 μL (1 U).

2. Add 5 μL cDNA per sample, as well as any negative or positive controls (*see* **Note 20**; Table 4).
3. Run in thermocycler with the following conditions (Tables 4 and 5) (*see* **Note 21**):

   a. Initial denaturation: 98°C for 30 s (one cycle).

   b. Denaturation: 98°C for 10 s (30–35 cycles).

   c. Annealing: X°C for 30 s (30–35 cycles) (*see* **Notes 22 and 23**).

   d. Extension: 72°C for X min (30–35 cycles) (*see* **Notes 22 and 23**).

   e. Final extension: 72°C for 10 min (one cycle).

   f. Hold: 4°C.

4. Confirm presence of desired bands in a 1% agarose gel (*see* **Note 24**). Examples of bands from two- and three-amplicon generation can be observed in Figs. 2b, c and 3b, c.
5. Afterwards, PCR cleanup should be performed before sending for nanopore sequencing (*see* **Note 25**).

Table 4
Conditions for PCR amplicon generation

<table>
<tr><th>Reagents</th><th>Volume (50 μL ×1)</th><th>Final conc.</th><th>Thermocycler</th></tr>
<tr><td>Water</td><td>Add to 50 μL</td><td>—</td><td rowspan="7">(1 cycle)
98°C: 30 s (initial denaturation)
98°C: 10 s (denaturation)
(30–35 cycles)
X°C: 30 s (annealing)
72°C: X min (extension)
(1 cycle)
72°C: 10 min (final extension)
4°C: hold</td></tr>
<tr><td>Phusion 5× HF buffer</td><td>10 μL</td><td>1×</td></tr>
<tr><td>dNTPs</td><td>—</td><td>0.2 mM</td></tr>
<tr><td>Forward primer</td><td>—</td><td>0.5 μM</td></tr>
<tr><td>Reverse primer</td><td>—</td><td>0.5 μM</td></tr>
<tr><td>Phusion™ DNA Polymerase</td><td>0.5 μL (2 U/μL)</td><td>0.02 U/μL</td></tr>
<tr><td>cDNA</td><td>Up to 5 μL</td><td>—</td></tr>
</table>

Table 5
Annealing temperature and extension times for PCR amplicon generation

Strategy	Amplicon	Primers	Annealing temp.	Extension time (m:s)
Two-amplicon generation	Amplicon 1 ~7598 bp	Primer 5′UTR-Fw Primer F1a-Rv	62.6°C	05:04
	Amplicon 2 ~6643 bp	Primer F2a-Fw Primer 3′ UTR-Rv	57.4°C	04:26
Three-amplicon generation	Amplicon 1 ~4204 bp	Primer 5′UTR-Fw Primer F1b-Rv	55.9°C	02:48
	Amplicon 2 ~4215 bp	Primer F2b-Fw Primer F2b-Rv	63.2°C	02:49
	Amplicon 3 ~4353 bp	Primer F3b-Fw Primer 3′ UTR-Rv	57.4°C	02:54

Amplicon sizes are estimated from LR2006_OPY1 and AF15561 genomes

4 Notes

1. Depending on the experiment, CHIKV samples may be derived from cells, tissues, or blood. This book chapter will focus on extraction of CHIKV from infected cell culture supernatants. However, these protocols can be applied to other sample types, including tissue homogenates prepared in PBS or TRIzol™, as well as serum, plasma, or whole blood.
2. TRIzol™ and chloroform reagents should be opened under a fume hood with appropriate personal protective equipment (PPE). TRIzol™ is corrosive, toxic, carcinogenic, and combustible, and chloroform is toxic, carcinogenic, and can irritate skin.
3. For the random-primer strategy, both Induro® and Maxima™ H were tested, and no differences in efficiency were found between the two reverse transcriptases.
4. dT(23)_GA CHIKV refers to an anchored primer (Table 2) designed to target the poly(A) tail and the first two bases preceding it in CHIKV, which are conserved within alphaviruses [23]. Anchored primers prevent annealing to internal poly(A) sites by specifically binding at the start of the poly(A) tail, right at the transition from the RNA sequence. This reduces the risk of truncated cDNA synthesis.

5. Phusion™ is a high-fidelity DNA polymerase that minimizes errors during PCR.
6. The sample: TRIzol™ can vary if TRIzol™ comprises more than 50% of the final volume. When processing solid samples such as tissues or cells, follow manufacturer protocols for optimization.
7. A master mix of isopropanol and Glycoblue™ can be prepared if processing multiple samples; ensure that the mix is homogenized thoroughly.
8. When extracting from viral stocks or supernatant, the upper aqueous phase will generally appear clear; however, in the case of tissues or cell pellets, it may be opaque and have a brown color.
9. Samples can be precipitated overnight at −20°C to maximize RNA yield.
10. If the Glycoblue™ is not visible, add another 1 μL and spin again.
11. Vortexing helps break down the pellet and solubilize salts and other contaminants.
12. The resuspension volume depends on how much RNA is expected from the extraction; if unsure, resuspend at a lower volume (i.e., 22 μL), quantify using nanodrop, and then add additional volume to reach desired concentration (i.e., 20–200 ng/μL).
13. Samples from tissues or cells will contain RNA from the host that copurifies with RNA from virus, so the measured concentration will be representative of both sources of RNA. RNA quality can be assessed using a TapeStation or a Bioanalyzer if necessary.
14. Inclusion of a non-template control, such as water, is recommended. For a true negative control, RNA extraction can be performed on the corresponding material (supernatant, cells, etc.) from an uninfected sample to capture any background amplification that could be misinterpreted as real data.
15. Both the random-primer and oligo(dT) strategies can be up- or down-scaled proportionally by adjusting reagent volumes.
16. Around 1 μg of RNA is recommended for 20 μL Maxima™ H reactions; up to 1 μg of RNA is recommended for 20 μL Induro® reaction.
17. The Induro® protocol calls for water addition in both Master Mix 1 and Master Mix 2 (for reaction volumes of 10 μL then 20 μL). However, our optimization found that adding the water from Master Mix 2 into Master Mix 1 instead (for reaction volumes of 15 μL then 20 μL) does not compromise the

final cDNA yield. This adjustment allows for more RNA to be added into the first master mix.

18. The impact of primer mismatch can be seen in Fig. 3c. Comparing the forward primer of amplicon 3, optimized for CHIKV IOL, to the sequences of the 181/25 (Asian strain), the presence of two mismatches likely contributed to a reduced yield during PCR amplification.
19. The PCR reaction described is for 50 μL reactions for PCR cleanup and sequencing. If a smaller analytical reaction is desired to optimize conditions for different cDNA templates, primers, or thermocycler settings, the reaction can be scaled down to 10 μL.
20. Inclusion of a non-template control for the PCR reaction, such as water, as well as the non-template control from the cDNA synthesis is recommended. For a positive control, plasmids known to contain the viral genome can be used (2 ng plasmid DNA per 10 μL PCR reaction).
21. Depending on the downstream applications, for example—deep sequencing, it is recommended that PCR be performed with as few as 25 cycles to minimize amplification-induced errors. However, when the initial template is limited, increasing the reaction to up to 35 cycles may be necessary, with careful consideration of the limitations associated with higher cycle numbers and inclusion of appropriate controls.
22. Annealing temperature depends on the Tm (melting temperature) of the primer sets used—this protocol employed the ThermoFisher Tm Calculator for Primers to determine annealing temperatures (Table 5). Extension time depends on the expected amplicon size—this protocol used 40 s per 1 kb as per manufacturer instructions (Table 5). Expected amplicon sizes and extension times described in Table 5 are based on CHIKV LR2006_OPY1 and 181/25 genomes. Adjust extension times as needed according to the expected amplicon sizes of the strain being tested, but do not exceed 1 min per 1 kb.
23. To amplify all fragments in a single thermal cycler run, use an annealing temperature gradient to accommodate different primer Tm values, and set the extension time based on the largest expected amplicon size.
24. If desired bands are absent in the positive control, this likely indicates a PCR fault. If desired bands appear in the positive control but are absent or faint in the cDNA sample, this could indicate low RNA template or a problem in the cDNA synthesis. If the viral sequence is known, verify minimal primer mismatches. In the case of weak amplification due to low starting template, design shorter amplicons with new primer sets to increase yield.

25. The PCR cleanup kit used for this book chapter was the Macherey-Nagel NucleoSpin Gel and PCR Clean-up. However, any PCR cleanup strategy may serve the same purpose.

Acknowledgments

We thank the members of the Noval lab and Dr. Franco Izzo for their critical feedback on this book chapter. This work was supported by the Icahn School of Medicine at Mount Sinai and the National Institutes of Health (NIH FIRST U54CA267776).

References

1. Bartholomeeusen K, Daniel M, LaBeaud DA et al (2023) Chikungunya fever. Nat Rev Dis Primer 9:17
2. Ribeiro Dos Santos G, Jawed F, Mukandavire C et al (2025) Global burden of chikungunya virus infections and the potential benefit of vaccination campaigns. Nat Med 31:2342–2349
3. Lumsden WH (1955) An epidemic of virus disease in Southern Province, Tanganyika Territory, in 1952-53. II. General description and epidemiology. Trans R Soc Trop Med Hyg 49:33–57
4. Weaver SC, Forrester NL (2015) Chikungunya: evolutionary history and recent epidemic spread. Antiviral Res 120:32–39
5. Parola P, De Lamballerie X, Jourdan J et al (2006) Novel Chikungunya virus variant in travelers returning from Indian Ocean Islands. Emerg Infect Dis 12:1493–1499
6. Tsetsarkin KA, Vanlandingham DL, McGee CE et al (2007) A single mutation in chikungunya virus affects vector specificity and epidemic potential. PLoS Pathog 3:e201
7. European Centre for Disease Prevention and Control (ECDC) Chikungunya virus disease worldwide overview. Stockholm: European Centre for Disease Prevention and Control; 2025 (Reviewed November 2025), Stockholm, Sweden. Available from: https://www.ecdc.europa.eu/en/chikungunya-monthly
8. Cotella JI, Sauce AL, Saldarriaga CI et al (2021) Chikungunya and the Heart. Cardiology 146:324–334
9. Brito C, Falcão MB, Albuquerque MdeFPMde et al (2025) Chikungunya: from hypothesis to evidence of increased severe disease and fatalities. Viruses 17:62
10. U.S. Centers for Disease Control and Prevention (2025) Chikungunya vaccines. (Reviewed November 2025)
11. Frolov I, Frolova EI (2022) Molecular virology of chikungunya virus. Curr Top Microbiol Immunol 435:1–31
12. Levi LI, Rezelj VV, Henrion-Lacritick A et al (2021) Defective viral genomes from chikungunya virus are broad-spectrum antivirals and prevent virus dissemination in mosquitoes. PLoS Pathog 17:e1009110
13. Poirier EZ, Mounce BC, Rozen-Gagnon K et al (2015) Low-fidelity polymerases of alphaviruses recombine at higher rates to overproduce defective interfering particles. J Virol 90:2446–2454
14. Dolan PT, Whitfield ZJ, Andino R (2018) Mechanisms and concepts in RNA virus population dynamics and evolution. Annu Rev Virol 5:69–92
15. De Souza WM, De Lima STS, Simões Mello LM et al (2023) Spatiotemporal dynamics and recurrence of chikungunya virus in Brazil: an epidemiological study. Lancet Microbe 4:e319–e329
16. Morley VJ, Noval MG, Chen R et al (2018) Chikungunya virus evolution following a large 3'UTR deletion results in host-specific molecular changes in protein-coding regions. Virus Evol 4:vey012
17. Stapleford KA, Coffey LL, Lay S et al (2014) Emergence and transmission of arbovirus evolutionary intermediates with epidemic potential. Cell Host Microbe 15:706–716
18. Stapleford KA, Moratorio G, Henningsson R et al (2016) Whole-genome sequencing analysis from the chikungunya virus Caribbean outbreak reveals novel evolutionary genomic elements. PLoS Negl Trop Dis 10:e0004402
19. Noval MG, Rodriguez-Rodriguez BA, Rangel MV et al (2019) Evolution-driven attenuation of alphaviruses highlights key glycoprotein determinants regulating viral infectivity and dissemination. Cell Rep 28:460–471. e5

20. Kaczmarek ME, Herzog NL, Noval MG et al (2020) Distinct New York City Aedes albopictus mosquito populations display differences in salivary gland protein D7 diversity and chikungunya virus replication. Viruses 12:698
21. Stoeck T, Katzenmeier SN, Breiner H-W et al (2024) Nanopore duplex sequencing as an alternative to Illumina MiSeq sequencing for eDNA-based biomonitoring of coastal aquaculture impacts. Metabarcoding Metagenomics 8:e121817
22. Kolmogorov M, Billingsley KJ, Mastoras M et al (2023) Scalable Nanopore sequencing of human genomes provides a comprehensive view of haplotype-resolved variation and methylation. Nat Methods 20:1483–1492
23. Barraza Sánchez JJ, Volpe S, Faraj SE et al (2025) Functional RNA elements in alphavirus genomes. Curr Microbiol 82:364
24. Frumence E, Piorkowski G, Traversier N et al (2025) Genomic insights into the re-emergence of chikungunya virus on Réunion Island, France, 2024 to 2025. Euro Surveill Bull Eur Sur Mal Transm Eur Commun Dis Bull 30:2500344
25. Hallengärd D, Kakoulidou M, Lulla A et al (2014) Novel attenuated Chikungunya vaccine candidates elicit protective immunity in C57BL/6 mice. J Virol 88:2858–2866
26. Gorchakov R, Wang E, Leal G et al (2012) Attenuation of chikungunya virus vaccine strain 181/clone 25 is determined by two amino acid substitutions in the E2 envelope glycoprotein. J Virol 86:6084–6096

Check for updates

Chapter 26

Protocol for Detecting and Sequencing Chikungunya Virus from Field-Collected Mosquitoes

Cecília Artico Banho, Olivia Borghi Nascimento, Lívia Sacchetto, Nikos Vasilakis and Mauricio Lacerda Nogueira

Abstract

Arboviral diseases represent a major public health challenge, especially in tropical regions where environmental conditions may favor the proliferation and spread of mosquito vectors. Thus, early and accurate detection of chikungunya virus (CHIKV) in mosquito populations can be a valuable tool for effective surveillance of circulating variants and for identifying new viral introductions. Given the challenges of detecting arboviruses in field-captured mosquitoes, we describe an integrated workflow for CHIKV molecular detection and whole-genome sequencing. This protocol includes mosquito homogenization using a bead-based mechanical disruptor, RNA extraction using TRIzol reagent with minor modifications, molecular screening using CHIKV-specific RT-qPCR, and whole-genome amplification followed by sequencing on Illumina platforms. Despite the protocol being optimized for individual mosquitoes, it results in high-quality RNA suitable for both entomological surveillance and genomic analysis. As this protocol allows recovery of complete CHIKV genomes from mosquito specimens, it can serve as a basis for genomic epidemiology studies, enabling monitoring of viral diversity and lineage dynamics, and facilitating early detection of emerging variants to support timely and targeted public health interventions in endemic and at-risk regions

Keywords Arbovirus, Chikungunya virus, Vector, Surveillance, RT-qPCR, Whole-genome sequencing

1 Introduction

Chikungunya virus (CHIKV) is an arthropod-borne virus (arbovirus) belonging to the genus *Alphavirus* within the family *Togaviridae* [1]. It is an emerging and re-emerging pathogen of global health importance, transmitted to humans primarily by *Aedes aegypti* and *Ae. albopictus* mosquitoes [2, 3]. CHIKV infection causes an acute febrile illness, typically characterized by high fever, maculopapular rash, severe arthralgia, and long-lasting sequelae, such as debilitating arthritis or more severe outcomes, such as neurological disorders [4–6], resulting in a significant social and economic burden in affected regions [7]. Early detection of CHIKV circulation in

Claudia Filomatori and Eliana F. Castro (eds.), *Chikungunya Virus*, Methods in Molecular Biology, vol. 3067, https://doi.org/10.1007/978-1-0716-5496-5_26,

mosquito populations is critical for guiding vector control and surveillance programs, as viral activity in vectors often precedes human outbreaks [8]. Molecular techniques such as reverse transcription polymerase chain reaction (RT-PCR) and next-generation sequencing (NGS) have demonstrated high sensitivity and specificity for CHIKV detection and genomic characterization directly from field-collected mosquitoes [8–11]. These methods allow rapid identification of infected vectors and genomic surveillance, which facilitates monitoring of viral evolution, lineage replacement, and the introduction of new strains into endemic susceptible areas.

This chapter presents a standardized protocol for the viral RNA extraction from mosquitoes, such as *Aedes* spp. and *Culex* spp., using TRIzol reagent, followed by molecular detection of CHIKV and genome sequencing of positive samples. Although the described procedure is optimized for field-collected specimens, it may be adaptable for laboratories using different arbovirus surveillance systems. Therefore, by providing a complete workflow for mosquito sample molecular analysis (Fig. 1), this protocol aims to support entomological surveillance programs and improve the early warning capacity for arboviral transmission in endemic and at-risk regions.

2 Materials

2.1 Mosquito Homogenization

1. Class II biological safety cabinet (laminar flow hood) for RNA handling and sample aliquoting.
2. Individual mosquitoes or pools of mosquito specimens, identified by species, sex, and collection site (Fig. 1, step 2).
3. 2-mL screw-cap conical microtubes, impact- and low-temperature–resistant, compatible with the L-BEADER® (Loccus, Cotial, SP, Brazil) mechanical disruptor or an equivalent homogenization device (e.g., TissueLyser II, Qiagen, Hilden, Germany).
4. Dulbecco's Modified Eagle Medium (DMEM) supplemented with 2% fetal bovine serum (FBS), 1% penicillin, and 1% amphotericin B.
5. Ice.
6. Sterile stainless-steel beads 5 mm diameter.
7. L-BEADER® mechanical disruptor or equivalent bead-beating homogenizer.
8. Refrigerated microcentrifuge.
9. Micropipettes with sterile, nuclease-free, aerosol-resistant filter tips.
10. Personal protective equipment (PPE): laboratory coat, safety glasses, and nitrile gloves.

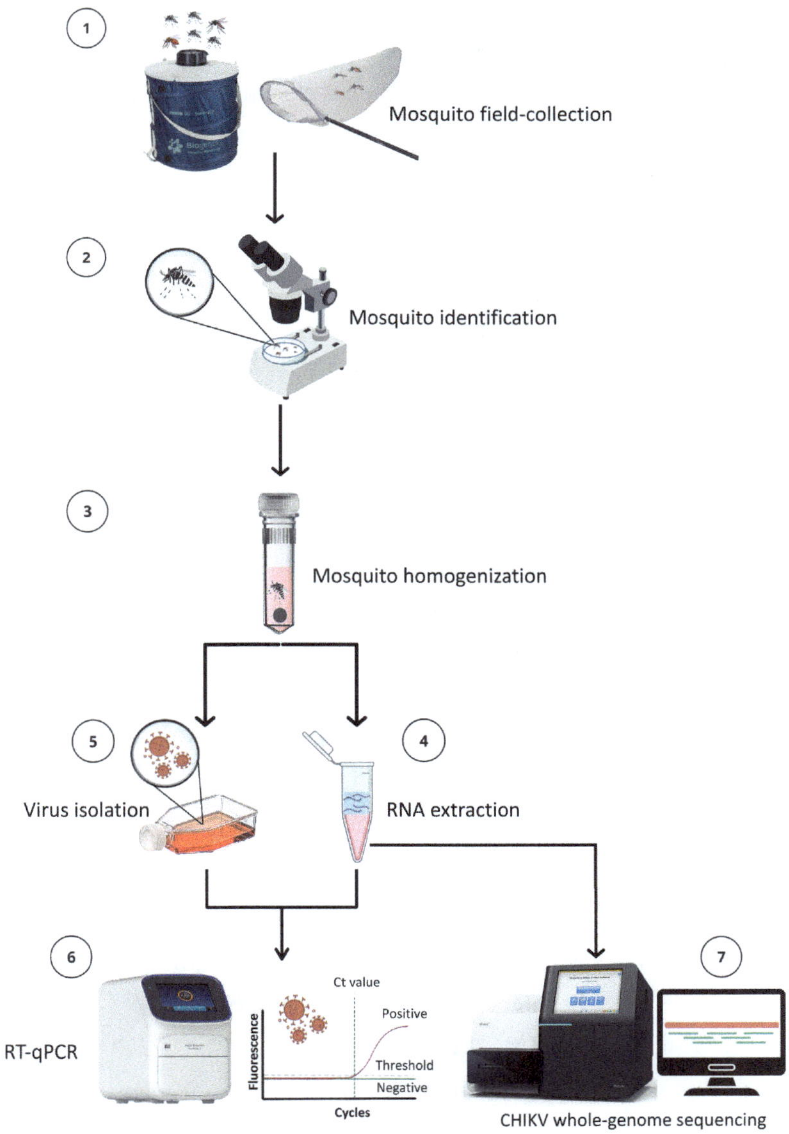

Fig. 1 Schematic workflow for mosquito sampling, processing, and arbovirus molecular screening. Figure created using Canva Pro (https://www.canva.com/design/DAG5_FW1bAw/QvJEXu0DvRK4y_lajpBgow/view?utm_content=DAG5_FW1bAw&utm_campaign=designshare&utm_medium=link2&utm_source=uniquelinks&utlId=h013ff80f23)

2.2 RNA Extraction

1. TRIzol® reagent or equivalent reagent containing phenol and guanidine isothiocyanate.
2. Chloroform.
3. Isopropanol.
4. 100% (v/v) and 75% (v/v) ethanol.
5. Nuclease-free water or DEPC-treated water.

6. Chemical fume hood, for all extraction steps involving TRIzol and chloroform reagents.
7. 1.5 mL RNase-free microcentrifuge tubes.
8. Refrigerated centrifuge.
9. Ice.
10. Micropipettes with sterile, nuclease-free, aerosol-resistant pipette tips.
11. Personal protective equipment (PPE): laboratory coat, nitrile gloves, safety glasses, and a PFF2/N95 (or equivalent) respirator when required.

2.3 RT-qPCR for CHIKV Detection

1. Biosafety cabinet (Class II laminar-flow hood) for stock reagent aliquoting and RT-qPCR mix preparation.
2. Lyophilized primers and probes (Table 1).
3. Benchtop microcentrifuge.
4. Nuclease-free water.
5. 0.6 mL RNase-free microcentrifuge tubes.
6. 1.5 mL RNase-free microcentrifuge tubes.
7. 96-well optical PCR plates, 0.2 mL low-profile PCR tubes, or other compatible reaction microtubes appropriate for the instrument used.
8. Adhesive optical sealing films for PCR plates.
9. GoTaq® Probe 1-Step RT-qPCR System (Promega, Madison, USA) or equivalent reagents.
10. Micropipettes dedicated to pre-amplification work.
11. Sterile, aerosol-resistant pipette tips.
12. PCR 96-well cooler.
13. RNA extracted from mosquito samples.

Table 1

Chikungunya virus-specific primers and probe sequences and their expected product sizes. Primer sequences were obtained from Lanciotti et al. [12]

Primer name	Genome position	Sequence 5'→ 3'
CHIKV_forward	874–894	AAAGGGCAAACTCAGCTTCAC
CHIKV_reverse	961–942	GCCTGGGCTCATCGTTATTC
CHIKV_probe	899–923	FAM-CGCTGTGATACAGTGGTTTCGTGTG-BBQ

14. Reconstituted primers, diluted to 10 μM.
15. Positive control RNA.
16. Plate spin or centrifuge.
17. Real-time PCR instrument.
18. Personal protective equipment (PPE): laboratory coat and nitrile gloves.

2.4 Multiplex PCR Method and Illumina Sequencing of CHIKV Genomes Directly from Mosquito Samples

1. Biosafety cabinet (Class II laminar-flow hood) for stockreagent aliquoting.
2. Nuclease-free water.
3. Micropipettes dedicated to pre-amplification work.
4. Sterile, aerosol-resistant pipette tips.
5. 1.5 mL RNase-free microcentrifuge tubes.
6. RNA extracted from mosquito samples.
7. 10 μM reconstituted primer pools.
8. Benchtop microcentrifuge.
9. Plate spin or centrifuge.
10. 0.2 mL 96-well optical PCR plates or other compatible reaction vessels appropriate for the instrument used.
11. Adhesive optical sealing films for PCR plates.
12. PCR 96-well cooler.
13. Micropipettes dedicated to post-amplification work.
14. Thermocycler.
15. Magnetic racks for tubes.
16. 96-well plate magnetic separation rack.
17. Illumina Microbial Amplicon Prep.
18. 4150 TapeStation system.
19. D1000 ScreenTape.
20. D1000 Reagents.
21. Quibit Fluorometer.
22. Qubit dsDNA HS Assay Kit.
23. Qubit Assay Tubes.
24. 96-well plate shaker.
25. MiSeq (Illumina, San Diego, CA, USA).
26. 300-cycle MiSeq Reagent Kit v2.
27. Personal protective equipment (PPE): laboratory coat and nitrile gloves.

3 Methods

As all procedures described here involve working with RNA, strict precautions are necessary to minimize contamination and RNase exposure. RNA work should be performed using RNase-free materials, and post-extraction steps should be conducted within a biosafety cabinet (Class II laminar-flow hood) to prevent RNA degradation and cross-contamination. Designate separate work areas and equipment for RNA extraction, pre-RT-qPCR, and post-amplification to minimize the contamination risk. Always use nuclease-free reagents, consumables, and aerosol-resistant pipette tips, and wear gloves, changing them between steps to avoid contamination.

Perform the RNA extraction procedure in a certified chemical fume hood, as both TRIzol reagent and chloroform contain toxic, corrosive, and volatile components. Wear appropriate personal protective equipment (PPE), including a laboratory coat, nitrile gloves, safety glasses, and a PFF2/N95 mask (or equivalent), when necessary. Wear disposable gloves when handling reagents and RNA samples to prevent RNase contamination from skin surfaces.

Chemical waste containing phenol or chloroform should be collected in designated chemical waste containers and disposed of in accordance with institutional biosafety and environmental safety protocols.

3.1 Sample Preparation: Mosquito Homogenization

1. Before starting, decontaminate all work surfaces, micropipettes, centrifuges, and the mechanical disruptor using RNase/DNase decontamination solutions, then clean them with 70% (v/v) ethanol to remove any remaining residues.
2. Perform all steps of this procedure inside a Class II Biological Safety Cabinet dedicated to sample handling.
3. Place the fresh or frozen mosquito samples (field-collected and morphologically identified specimens) in 2-mL screw-cap conical microcentrifuge tubes, resistant to low temperatures and compatible with the mechanical disruptor, and keep them on ice (*see* **Note 1**) (Fig. 1, step 3).
4. Add one sterile 5-mm stainless-steel bead to each screw-cap microtube (Fig. 1, step 3).
5. Add 400 μL of DMEM supplemented with 2% fetal bovine serum (FBS), 1% penicillin, and 1% amphotericin B to each microtube and secure the caps tightly. Change tips between each sample to avoid cross-contamination (*see* **Note 2**) (Fig. 1, step 3).
6. Remove the tubes from the ice and place them in the L-BEADER® mechanical disruptor. Homogenize at 3000 rpm for 30 s, performing three cycles with 30-s intervals between cycles (*see* **Note 3**).

7. Remove the tubes from the disruptor and place them back on ice.
8. Centrifuge the homogenates at 8400 × *g* for 10 min at 4°C.

3.2 RNA Extraction

This RNA extraction protocol is adapted from the TRIzol™ Reagent user guide, with minor modifications. Perform RNA extraction inside a certified chemical fume hood and wear appropriate personal protective equipment (PPE) to ensure safety.

1. Transfer 150 µL of the clarified mosquito homogenate to a new, properly labeled 1.5-mL snap-cap microtube. Only the supernatant should be collected, avoiding any chitin residues or debris. Change pipette tips between each sample to prevent cross-contamination.
2. Add a negative control during RNA extraction. For this, add 150 µL DMEM, supplemented with 2% FBS, 1% penicillin, and 1% amphotericin B (the same solution used in the mosquito homogenization procedure) to a properly labeled1.5-mL snap-cap microtube.
3. Store the remaining mosquito homogenate at −80 °C.
4. Add 150 µL of TRIzol® reagent or equivalent and mix thoroughly by pipetting up and down several times to ensure complete homogenization (*see* **Note 4**).
5. Incubate the mixture at room temperature for 5 min.
6. Add 75 µL of chloroform, securely cap the tube, and invert 30 times to mix.
7. Incubate for 3 min at room temperature.
8. Centrifuge the samples for 15 min at 12,000 × *g* at 4°C. The mixture will separate into three phases: a lower phenol–chloroform phase, an interphase, and a clear upper aqueous phase (Fig. 1, step 4).
9. Carefully transfer the upper aqueous phase (~200 µL), which contains the RNA, to a new, properly labeled 1.5-mL microtube by angling the tube at 45° and aspirating without disturbing the interphase. The remaining phases should be discarded into designated chemical waste containers, in accordance with institutional biosafety and environmental disposal guidelines. Alternatively, the phenol phase and interphase may be stored for subsequent extraction of DNA or proteins, if required.
10. Add an equal volume of isopropanol (~200 µL) to the recovered aqueous phase.
11. Mix the aqueous phase by inverting 30 times.
12. Incubate for 10 min at room temperature (15–30°C).

13. Centrifuge for 10 min at 12,000 × *g* at 4°C. A small gel-like or white RNA pellet will form at the bottom of the tube.
14. Carefully discard the supernatant by inverting the tube into an appropriate vessel, without disturbing the pellet.
15. Prepare 75% (v/v) fresh cold ethanol.
16. Add 1 mL of ice-cold 75% (v/v) ethanol and resuspend the pellet by briefly vortexing.
17. Centrifuge for 5 min at 7500 × *g*, 4°C.
18. Discard the supernatant completely by carefully inverting the tube into an appropriate vessel.
19. Air-dry the RNA pellet for 10–40 min at room temperature.
20. Do not dry the RNA using a vacuum centrifuge, as this may over-dry the pellet and reduce solubility.
21. Resuspend the RNA pellet in 30 μL of nuclease-free or DEPC-treated water.
22. Transfer 9 μL of the resuspended RNA to a new, properly labeled 0.6-mL microtube (*see* **Note 5**).
23. Proceed immediately to downstream applications or store the remaining RNA at −80°C until molecular analysis (each sample will yield two microtubes, one containing 9 μL and the other 21 μL of the resuspended RNA).
24. Incubate the eluted RNA in a water bath or heat block set at 56°C for 15 min.

3.3 Molecular Detection of Chikungunya Virus Through RT-qPCR

3.3.1 Primer Reconstitution

Perform all steps of this procedure inside a Class II Biological Safety Cabinet dedicated to pre-amplification workflows to minimize the risk of contamination.

1. Briefly centrifuge the lyophilized primers (a short spin) at room temperature before opening the tubes to ensure that all DNA is at the bottom.
2. The stock solutions of the forward primer, reverse primer, and probe should be prepared by adding the required volume of nuclease-free water to achieve a final concentration of 100 μM.
3. Homogenize the eluted primers by vortexing to ensure complete DNA reconstitution. Briefly centrifuge (short spin) before opening the reconstituted primer stock solutions. Wait at least 10 min before proceeding to the next steps.
4. Prepare small working aliquots by diluting the stock primer solution (10 μM), in order to avoid repeated freeze–thaw cycles. In a properly labeled 0.6-mL snap-cap microtube, add

10 μL of the previously reconstituted stock solution. Then add 90 μL of nuclease-free water to each small working aliquot to prepare a 10 μM solution. Change pipette tips between each aliquot to prevent cross-contamination.

5. Store the working dilution primer aliquots at −20°C.

3.3.2 Real-Time RT-PCR Reaction

Perform all steps of this procedure inside a Class II Biological Safety Cabinet dedicated to pre-amplification workflows to minimize the risk of contamination.

1. Prepare the reaction master mix on ice for the required number of reactions:

Component	Volume for one reaction	Final concentration
Nuclease-free water	1.0	–
GoTaq® Probe qPCRMaster Mix with dUTP	5	1×
Forward primer (workingsolution at 10 μM)	0.5	500 nM
Reverse primer (workingsolution at 10 μM)	0.5	500 nM
Probe (workingsolution at 10 μ M)	0.2	200 nM
GoScript™ RT Mix for1-Step RT-qPCR	0.3	–
Template RNA	2.5	–
Total volume	10	–

2. Briefly centrifuge (short spin) the PCR reagents before opening to minimize aerosol generation.
3. It is recommended to add the PCR reagents to a new, properly labeled, nuclease-free 1.5-mL snap-cap microtube in the following order: nuclease-free water; GoTaq® Probe 1-Step RT-qPCR System master mix or equivalent reagents; forward primer (working solution at 10 μM); reverse primer (working solution at 10 μM); probe (working solution at 10 μM); and the RT enzyme (*see* **Note 6**).
4. Mix by vortexing the microtube, then briefly centrifuge before proceeding to the next step.
5. Add 7.5 μL of the reaction master mix into each well of a 96-well optical PCR plate or into 0.2-mL low-profile PCR tubes (depending on the number of samples). Keep the PCR plate or tubes chilled on a 96-well PCR cooler or on ice until they are placed in the thermal cycler.

6. In a designated work area, outside the Class II Biological Safety Cabinet dedicated to pre-amplification workflows, add the RNA samples to the reaction wells of the plate or tubes. Add the negative extraction control, the no-template control (nuclease-free water), and the positive control RNA. RNA samples should be kept on ice if used immediately after extraction; if stored at −80°C, they should be thawed on ice before use.
7. Seal the 96-well optical PCR plate and briefly centrifuge it (1000 × *g* for 1 min). Inspect the plate for air bubbles before placing it in the real-time PCR instrument, as they may interfere with RT-qPCR performance.
8. Place the plate in the real-time PCR instrument and run the pre-programmed (Fig. 1, step 6) standard cycling conditions for molecular screening of chikungunya virus in mosquito samples. Cycling conditions are based on the manufacturer's instructions for the GoTaq® Probe 1-Step RT-qPCR System master mix (Promega, Madison, USA). If other RT-qPCR master mix reagents are used, cycling parameters should be optimized accordingly.

Step	Cycles	Temperature
Reverse transcription	1	45°C
Reverse transcriptase inactivation and GoTaq DNA polymerase activation	1	95°C
Denaturation	40	95°C
Annealing and extension		60°C

3.3.3 Real Time RT-PCR Result Analysis

1. Results are reported as cycle threshold (Ct) values. The Ct represents the number of PCR cycles required for the fluorescence signal to increase exponentially and cross the detection threshold (i.e., the background level). Check for amplification in the positive control and the absence of contamination in the negative controls to ensure that the reaction conditions were appropriate. Positive samples are expected to show Ct values ≤ 37, according to Lanciotti et al. [12] (Fig. 1, step 6).

3.4 Multiplex PCR Method and Illumina Sequencing of CHIKV Genomes Directly from Field-Collected Mosquito Samples

3.4.1 Primer Pools Preparation

Perform all steps of this procedure inside a Class II Biological Safety Cabinet dedicated to pre-amplification workflows to minimize the risk of contamination.

1. Briefly centrifuge the lyophilized primers (a short spin) at room temperature before opening the tubes to ensure the DNA pellet is at the bottom of each tube.
2. Prepare the stock solutions for each primer by adding the appropriate volume of nuclease-free water to obtain a final concentration of 100 μM. If the primers were ordered pre-diluted to 100 μM, proceed to the next step.
3. Homogenize the eluted primers by vortexing to ensure complete DNA reconstitution. Briefly centrifuge the tubes (short spin) before opening the reconstituted primer stock solutions.
4. Label two 1.5 mL RNase-free microcentrifuge tubes as indicated in Table 2, "Pool 1" and "Pool 2." The primers were designed by Quick et al. [13] for adjacent genomic regions and are distributed into alternating pools so that amplicons overlap between pools but not within each pool. To prepare the primer pool stocks, add equal volumes (e.g., 10 μL) of each 100 μM primer stock to the corresponding tube, ensuring that both the forward and reverse primers for alternating regions are combined in the same pool. For example, "Pool 1" should contain primers CHIK_400_1_LEFT_3, CHIK_400_1_RIGHT_3, CHIK_400_3_LEFT_0, CHIK_400_3_RIGHT_0, CHIK_400_5_LEFT_0, CHIK_400_5_RIGHT_0, and so on.

   Similarly, "Pool 2" should contain primers CHIK_400_2_LEFT_0, CHIK_400_2_RIGHT_0, CHIK_400_4_LEFT_4, CHIK_400_4_RIGHT_4, CHIK_400_6_LEFT_4, CHIK_400_6_RIGHT_4, and so on. Always change pipette tips between each aliquot to prevent cross-contamination.
5. Homogenize the primer pool stocks by vortexing, then perform a brief centrifugation (short spin).
6. In a new, properly labeled 1.5-mL snap-cap microtube, dilute each stock primer pool 1:10 with nuclease-free water to obtain a working concentration of 10 μM. For example, add 10 μL of the reconstituted 100 μM "Pool 1" stock solution to the tube, then add 90 μL of nuclease-free water to prepare a 10 μM working aliquot.
7. Store the working dilution primer aliquots at −20°C.

3.4.2 CHIKV Whole-Genome Sequencing

1. Before starting the library preparation, determine the number of samples per flow cell. This calculation depends on the MiSeq Reagent Kit selected (e.g., v2 Nano, v2 Micro, v2, or v3 Reagents), the viral genome size, and the desired coverage depth for genome assembly and downstream analyses. Perform the calculation using the Illumina Sequencing Coverage Calculator, available at https://support.illumina.com/downloads/sequencing_coverage_calculator.html. For example, given that the CHIKV genome is approximately

Table 2
Primer sets used for whole-genome CHIKV amplification, obtained from Quick et al. [13]

Primer name	Sequence 5′→ 3′	Primer pool[a]
CHIK_400_1_LEFT_3	CTTTTTGAAGGCCCTGCAACGT	1
CHIK_400_1_RIGHT_3	ACGGCCATCACCTCTTGTAAGT	1
CHIK_400_2_LEFT_0	TGTCGGACAGGAAGTACCACTG	2
CHIK_400_2_RIGHT_0	CTCATCCGCCCAATTTGTCGAG	2
CHIK_400_3_LEFT_0	GTACTGGGTAGGGTTCGACACA	1
CHIK_400_3_RIGHT_0	CACACGAAACCACTGTGTCACA	1
CHIK_400_4_LEFT_4	TCGGTGTTCCATCTAAAGGGCA	2
CHIK_400_4_RIGHT_4	CTTCATGGTGTTCGTGTTCCGT	2
CHIK_400_5_LEFT_0	GGTGGGGCTGAACCAGAGAATA	1
CHIK_400_5_RIGHT_0	TACAAAGCTGTCGAATTCGGCC	1
CHIK_400_6_LEFT_4	AAGCAGAAAACACACACGGTCT	2
CHIK_400_6_RIGHT_4	GCGCCTCTCGGAGTCTCTATTA	2
CHIK_400_7_LEFT_0	TACAGGCAGCACAGGAAGATGT	1
CHIK_400_7_RIGHT_0	TGGAAGTCTTCGGGCGAAATTG	1
CHIK_400_8_LEFT_1	GGAGCAAGTGAAGACGTGTACG	2
CHIK_400_8_RIGHT_1	TATATGGACAAGCGGGGCGAAT	2
CHIK_400_9_LEFT_0	TGCAAGAAGGAAGAAGCTGCAG	1
CHIK_400_9_RIGHT_0	AAGGCGATCAAGGCAAGTAACG	1
CHIK_400_10_LEFT_0	ATCTGCACGTACGGTTGATTCG	2
CHIK_400_10_RIGHT_0	GGTTTTGTTGAGCCCGTAGTGT	2
CHIK_400_11_LEFT_0	GGCAAAATGCGCACTACGAATG	1
CHIK_400_11_RIGHT_0	CCTTCCGTACGCGTTAGGAGTA	1
CHIK_400_12_LEFT_1	AGGAGTTTACGCAGTTAGGCAAA	2
CHIK_400_12_RIGHT_1	TTCATTCAGGGCTACTTCGGGT	2
CHIK_400_13_LEFT_0	TGGTCCCTATCCTCGAAACAGC	1
CHIK_400_13_RIGHT_0	TCCTGGTAGTCACGCAGATCTG	1
CHIK_400_14_LEFT_0	TGAGGCAGCGTCCATTCTAGAA	2
CHIK_400_14_RIGHT_0	ATACCTACCAAGTGTTGCCGGT	2
CHIK_400_15_LEFT_4	ACCTTGCACTGCCTACTAAGAGA	1
CHIK_400_15_RIGHT_4	ACATGGTGGTTTCAATGCTCTGG	1
CHIK_400_16_LEFT_0	TCAGAGCATACGGTTACGCAGA	2
CHIK_400_16_RIGHT_0	TTTGCGGTTCCTACTGGTGTTG	2
CHIK_400_17_LEFT_0	GACGGTGTTTGCAAGGCAGTAT	1
CHIK_400_17_RIGHT_0	TCCATGGCTGTAAAGAGGTGGT	1
CHIK_400_18_LEFT_0	ACCTCTCCTCTCCACAGGTGTA	2
CHIK_400_18_RIGHT_0	GCCTCTATTTGCTTTGGCCACA	2
CHIK_400_19_LEFT_0	AAGGGACACGTTTTCACCAGAC	1
CHIK_400_19_RIGHT_0	GGAGCATTTGACTTTTTGCACTCC	1
CHIK_400_20_LEFT_0	CGACTTCGCATGAACCATGTCA	2
CHIK_400_20_RIGHT_0	GTATGTATCGCCCCGTCGTCTA	2
CHIK_400_21_LEFT_0	AGTTTGATCTAAGCGCCGATGG	1

Table 2 (continued)

Primer name	Sequence 5′→ 3′	Primer pool[a]
CHIK_400_21_RIGHT_0	AAATGTGATGGGGAACGTCTCG	1
CHIK_400_22_LEFT_0	CGCGTGACACAGCTATTTCCTT	2
CHIK_400_22_RIGHT_0	TGGCGTACCGACTTCTGTTGTA	2
CHIK_400_23_LEFT_0	ACACGGACGACGAGTTATGACT	1
CHIK_400_23_RIGHT_0	GACAATCGGACGTTGATCGGAG	1
CHIK_400_24_LEFT_0	TAATGGCAGAGACCCCGAAAGT	2
CHIK_400_24_RIGHT_0	TGTGTTCTGGAATGGGGAAGGT	2
CHIK_400_25_LEFT_0	CGAGCGACATTCAATCCGTCAA	1
CHIK_400_25_RIGHT_0	CGGCAGCAGATTATGGGTTCTT	1
CHIK_400_26_LEFT_3	GAAGAATTTGCTGCCAGCCCTA	2
CHIK_400_26_RIGHT_3	GTGCGGCTATAATGGCATCGAA	2
CHIK_400_27_LEFT_0	TTAGGAGATTGAACGCCGTCCT	1
CHIK_400_27_RIGHT_0	GTCAGACGATCTTCCAACACCC	1
CHIK_400_28_LEFT_2	GACAGGTACGCGCTTCAAGTTC	2
CHIK_400_28_RIGHT_2	CGCTAACGGTTTGCCCAGTTTA	2
CHIK_400_29_LEFT_0	AGCTCCCTACTTTTGTGGAGGG	1
CHIK_400_29_RIGHT_0	GACCGCCGTACAAGGTTATGAC	1
CHIK_400_30_LEFT_0	TAATGTCCATGGCCACCTTTGC	2
CHIK_400_30_RIGHT_0	TATTCTTCCGATTCTTGCGCGG	2
CHIK_400_31_LEFT_0	CAACTTGCCCAGCTGATCTCAG	1
CHIK_400_31_RIGHT_0	ATCGATGGTCCCCTTTACGTGT	1
CHIK_400_32_LEFT_0	TAAGGTAACAGGTTACGCGTGC	2
CHIK_400_32_RIGHT_0	TACGGGCTCCTTCATTAGCTCC	2
CHIK_400_33_LEFT_2	GACCGATCTTCGACAACAAGGG	1
CHIK_400_33_RIGHT_2	GCGTCGGGGAGAACATGTTAAG	1
CHIK_400_34_LEFT_0	GAGAAAACCTTGCGCATGCTTG	2
CHIK_400_34_RIGHT_0	CTGCTGGCATGTGATTGTCCAT	2
CHIK_400_35_LEFT_0	CCAGGTTTCCTTGCAAATCGGA	1
CHIK_400_35_RIGHT_0	GTACGTGCTGCAAGGTAGTTCC	1
CHIK_400_36_LEFT_1	CATGTACGCACCCATTTCACCA	2
CHIK_400_36_RIGHT_1	GGTCCCCGAATTCAGCATTACG	2
CHIK_400_37_LEFT_0	CTGCAAGGTCGATCAATGCCAT	1
CHIK_400_37_RIGHT_0	TCATTGTTACCCCACGTGACCT	1
CHIK_400_38_LEFT_0	AGTGGGTGACGCATAAGAAGGA	2
CHIK_400_38_RIGHT_0	TGGTGTCAGTTCGTACGGTGTA	2
CHIK_400_39_LEFT_0	ATACTCCTGTCGATGGTGGGTG	1
CHIK_400_39_RIGHT_0	CGGGATCACTGTTACGTGTTCG	1
CHIK_400_40_LEFT_0	TTTTTAGCCGTACTGAGCGTCG	2
CHIK_400_40_RIGHT_0	ATTTTCGGTGTCGCAGAAGCAG	2
CHIK_400_41_LEFT_0	GCGGTACAGCAGAGTGTAAGGA	1
CHIK_400_41_RIGHT_0	GTCGCCTTTGTACACCACGATT	1

Table 2 (continued)

Primer name	Sequence 5′→ 3′	Primer pool[a]
CHIK_400_42_LEFT_0	CCATGCCGTCACAGTTAAGGAC	2
CHIK_400_42_RIGHT_0	AGTTCATCGCTCTTACCGGGTT	2
CHIK_400_43_LEFT_0	ATTGGCTAAAAGAACGAGGGGC	1
CHIK_400_43_RIGHT_0	TTCCCGAATAGTGACGGCGTTA	1
CHIK_400_44_LEFT_0	GCACCCATTCCTCAGACTTTGG	2
CHIK_400_44_RIGHT_0	CCTGCTAAACGACACGCATAGC	2

[a]**Primer pool:** Forward and reverse primer sequences designed for adjacent CHIKV genomic regions, which are distributed into alternating pools so that amplicons overlap between pools but not within each pool. The primers are pooled because they share similar characteristics, such as melting temperatures, to reduce the number of PCR reactions required for whole-genome amplification.

11,000 base pairs, if a mean coverage depth of 5000× is required, include approximately five samples per flow cell using the v2 Nano Reagents, 21 samples per flow cell using the v2 Micro Reagents, 80 samples per flow cell using the v2 Reagents, and 133 samples per flow cell using the v3 Reagents. It is strongly recommended to include a negative control to facilitate detection of possible cross-contamination related to primer pool handling rather than to sample-specific issues.

2. Conduct library preparation in a designated work area exclusively for sequencing (post-amplification steps). This area must contain separate materials, such as dedicated pipettes, and specific equipment, to avoid contamination with PCR products.
3. Use the 9 μL RNA aliquots of positive samples, stored at −80 °C and not previously thawed, for sequencing.
4. Perform library preparation using the Illumina Microbial Amplicon Prep Kit according to the manufacturer's instructions. As a modification to the standard protocol, perform the amplicon PCR at 63°C. During the amplicon indexing step, which amplifies the tagmented amplicons using a PCR program, adjust the number of cycles from 7 to 10 (*see* **Note 6**).
5. Quantify the library using the Qubit dsDNA HS Assay Kit and measure the concentration of the pooled DNA library, following the manufacturer's instructions. Perform quantification on a Qubit Fluorometer 3.0 or later.
6. Perform library quality control by assessing the size distribution of the previously quantified pooled library using the D1000 ScreenTape and D1000 Reagents on the 4150 TapeStation system, following the manufacturer's instructions.
7. Perform library denaturation and dilution according to the manufacturer's instructions for the MiSeq system (Fig. 1, step 7).
8. Prior to starting the sequencing run, generate the SampleSheet.csv file in the Illumina Local Run Manager using the required

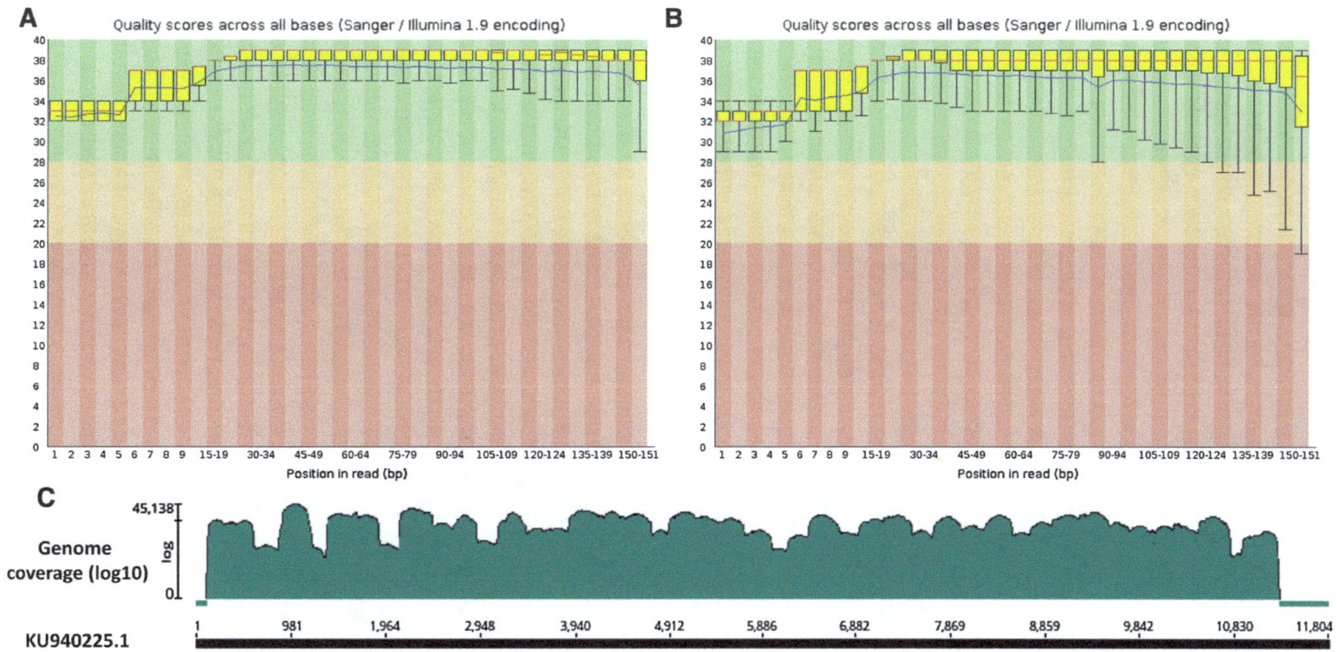

Fig. 2 CHIKV whole-genome sequencing and assembly. (**A, B**) Quality check of raw reads visualized using fastQC v. 0.12.1 software (https://www.bioinformatics.babraham.ac.uk/projects/fastqc/). (**C**) Genome assembly coverage depth of the CHIKV genome derived from *Aedes aegypti* mosquitoes, collected in São José do Rio Preto, São Paulo, Brazil. The sequence was generated according to the protocol previously described and assembled using the Geneious Prime v. 2025.2 (https://www.geneious.com/updates/geneious-prime-2025-2)

sample and index information. Proceed with the instrument setup and start the run according to the manufacturer's recommended guidelines.

9. Base calling and demultiplexing are automatically performed by the MiSeq system using the sample and index information specified in the SampleSheet.csv file.
10. Proceed with the appropriate bioinformatic pipelines for quality checking, genome assembly, and downstream analyses (Fig. 2).

4 Notes

1. After field collection, mosquito identification and handling (species and sex determination) should be performed on a chilled surface, such as a cold plate or a Petri dish placed on ice, to preserve the integrity of any viral particles present in the specimens. Following identification, mosquitoes should be transferred individually or in pools of up to ten individuals (grouped by species, sex, date, and collection site) into 2-mL screw-cap conical microcentrifuge tubes compatible with the L-BEADER® mechanical disruptor or an equivalent device suitable for mosquito grinding. If the samples are stored in snap-cap microtubes, transfer them to screw-cap microtubes before proceeding (Fig. 1, step 2). The DMEM supplemented with 2% fetal bovine serum (FBS), 1% penicillin, and 1% amphotericin B is used for mosquito maceration because, in addition to molecular diagnosis, this protocol also aims to enable viral isolation from positive samples. Homogenization in DMEM supplemented with antibiotics and antifungal agents is compatible with virus isolation, as it does not contain inhibitors that would impair viral viability. Alternatively, this solution may be replaced with RNase-free 1× phosphate-buffered saline (PBS) supplemented with 2% penicillin and 1% amphotericin B, Minimum Essential Medium (MEM), or L-15 cell culture medium (supplemented as required) when viral isolation is one of the experimental objectives (Fig. 1, step 5). If the goal is exclusively molecular diagnosis, the mosquito grinding step can be optimized by substituting DMEM with TRIzol, which provides immediate viral inactivation and efficient preservation of RNA, preventing mRNA degradation.
2. If a mechanical disruptor, such as the L-BEADER®, is unavailable, several alternative methods can be used to achieve efficient mosquito maceration for downstream RNA extraction or viral isolation. The choice of method should consider sample

throughput, biosafety, RNA integrity, and whether viral viability must be preserved. Regardless of the chosen alternative, you should always perform the homogenization step on ice, using RNase-free materials and solutions suitable for the experimental objective (e.g., DMEM, PBS 1×, MEM, or L-15 medium for viral isolation, or TRIzol for RNA-only workflows).

3. TRIzol™ reagent may be substituted with other phenol–guanidinium thiocyanate-based extraction reagents, such as QIAzol® or TRI Reagent®.
4. This step is optional. A 9 μL RNA aliquot is prepared to avoid repeated freeze–thaw cycles during CHIKV (or arbovirus) screening. Specifically, if a sample tests positive in the initial RT-qPCR screening, you can use this aliquot directly in the NGS workflow. The Illumina Microbial Amplicon Prep Kit used in this protocol requires 8.5 μL of RNA as input for library preparation, which justifies the aliquot volume. If you choose to use another library preparation kit, you can adapt the volume as required.
5. If you choose to use the GoTaq® Probe 1-Step RT-qPCR System master mix, carefully follow the manufacturer's instructions, as this mix does not include a reference dye. However, a tube of carboxy-X-rhodamine (CXR) reference dye is provided, allowing its addition if desired. Incorporating the reference dye enhances the performance of the GoTaq® Probe 1-Step RT-qPCR System master mix when used with real-time PCR instruments that support dye normalization. The required volume of CXR depends on the RT-PCR instrument, as some platforms require low dye concentrations and others require high concentrations. Please check the manufacturer's guidelines for specific recommendations. Alternatively, the GoTaq® Probe 1-Step RT-qPCR System master mix may be replaced with an equivalent reagent, which will require adjustments to cycling conditions.
6. If more than 48 samples are being sequenced, the indexes provided in the kit should be replaced with an alternative index set, such as the IDT-ILMN PCR Index Set 1-4 (384 IDX), which allows multiplexing up to 384 samples.

Acknowledgments

We would like to thank the Multiuser Laboratory (LMU) at FAMERP, Brazil, for the use of the Illumina MiSeq system. This work received support from the Fundação de Amparo à Pesquisa do Estado de São Paulo (FAPESP, grant numbers 2022/03645-1 to MLN, 2023/14670-0, and to CAB). This study was funded in part by the INCT Viral Genomic Surveillance, the One Health grant 405786/2022-0, and by the Centers for Research in Emerging Infectious Diseases (CREID), via the "Coordinating Research on Emerging Arboviral Threats Encompassing the Neotropics (CREATE-NEO)" grant U01AI151807 awarded to NV by the National Institutes of Health (NIH/USA). The funders had no role in the study design, data collection, analysis, interpretation, manuscript writing, or the decision to publish the results. Grammarly AI was used solely for language editing (grammar, spelling, and punctuation). The tool did not generate, modify, or interpret any scientific content; all analyses, interpretations, and conclusions are the authors' own.

Data Availability All the chikungunya virus raw data (FASTq files) generated using this protocol are available in the SRA-NCBI database (https://www.ncbi.nlm.nih.gov/sra/PRJNA1338711).

References

1. Griffin DE, Weaver SC (2021) Alphaviruses. In: Howley PM, Knipe DM (eds) Fields virology, vol I, 7th edn. Wolters Kluwer, pp 458–90
2. Luhken R et al (2024) High vector competence for chikungunya virus but heavily reduced locomotor activity of Aedes albopictus from Germany at low temperatures. Parasit Vectors 17(1):502. https://doi.org/10.1186/s13071-024-06594-x
3. Vega-Rua A et al (2013) High efficiency of temperate Aedes albopictus to transmit chikungunya and dengue viruses in the Southeast of France. PLoS One 8(3):e59716. https://doi.org/10.1371/journal.pone.0059716
4. da Costa VG et al (2024) A meta-analysis of chikungunya virus in neurological disorders. Infect Dis Now 54(5):104938. https://doi.org/10.1016/j.idnow.2024.104938
5. Khongwichit S, Chansaenroj J, Chirathaworn C, Poovorawan Y (2021) Chikungunya virus infection: molecular biology, clinical characteristics, and epidemiology in Asian countries. J Biomed Sci 28(1):84. https://doi.org/10.1186/s12929-021-00778-8
6. Freppel W, Silva LA, Stapleford KA, Herrero LJ (2024) Pathogenicity and virulence of chikungunya virus. Virulence 15(1):2396484. https://doi.org/10.1080/21505594.2024.2396484
7. Ng LC, Hapuarachchi HC (2010) Tracing the path of chikungunya virus-evolution and adaptation. Infect Genet Evol 10(7):876–885. https://doi.org/10.1016/j.meegid.2010.07.012
8. Banho CA et al (2025) Vector surveillance during a major chikungunya outbreak in northwestern Sao Paulo state, Brazil. Npj Viruses 3(1):78. https://doi.org/10.1038/s44298-025-00161-6
9. Nath A et al (2023) Genomic characterization of mosquito isolates of chikungunya virus (outbreak strains 2022) using next-generation sequencing. Vector Borne Zoonotic Dis 23(6):341–349. https://doi.org/10.1089/vbz.2022.0100
10. Zhou X et al (2025) First whole-genome chikungunya virus sequence detected in mosquitoes during the 2025 Foshan outbreak: evidence of field vector infection and transmission potential in China. Infect Dis Poverty 14(1):108. https://doi.org/10.1186/s40249-025-01383-9
11. Ramirez AL, van den Hurk AF, Meyer DB, Ritchie SA (2018) Searching for the proverbial needle in a haystack: advances in mosquito-borne arbovirus surveillance. Parasit Vectors 11(1):320. https://doi.org/10.1186/s13071-018-2901-x
12. Lanciotti RS et al (2007) Chikungunya virus in US travelers returning from India, 2006. Emerg Infect Dis 13(5):764–767. https://doi.org/10.3201/eid1305.070015
13. Quick J et al (2017) Multiplex PCR method for MinION and Illumina sequencing of Zika and other virus genomes directly from clinical samples. Nat Protoc 12(6):1261–1276. https://doi.org/10.1038/nprot.2017.066

Index

Claudia Filomatori and Eliana F. Castro (eds.), *Chikungunya Virus*, Methods in Molecular Biology, vol. 3067, https://doi.org/10.1007/978-1-0716-5496-5,

FSC
www.fsc.org
MIX
Papier | Fördert gute Waldnutzung
FSC® C083411

Kolibri 360 GmbH
Ferdinand-Jühlke-Straße 7
99095 Erfurt, Deutschland
produktsicherheit@kolibri360.de